AF356443

Interaction between Waves and Maritime Structures

Interaction between Waves and Maritime Structures

Editors

Mariano Buccino
Luca Martinelli

MDPI • Basel • Beijing • Wuhan • Barcelona • Belgrade • Manchester • Tokyo • Cluj • Tianjin

Editors
Mariano Buccino
University of Naples "Federico II"
Italy

Luca Martinelli
Padova University
Italy

Editorial Office
MDPI
St. Alban-Anlage 66
4052 Basel, Switzerland

This is a reprint of articles from the Special Issue published online in the open access journal *Water* (ISSN 2073-4441) (available at: https://www.mdpi.com/journal/water/special_issues/Waves_Maritime_Structures).

For citation purposes, cite each article independently as indicated on the article page online and as indicated below:

LastName, A.A.; LastName, B.B.; LastName, C.C. Article Title. *Journal Name* **Year**, *Volume Number*, Page Range.

ISBN 978-3-0365-0624-1 (Hbk)
ISBN 978-3-0365-0625-8 (PDF)

Cover image courtesy of Luca Martinelli.

Contents

About the Editors . . . . **vii**

Mariano Buccino and Luca Martinelli
Interaction between Waves and Maritime Structures
Reprinted from: *Water* **2020**, *12*, 3472, doi:10.3390/w12123472 . . . . **1**

Mariano Buccino, Gianluigi Di Paola, Margherita C. Ciccaglione, Giuseppe Del Giudice and Carmen M. Rosskopf
A Medium-Term Study of Molise Coast Evolution Based on the One-Line Equation and "Equivalent Wave" Concept
Reprinted from: *Water* **2020**, *12*, 2831, doi:10.3390/w12102831 . . . . **7**

Roman Gabl, Thomas Davey and David M. Ingram
Roll Motion of a Water Filled Floating Cylinder—Additional Experimental Verification
Reprinted from: *Water* **2020**, *12*, 2219, doi:10.3390/w12082219 . . . . **29**

Dimitris Stagonas, Rajendran Ravindar, Venkatachalam Sriram and Stefan Schimmels
Experimental Evidence of the Influence of Recurves on Wave Loads at Vertical Seawalls
Reprinted from: *Water* **2020**, *12*, 889, doi:10.3390/w12030889 . . . . **49**

Der-Chang Lo, Keh-Han Wang and Tai-Wen Hsu
Two-Dimensional Free-Surface Flow Modeling for Wave-Structure Interactions and Induced Motions of Floating Bodies
Reprinted from: *Water* **2020**, *12*, 543, doi:10.3390/w12020543 . . . . **63**

Piero Ruol, Chiara Favaretto, Matteo Volpato and Luca Martinelli

Reprinted from: *Water* **2020**, *12*, 427, doi:10.3390/w12020427 . . . . **93**

Jassiel V. Hernández-Fontes, Paulo de Tarso T. Esperança, Juan F. Bárcenas Graniel, Sergio H. Sphaier and Rodolfo Silva
Green Water on A Fixed Structure Due to Incident Bores: Guidelines and Database for Model Validations Regarding Flow Evolution
Reprinted from: *Water* **2019**, *11*, 2584, doi:10.3390/w11122584 . . . . **109**

Roman Gabl, Thomas Davey, Edward Nixon, Jeffrey Steynor and David M. Ingram
.Comparison of a Floating Cylinder with Solid and Water Ballast
Reprinted from: *Water* **2019**, *11*, 2487, doi:10.3390/w11122487 . . . . **135**

Sara Mizar Formentin, Maria Gabriella Gaeta, Giuseppina Palma, Barbara Zanuttigh and Massimo Guerrero
Flow Depths and Velocities across a Smooth Dike Crest
Reprinted from: *Water* **2019**, *11*, 2197, doi:10.3390/w11102197 . . . . **153**

Dong-Liang Zhang, Chun-Wei Bi, Guan-Ye Wu, Sheng-Xiao Zhao and Guo-Hai Dong
Laboratory Experimental Investigation on the Hydrodynamic Responses of an Extra-Large Electrical Platform in Wave and Storm Conditions
Reprinted from: *Water* **2019**, *11*, 2042, doi:10.3390/w11102042 . . . . **185**

José Sande, Andrés Figuero, Javier Tarrío-Saavedra, Enrique Peña, Alberto Alvarellos and Juan Ramón Rabuñal
Application of an Analytic Methodology to Estimate the Movements of Moored Vessels Based on Forecast Data
Reprinted from: *Water* **2019**, *11*, 1841, doi:10.3390/w11091841 . . . . **205**

Enrique Maciñeira, Enrique Peña, José Sande and Andrés Figuero
Dynamic Calculation of Breakwater Crown Walls under Wave Action: Influence of Soil Mechanics and Shape of the Loading State
Reprinted from: *Water* **2019**, *11*, 1149, doi:10.3390/w11061149 . . . . **225**

Mads Røge Eldrup, Thomas Lykke Andersen and Hans Falk Burcharth
Stability of Rubble Mound Breakwaters—A Study of the Notional Permeability Factor, Based on Physical Model Tests
Reprinted from: *Water* **2019**, *11*, 934, doi:10.3390/w11050934 . . . . **255**

Mads Røge Eldrup, Thomas Lykke Andersen and Hans Falk Burcharth
Correction: Eldrup, M.R., et al. Stability of Rubble Mound Breakwaters—A Study of the Notional Permeability Factor, based on Physical Model Tests. *Water* 2019, *11*, 934
Reprinted from: *Water* **2019**, *11*, 2483, doi:10.3390/w11122483 . . . . **273**

About the Editors

Mariano Buccino (MB) is an Associate Professor at the Department of Civil, Architectural and Environmental Engineering of University of Napoli Federico II. Since the early 2000s, he has been teaching Maritime Structures Shore Protection Measures and Coastal Engineering, first as an Assistant Professor and then as an Associate Professor. During this period, he tutored seven PhD. Programs and has headed a number of Field and Laboratory projects on account of both the University of Napoli Federico II and the C.U.G.RI. University Centre for Research on Major Hazards. MB is a member of the groups of Italian Maritimes Academy, and of the Italian Group of Hydraulics. He is on the Editorial Board of Energies and *Journal for Marine Science and Engineering*, and has guest-edited Special Issues of both Water and *Sustainability*.

Luca Martinelli (LM) is an Associate Professor at the University of Padova, ICEA Department, in the sector related to Coastal Engineering, in the research group founded by Prof. Guido Ferro in 1927 and now led by Prof. Piero Ruol. LM's main qualifications are related to physical model testing in the field of maritime constructions (including test design, data analysis and interpretation, even through numerical modelling). LM is now the Director of the Maritime Laboratory of Padova University, member of the Doctorate Council of the Department, member of CONISMA (Consorzio Internazionale Interuniversitario per le Scienze del Mare), ISOPE (International Society of Offshore and Polar Engineers), of the groups of Italian Maritimes Academy, and of the Italian Group of Hydraulics. He is a member of the editorial board for the *Journal for Marine Science and Engineering*, and Guest Editor of two Special Issues of *Water*.

Editorial

Interaction between Waves and Maritime Structures

Mariano Buccino [1,*] **and Luca Martinelli** [2]

[1] Department of Civil, Architectural and Environmental Engineering, University of Napoli Federico II,
    Via Claudio 21, 80125 Napoli, Italy
[2] Department of Civil, Environmental and Architectural Engineering, University of Padova,
    Via Ognissanti, 39, 35129 Padova, Italy; luca.martinelli@unipd.it
*   Correspondence: buccino@unina.it

Received: 1 December 2020; Accepted: 8 December 2020; Published: 10 December 2020

---

1. Introduction

Understanding the interaction between waves and maritime structures (IWMS) has been a primary concern for humans since ancient times, when they started sailing oceans and defending land from flooding and erosion.

However, understanding of this interaction started to truly develop only in the first half of twentieth century, when maritime engineering was recognized as a new discipline [1]. Since then, a remarkable body of research has been produced, both of a theoretical and, far more frequently, an experimental nature. Several water wave theories were developed in the nineteenth century (e.g., [2]) partly based on results from other branches of physics, such as thermodynamics [3], optics [4], and electromagnetism [5]. However, IWMS could not be described by theory alone. Starting from the early 1940s, the emerging "art and science of physical modeling" [6] had a tremendous impact on IWMS, putting a new design method at engineers' disposal. For example, it was thought that the stability of rubble mound breakwaters was a question of trial and error [7] for many years, until intensive laboratory testing (1942–1950) allowed Hudson [8], based on previous research by Iribarren [9], to propose a stability formula, which was widely employed in design practice until the early 1990s.

In recent years, the experimental approach to IWMS has taken further advantage of laboratory data by categorizing information into wide databases. This was the case, for example, for the 10,000 wave overtopping data collected within the framework of the European program CLASH (Crest Level Assessment of Coastal Structures by Full-scale Monitoring, Neural Network Prediction and Hazard Analysis on Permissible Wave Overtopping, [www.clash-eu.org]), which were then enlarged to 17,000 under conditions set out in EurOtop 2018 [www.overtopping-manual.com].

Clearly, large databases ease the analysis of relationships between predictive quantities and output variables and reduce the bias from laboratory/scale effects via a critical comparison of various datasets. Moreover, advanced numerical modeling techniques, such as computational fluid dynamics (CFD) or smooth particle hydrodynamics (SPH) have emerged as powerful experimental tools that may serve to either deepen laboratory findings or approach new or little-understood research topics [10–14].

2. Structure of the Special Issue

The arguments of the papers gathered within this Special Issue can be categorized into:

1. Structural performance, concerning the stability/integrity of the structure (or any part thereof);
2. Functional performance, i.e., related to the capability of the structure to meet the scope for which it has been designed.

In turn, Item 2 may be subdivided into:

-   Hydraulic response, such as wave reflection, wave transmission, wave overtopping, etc.;

- Morphodynamic response, which is the shoreline and beach response to a structure placement;
- Floating body response, such as the oscillation of floating bodies or moored vessels.

This Special Issue includes 12 articles, four of which focus on structural performance and eight on functional response. The following section provides a brief summary of each of the contributions.

3. Article Overview

3.1. Contributions on Structural Processes

Two of the articles on structural performance deal with mound breakwaters; one addresses vertical seawalls and one analyzes the response of a complex offshore structure.

3.1.1. Rubble Mound Breakwaters: Effect of Permeability and Dynamics of Crown Walls

Eldrup et al. [15] deals with the role of notational permeability factor (NPF) in the hydraulic stability of rubble mound breakwaters. NPF is a leading variable of the widely employed van der Meer stability formulae [16] and has been introduced to globally account for the water seepage at a given cross-section material assemblage (including armor, underlayers, and core).

However, the original work by van der Meer indicated only four typical values of NPF, thereby creating some uncertainty in the engineering applications. Based on the analysis of new physical model tests conducted at Aalborg University (Denmark), the authors in [15] provide NPF values for seven different layer compositions and propose a pragmatic empirical method to help estimate the permeability parameter in non-tested situations.

The research of Macineira et al. [17] examines the dynamics of breakwater crown walls under wave actions. The proposed approach considers only two degrees of freedom, corresponding to the horizontal translation of the wall (normal to the breakwater axis) and the rotation in the cross-section plane. Stiffness is estimated as a function of the instantaneous wave force by assimilating the soil to an elastic medium, and damping effects are modeled as delayed soil reactions to the foundation movements. These assumptions result in simplified equations of motion, which can be iteratively integrated in the time domain. The article presents an application to the main breakwater of La Coruña, Punta Langosteira Port (Spain). A total of 752 simulations are carried out, in which both structure and soil characteristics are varied. Moreover, three loading types are considered, including permanent, sinusoidal, and impulsive chronograms.

3.1.2. Large Scale Experiments on Wave Loadings at Vertical Seawalls with Curved Outer Profiles

The study of wave loadings acting on the outer face of monolithic breakwaters and seawalls drew scientists' attention for many years, especially in the second half of the twentieth century. Stagonas et al. [18] concentrate on the effect of the wall curvature, which, despite having long been researched with respect to wave overtopping, is still a relatively unexplored facet of wave loading analysis. The authors discuss the results of regular wave large-scale physical model tests conducted on seawalls having an arch shape profile at their top. The angle in the center of the arch (α_e) varies between 48° and 90°; wave attacks are designed to induce both pulsating and impulsive pressures.

The research shows that the effect of α_e is negligible under quasi-static conditions while, in marked contrast, the mean of the maximum impulsive pressure and force peaks increases correspondingly by a factor of two. It is also shown that the deflection of the run-up wedge back to the incoming wave field has almost no effect on incident waves.

3.1.3. An Experimental Study on Dynamic Behavior of a Large Offshore Platform

Zhang et al. [19] performed physical model tests at the State Key Laboratory of Coastal and Offshore Engineering of Dalian University of Technology (China), with the purpose of investigating the dynamic response of a 10,000 ton offshore electrical platform.

First, the authors tackle the problem of scaling down the structural model, which is made of plexiglass; a hydroelastic similarity criterion is used, which guarantees both the hydraulic Froude condition and the similitude of elastic deformations.

The experiments are conducted in a basin, where irregular waves, currents, and wind (with three different angles of attack) are employed as climatic forcings. In addition, a set of regular wave tests are specifically conducted to gain a deeper insight into the platform hydrodynamic response.

Results highlight the role of wave slamming in amplifying both strains and accelerations and show that the maximum strain value under combined wave-current-wind conditions may be four times larger than that attained with waves alone.

3.2. Contributions on Functional Processes

Lo et al. [20] present a Navier–Stokes equations solver for two-phase flows, which uses the level set (LS) method [21] to track the interface between the two fluids and introduces a force density source term to account for the effects of immersed bodies. The force density term is then separately calculated via the immerse body (IB) concept [22]. The resulting Eulerian Cartesian/Lagrangian grid system does not require re-meshing procedures that involve coupled fluid-body interactions. The solver accuracy is checked against a number of cases, among which the most relevant to the scope of the present issue are those of wave propagation across a submerged barrier and the interaction of a floating body with a wave group. In these applications, the combined LS/IB approach is compared to the results of 2D laboratory experiments conducted by Beij and Battjes [23] and Hadzic et al. [24]. The comparisons appear encouraging.

3.2.1. Experimental Studies on Wave Overtopping

Ruol et al. [25] analyze the role of wave overtopping in the flooding of Piazza San Marco, Venice (Italy), which is one of the most stunning and architecturally valuable squares in the world. The authors reason that despite the Mo.S.E. (MOdulo Sperimentale Elettromeccanico) gates that protect the lagoon from flooding [www.mosevenezia.eu], Piazza San Marco could nonetheless be inundated, as waves generated by boats or SE winds may come over top of the Riva San Marco quay. Hence, to investigate this process, specific hydraulic model tests are carried out at the wave flume of the University of Padova (Italy). The experiments are necessary as the quay has a layout with a very mild slope that has been never been investigated before. Altogether, three different sections of the structure are considered under 10 random wave attacks and three water levels. Besides providing information on the risk of flooding at the specific area, the research adds to the database on wave overtopping with new accurate measurements that may aid both conceptual design and numerical modeling.

Formentin et al. [26] research the depth (h) and velocity (u) of the overtopping flow, which prove to affect a number of engineering phenomena, such as scour at the landward toe of breakwaters and wave overtopping propagation. In the article, impermeable trapezoidal dikes are considered, with a special focus on low-crested and submerged structures, which have been little researched thus far. Based on the analysis of 94 numerical flume random wave experiments and 60 2D hydraulic model tests, the dynamics of the overtopping flow along the dike crest are investigated in depth. Moreover, two new formulations for predicting h and u at the seaward edge of the structure crown are proposed. The new formulae keep the mathematical structure of the existing predictive tools [27], but modify the dependence on wave run-up and crest freeboard; moreover, new coefficients are introduced, which are functions of the dike offshore slope.

The article by Hernàndez-Fontes et al. [28] tackles the problem of the occurrence of "green water" events on a horizontal deck, which consist of a compact mass of water overtopping the structure, due, for example, to large wave–ship relative movements [29]. The authors carry out physical model experiments in a 1.95m long and 0.47m wide tank, where isolated bores are generated via a "wet dam–break" approach. This approach creates moving hydraulic jumps in the facility by removing a gate that separates two different water volumes. The experiments have a duration of nearly 3 s, and bore characteristics and green water flow are analyzed through a high-speed digital camera. The authors provide details on the air pocket trapped between the overtopping bore and the horizontal deck (air cavity), establishing an effective analogy with the kinematics of "flip-through" waves, which have been extensively investigated for vertical face breakwaters [30]. The time evolution of water elevation at the structure is also analyzed, and the backflow that follows the "green water" events is accurately studied.

3.2.2. Morphodynamic Response: Medium-Term Shoreline Analysis at a "Highly Structured" Coast

Buccino et al. [31] analyze the evolution of the Molise coast (South Italy) in the period 2004–2016. Using the linear regression rate concept [32] and numerical simulations, the authors assess the impact of a number of structural systems for shore erosion control, most of which include submerged/low-crested breakwaters [33,34] and groins [35]. The authors find out that most of erosion areas result from the existence of a net NW–SE littoral drift, which is blocked by the structures. Moreover, they show that, in spite of the inherent bimodality of the local wave climate, the shoreline evolution trend can be reasonably reproduced numerically via a single "equivalent" wave component from 10° N. However, it is argued that bimodality might have enhanced the erosion process, especially where structural measures alternate with undefended shoreline segments. This is because the reaches of coast between the structures would tend to experience erosion regardless of the wave direction.

3.2.3. Mechanical Response: Oscillation of a Ballasted Cylinder and Movements of Moored Vessels

Gabl et al. [36] notice that understanding the motion of water-filled bodies under wave excitation is of great interest to many fields of engineering, such as the design of wave energy converters (WECs) and ships transporting liquefied natural gas (LNG). However, they point out that most of the existing experimental studies consider only small motions, while the little research including the effect of large movements (mainly of a numerical nature) is conducted on specific (ship) geometries.

With the purpose of filling this gap, the article discusses the results of physical model tests carried out at the wave basin of the University of Edinburgh (UK). The experiments are conducted on a simple open-topped hollow cylinder, which is ballasted with either water or solid material to isolate and better comprehend the role of sloshing. Using constant amplitude sinusoidal waves with a frequency varying between 0.3 and 1.1 Hz, the authors find impacting waves with up to 20° pitch angles, whereas roll motion is generally small.

Nevertheless, under a specific frequency band, a sudden rotation switch from pitch to roll has been observed; this unexpected behavior has encouraged Gable et al. to conduct supplementary tests, the results of which are discussed in [37]. Here, variables that may affect the rotation switch are investigated, including position of the cylinder within the test tank, observation time, wave amplitude, wave direction, and mooring system.

The rationale of Sande et al. [38] might be well summarized by the maxim reported at the beginning of the article: "the better the vessels in port, the greater the economic returns". The authors argue that the "degree of comfort" of moored vessels depends to a large extent on the movements they experience during working operations. In addition, they claim that neither physical nor numerical models allow for the accurate reproduction of the effects on the cargo configuration of a continuous change in loading conditions.

Thus, a new method is developed, which relates moored vessels' degrees of freedom to ship dimensions and climatic forecast data. The approach uses simple transfer functions based on multivariate linear regression analysis and has been successfully validated against 27 vessels (15 bulk carrier and 12 general cargo) moored at the facilities of the outer port of Punta Langosteira, La Coruña (Spain).

Funding: This research received no external funding.

Acknowledgments: The editors wish to gratefully acknowledge all the researchers who contributed to this Special Issue as well as the reviewers who provided stimulating suggestions to improve the quality of the manuscripts.

Conflicts of Interest: The authors declare no conflict of interest.

References

1. Goda, Y. Overview on the applications of random wave concept in coastal engineering. *Proc. Jpn. Acad. Ser. B Phys. Biol. Sci.* **2008**, *84*, 374–385. [CrossRef] [PubMed]
2. Stokes, G.G. On the theory of oscillatory waves. *Trans. Camb. Philos. Soc.* **1847**, *VIII*, 197–237.
3. Pelnard-Considère, R. Essai de Theorie de l'Evolution des Forms de Rivages en Plage de Sable et de Galets. *Journées de l'hydraulique* **1959**, *4*, 289–298.
4. Penney, W.G.; Price, A.T.; Price, A.T. The diffraction theory of sea waves and the shelter afforded by breakwaters. *Philos. Trans. R. Soc.* **1952**, *244*, 236–253. [CrossRef]
5. Mei, C.C.; Black, J.L. Scattering of surface waves by rectangular obstacles in waters of finite depth. *J. Fluid Mech.* **1969**, *38*, 374–385. [CrossRef]
6. Hughes, S.A. *Physical Models and Laboratory Techniques in Coastal Engineering*; World Scientific Publishing: Singapore, November 1993; p. 558.
7. D'Angremond, K.; van Roode, F.; Verhagen, H.J. *Breakwaters and Closure Dams*, 2nd ed.; VSSD: Delft, The Netherlands, 2008; p. 347.
8. Hudson, R.Y. Laboratory investigation of rubble-mound breakwaters. *J. Waterw. Harb. Div. Asce.* **1959**, *2171*, 611–635.
9. Iribarren, R. A formula for the calculation of rock fill dykes (Translation of "Una Fórmula Para El Cálculo De Los Diques De Escollera" Rev. De Obras Publica. 1938). *Bull. Beach Eros. Board* **1949**, *3*, 1–15.
10. Kamath, A. CFD Based Investigation of Wave-Structure Interaction and Hydrodynamics of an Oscillating Water Column Device. Ph.D. Thesis, Norwegian University of Science and Technology, Trondheim, Norway, December 2015.
11. Miquel, A.M.; Kamath, A.; Chella, M.A.; Archetti, R.; Bihs, H. Analysis of different methods for wave generation and absorption in a CFD-based numerical wave tank. *J. Mar. Sci. Eng.* **2018**, *6*, 73. [CrossRef]
12. Buccino, M.; Daliri, M.; Dentale, F.; Di Leo, A.; Calabrese, M. CFD experiments on a low crested sloping top caisson breakwater. Part 1. nature of loadings and global stability. *J. Ocean Eng.* **2019**, *182*, 259–282. [CrossRef]
13. Gotoh, H.; Khayyer, A. On the state-of-the-art of particle methods for coastal and ocean engineering, Coastal Engineering Journal. *Coast. Eng. J.* **2017**, *60*, 79–103. [CrossRef]
14. Meringolo, D.D.; Aristodemo, F.; Veltri, P. SPH numerical modeling of wave–perforated breakwater interaction. *Coast. Eng.* **2015**, *101*, 48–68. [CrossRef]
15. Eldrup, M.R.; Andersen, T.L.; Burcharth, H.F. Stability of Rubble Mound Breakwaters—A Study of the Notional Permeability Factor, Based on Physical Model Tests. *Water* **2019**, *11*, 934. [CrossRef]
16. Van der Meer, J.W. Conceptual Design of Rubble Mound Breakwaters. In Proceedings of the 23rd International Conference of Engineering, Venice, Italy, 4–9 October 1992.
17. Maciñeira, E.; Peña, E.; Sande, J.; Figuero, A. Dynamic Calculation of Breakwater Crown Walls under Wave Action: Influence of Soil Mechanics and Shape of the Loading State. *Water* **2019**, *11*, 1149. [CrossRef]
18. Stagonas, D.; Ravindar, R.; Sriram, V.; Schimmels, S. Experimental Evidence of the Influence of Recurves on Wave Loads at Vertical Seawalls. *Water* **2020**, *12*, 889. [CrossRef]
19. Zhang, D.-L.; Bi, C.-W.; Wu, G.-Y.; Zhao, S.-X.; Dong, G.-H. Laboratory Experimental Investigation on the Hydrodynamic Responses of an Extra-Large Electrical Platform in Wave and Storm Conditions. *Water* **2019**, *11*, 2042. [CrossRef]
20. Lo, D.-C.; Wang, K.-H.; Hsu, T.-W. Two-Dimensional Free-Surface Flow Modeling for Wave-Structure Interactions and Induced Motions of Floating Bodies. *Water* **2020**, *12*, 543. [CrossRef]
21. Osher, S.; Fedkiw, R.P. *Level Set Methods and Dynamic Implicit Surfaces*; Springer: Berlin, Germany, 2002.
22. Peskin, C.S. Flow patterns around heart valves: A numerical method. *J. Comput. Phys.* **1972**, *10*, 252–271. [CrossRef]
23. Beji, S.; Battjes, J.A. Experimental investigation of wave propagation over a bar. *Coast. Eng.* **1993**, *19*, 151–162. [CrossRef]
24. Hadzic, I.; Hennig, J.; Peric, M.; Xing-Kaeding, Y. Computation of flow induced motion of floating bodies. *Appl. Math. Model.* **2005**, *29*, 1196–1210. [CrossRef]

25. Ruol, P.; Favaretto, C.; Volpato, M.; Martinelli, L. Flooding of Piazza San Marco (Venice): Physical Model Tests to Evaluate the Overtopping Discharge. *Water* **2020**, *12*, 427. [CrossRef]

26. Formentin, S.M.; Gaeta, M.G.; Palma, G.; Zanuttigh, B.; Guerrero, M. Flow Depths and Velocities across a Smooth Dike Crest. *Water* **2019**, *11*, 2197. [CrossRef]

27. Van Gent, M.R. Wave overtopping events at dikes. In Proceedings of the 28th ICCE 2002, Wales, UK, 7–12 July 2002; Volume 2, pp. 2203–2215.

28. Hernández-Fontes, J.V.; Esperança, P.T.T.; Graniel, J.F.B.; Sphaier, S.H.; Silva, R. Green Water on A Fixed Structure Due to Incident Bores: Guidelines and Database for Model Validations Regarding Flow Evolution. *Water* **2019**, *11*, 2584. [CrossRef]

29. Greco, M. *A Two-Dimensional Study of Green-Water Loading*; Norwegian University of Science and Technology: Trondheim, Norway, 2001.

30. Hattori, M.; Arami, A. Impact breaking wave pressures on vertical walls. In Proceedings of the 23rd Conference on Coastal Engineering, Venice, Italy, 4–9 October 1992; ASCE: Reston, VA, USA, 1992; pp. 1785–1799.

31. Buccino, M.; Di Paola, G.; Ciccaglione, M.C.; Del Giudice, G.; Rosskopf, C.M. A Medium-Term Study of Molise Coast Evolution Based on the One-Line Equation and "Equivalent Wave" Concept. *Water* **2020**, *12*, 2831. [CrossRef]

32. Douglas, B.C.; Crowell, M. Long-term shoreline position prediction and error propagation. *J. Coast. Res.* **2000**, *16*, 145–152.

33. Van der Meer, J.W.; Briganti, R.; Zanuttigh, B.; Wang, B. Wave transmission and reflection at low-crested. structures: Design formulae, oblique wave attack and spectral change. *Coast. Eng.* **2005**, *52*, 915–929. [CrossRef]

34. Buccino, M.; del Vita, I.; Calabrese, M. Engineering modeling of wave transmission of reef balls. *J. Waterway Port. Coast. Ocean Eng.* **2014**, *140*, 04014010. [CrossRef]

35. Kraus, N.C.; Rankin, K.L. *Functioning and Design of Coastal Groins: The Interaction of Groins and the Beach Processes and Planning*; Coastal Education and Research Foundation: West Palm Beach, FL, USA, 2004.

36. Gabl, R.; Davey, T.; Nixon, E.; Steynor, J.; Ingram, D.M. Comparison of a Floating Cylinder with Solid and Water Ballast. *Water* **2019**, *11*, 2487. [CrossRef]

37. Gabl, R.; Davey, T.; Ingram, D.M. Roll Motion of a Water Filled Floating Cylinder—Additional Experimental Verification. *Water* **2020**, *12*, 2219. [CrossRef]

38. Sande, J.; Figuero, A.; Tarrío-Saavedra, J.; Peña, E.; Alvarellos, A.; Rabuñal, J.R. Application of an Analytic Methodology to Estimate the Movements of Moored Vessels Based on Forecast Data. *Water* **2019**, *11*, 1841. [CrossRef]

Publisher's Note: MDPI stays neutral with regard to jurisdictional claims in published maps and institutional affiliations.

 *water*

Article

A Medium-Term Study of Molise Coast Evolution Based on the One-Line Equation and "Equivalent Wave" Concept

Mariano Buccino [1,*], Gianluigi Di Paola [2], Margherita C. Ciccaglione [1], Giuseppe Del Giudice [1] and Carmen M. Rosskopf [2]

[1] Department of Civil and Environmental Engineering, University of Naples Federico II Via Claufio 21, 80125 Naples, Italy; margheritacarmen.ciccaglione@unina.it (M.C.C.); Giuseppe.delgiudice@unina.it (G.D.G.)

[2] Department of Biosciences and Territory, University of Molise, Contrada Fonte Lappone, 86090 Pesche (IS), Italy; gianluigi.dipaola@unimol.it (G.D.P.); rosskopf@unimol.it (C.M.R.)

* Correspondence: buccino@unina.it

Received: 18 June 2020; Accepted: 5 October 2020; Published: 12 October 2020

Abstract: The Molise region (southern Italy) fronts the Adriatic Sea for nearly 36 km and has been suffering from erosion since the mid-20th century. In this article, an in-depth analysis has been conducted in the time-frame 2004–2016, with the purpose of discussing the most recent shoreline evolution trends and individuating the climate forcings that best correlate with them. The results of the study show that an intense erosion process took place between 2011 and 2016, both at the northern and southern parts of the coast. This shoreline retreat is at a large extent a downdrift effect of hard protection systems. Both the direct observation of the coast and numerical simulations, performed with the software GENESIS, indicate that the shoreline response is significantly influenced by wave attacks from approximately 10° N; however, the bimodality that characterizes the Molise coast wave climate may have played an important role in the beach dynamics, especially where structural systems alternate to unprotected shore segments.

Keywords: shoreline evolution; littoral drift; equivalent wave; one-line equation; coastal defenses; structure response

---

1. Nature of the Problem

The study of shoreline change, and the prediction of its future development, are essential for integrated coastal zone management. Nowadays, many shorelines of the world are suffering from a deficit of sand, which may lead to a gradual or fast coastline retreat; of course, this represents a leading concern to coastal scientists and planners.

In principles, beach erosion may result from either natural or anthropogenic causes; natural changes are generally difficult to interpret, and can be often attributed to a simple long-term fluctuation of the littoral system, as well as to the rise of the sea level [1]. On the other hand, anthropogenic causes are easily individuated, and include a reduction in the natural sand supply due to up river reservoir construction, and the presence of ports or navigational entrances, the structures of which (e.g., jetties) tend to interfere with longshore sediment transport.

Various approaches exist for beach stabilization, that can be broadly classified as structural (or hard) and non-structural (or soft). Purely non-structural approaches are limited to beach nourishment, which consists of introducing a given amount of good quality sand in the nearshore, so to counteract the misbalance between sediment inputs and outputs.

On the other hand, hard approaches include a number of variants, such as bank protection (essentially revetments and seawall), detached breakwaters and groins; structures interact with incoming waves via a number of complex hydrodynamic processes, including wave transmission, wave set-up, wave reflection and wave overtopping, which makes the prediction of shoreline response rather uncertain [2]. For this reason, all those processes have been intensively researched in recent decades, mostly through physical and numerical experiments [3–13]. However, a common feature of structural approaches is that they do not supply new sand to the littoral system, and accordingly if they succeed in locally widening the beach, they also inevitably induce erosion on the adjacent stretches of coast.

Thus, unsurprisingly, a number of literature studies document the negligible or deleterious effects of structural solutions in the long term; this especially occurs when structures are built in emergency conditions, as argued by Masria et al. [14] for the Mediterranean coasts. Among the other examples, Dean et al. [1] discussed the case of Palm Beach, Florida, where nearly 82,000 m^3 of sand were lost (over three years) after the placement of an experimental proprietary submerged breakwater of 1260 m in length.

The uncertainty related to structural solutions have sometimes led governments to give up on them, leaving beach nourishments as the sole alternative for beach erosion control. This is the case, for example, of the Romagna Riviera (Italy, North Adriatic Sea), where after having used breakwaters and groins with mixed success between late 1950s and late 1970s, the 1981 Regional Coastal Plan eventually prohibited the construction of new structures; since then, over one million cubic meters of sand has been used for repeated beach nourishment interventions [15].

2. Research Aims and Outline

This article deals with the case of the Molise coast, Southern Italy, which fronts the Adriatic Sea and is located nearly 300 km south of the Romagna Riviera.

In the frame of a collaboration between the University of Molise and the University of Napoli "Federico II", a shoreline change study was carried out, aiming to:

1.   Analyze the most recent trends of Molise coast evolution;
2.   Investigate the possible relationships between wave directions and shoreline response.

The reference time window is the period 2004–2016, which updates previous literature studies [16]; according to the definition originally introduced by Crowell et al. [17], the present analysis can be conventionally referred to as a "Mid-term study".

Differently from the approach followed in early research e.g., [16,18,19], in this paper attention was drawn to selected "hot spot" areas, and for each of them, the possibility of generating forcings are discussed. Then, a hypothesis was formulated about the existence of a dominant wave attack for littoral transport, which we will refer to as "equivalent wave" (EW), and its degree of correlation with the coastline evolution is examined.

In this regard, it is worth highlighting that the "Equivalent Wave Concept" is widely (and trustily) used in the field of practical coastal engineering, in spite of it lacking a firm theoretical basis. Thus, present research basically represents the first systematic attempt at investigating the explanatory power of the EW concept. In the following, this task is accomplished either numerically, employing the one line model GENESIS [15], or via physical considerations.

The results presented below apply to the whole Molise coast and are mainly of a qualitative nature; a quantitative comparison for a more restricted reach, is instead discussed in a further study submitted to this Special Issue [20].

The paper is organized as follows. After an overview of the Molise coast evolution and the related literature (Section 3), the GENESIS model is briefly reviewed in Section 4. Then, the wave climate in the study area is examined in Section 5 and the shoreline change process during the period 2004–2016 is illustrated in Section 6. Finally, the "equivalent wave" concept and its possible applications are dealt with in Section 7.

3. Study Area and Early Literature

The Molise coast extends for 36 km and is prevalently exposed to waves coming from the northern quadrants (Figure 1). The rocky cliff of Termoli promontory splits the coastline into two reaches. One stretches from the harbor of Marina Sveva to the promontory of Termoli; the second reach extends between the Termoli harbor and the Saccione stream mouth and includes the harbor of Marina di Santa Cristina (Figure 1).

Molise beaches are composed of medium sand (average diameter 0.26 mm) and are characterized by a very gentle foreshore (slope less than 1% between 0 and 10 m below the Mean Water Level, MWL). The average berm height is 2.0 m above the MWL [16,18].

Starting from the mid-20th century, a significant erosion has been taking place, which has been recognized to be caused to a large extent by the restrictions imposed to the river flows and the consequent decrease in sediment delivery to the coast [16,18,19]. In particular, the S. Salvo check-dam, built across the Trigno River during the period 1954–1977, and the Ponteliscione dam, built across the Biferno River between 1965 and 1977, have strongly contributed to channel incision and narrowing [21] and deprived the coastal sedimentary budget of significant volumes; De Vincenzo et al. [22,23] estimated that from 1965 to 2007, nearly 4.4×10^6 m^3 of sediments have accumulated within the Ponteliscione dam reservoir.

The response to progressive shoreline erosion was of a fully "structural" nature and nowadays nearly 62% of the coast is covered by hard protection measures; structures include mainly segmented systems of detached breakwaters, either submerged or emerged, and groins.

The evolution of the Molise coast has been intensively investigated in recent decades. Among the most relevant research, Aucelli et al. [24] analyzed the 1954–2000 shoreline changes of a 10 km long reach astride the Termoli promontory. The authors performed an accurate analysis of both wind and wave climates and concluded the main wave forcings to come from either the NW or NE. The maximum erosion risk was localized in the "densely structured" area neighboring the Biferno River mouth. Finally, based on the shoreline orientation of the examined coastal segment, a net NW to SE littoral drift was postulated.

Recently, Rosskopf et al. [16] extended the analysis to the whole Molise coastline and widened the reference time interval to 1954–2014. The coast was partitioned into nine physiographically homogeneous sub-reaches, the length of which ranged between 2 and 7 km; then, the average rate of shoreline change was calculated for the different time windows.

The authors warned that in spite of the substantial stability observed during the period 2004–2014, a new erosion process might be arising either at the northern or the southern part of the coast. Although a prevailing NW to SE littoral drift was inferred, a number of possible inversion areas were indicated, such as at the Trigno river mouth or right at the south of the Termoli harbor. Finally, a more in-depth knowledge of wave forcing was auspicated for future research work.

4. Overview of GENESIS Model

GENESIS (Generalized Model for Simulating Shoreline Changes) is a numerical modeling system, simulating shoreline evolution by spatial and temporal differences in longshore sand transport [25]. Despite being known worldwide, some of its characteristics are outlined below to facilitate the comprehension of Section 6.

GENESIS integrates the one-line contour equation (OCL), which expresses the conservation of sand volume; the leading hypotheses adopted are that the beach profile moves in parallel to itself and littoral transport is included between the berm height, D_B, and the depth of the closure, D_c [26]. The OCL reads:

$$\frac{\partial y}{\partial t} = \frac{1}{D_B + D_c} \frac{\partial Q}{\partial t} = 0 \tag{1}$$

where beside the already introduced quantities:

- $y(x,t)$ is the cross-shore coordinate of the shoreline at the alongshore position x and time t;
- Q is the littoral drift rate.

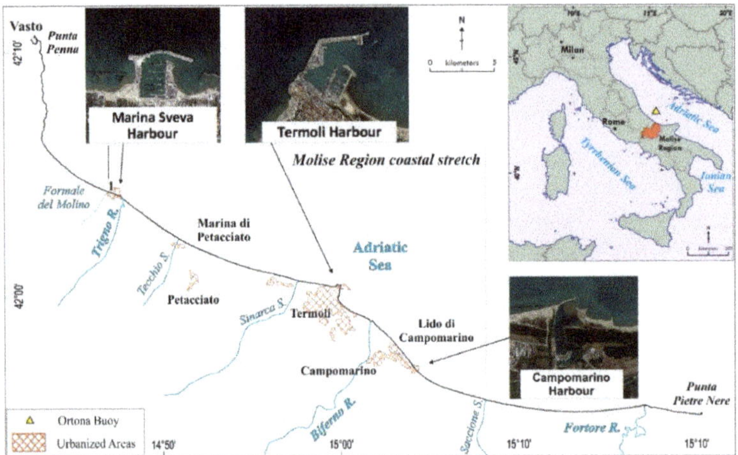

Figure 1. Molise coast with the main harbor location.

As is widely known, the beach change process is influenced by many environmental factors; these include wave propagation and breaking, nearshore currents activated by the release of wave momentum in the surf-zone and, finally, sediment transport. However, in the model the horizontal circulation that actually moves sediments is not simulated; the littoral drift rate is then empirically related to wave and sand characteristics, according to the formula:

$$Q = \left(H_s^2 \cdot c_g\right)_b \cdot \left[\frac{K_1}{C_1} \cdot \sin 2\theta + \frac{K_2}{C_2 tan\beta} \cdot \frac{\partial H_s}{\partial x}\right]_b \tag{2}$$

in which:

- H_s is the significant wave height;
- c_g is the group celerity;
- The subscript "*b*" denotes incipient breaking conditions;
- θ is the angle between the wave front and the shore;
- $tan\beta$ is the average beach slope between the shoreline, down to a depth $D = 2H$;
- K_1 and K_2 are transport coefficients;
- C_1 and C_2 are coefficients related to the sediment properties.

The first term at the right-hand side of Equation (2) corresponds to the well known Coastal Engineering Research Center (CERC) formula [27], and accounts for longshore sand transport produced by obliquely incident breaking waves. The second term simulates the effect of an alongshore gradient in breaking wave height, which may take place for a considerable length of beach in the vicinity of diffractive structures.

Breakwaters (both transmissive and non-transmissive), jetties and groins are easily implemented in GENESIS, but only structures located seaward the breaker-line are assumed to produce diffraction. Diffractive structures segment the calculation domain into *energy windows*, penetrable by waves; accordingly, the shoreline is divided into several "*sand transport calculation domains*", communicating with each other via sand bypassing and transmission.

As far as lateral boundary conditions are concerned, three constraint types can be used, and namely:

- *"Pinned"* if a portion of the beach is assumed to not move appreciably in time;
- *"Gate"* if the movement of sand alongshore is interrupted, partially or completely;
- *"Moving"* if the modeler specifies the rate of shoreline change on an end of the calculation grid.

5. Wave Climate Analysis

Data acquired by a directional wave buoy located in the offshore waters fronting the town of Ortona, 56 km north the mouth of Trigno river, were used to infer the wave climate at Molise coast (Figure 2). The device was anchored at a depth of 70 m below the low tide level, at a latitude of 42°24′54.0″ N and a longitude of 14°30′20.99″ E.

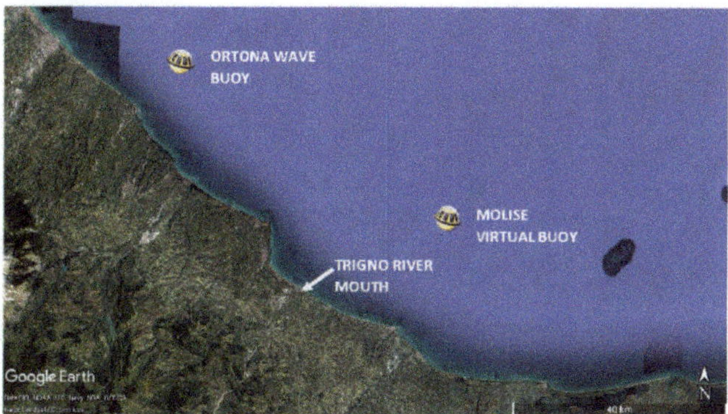

Figure 2. Ortona wave buoy and "virtual" wave buoy at the Molise coast.

Significant wave height, H_s, peak period, T_p, and azimuth of the mean wave direction, α, were recorded during the period 1989–2012, at an average interval of 3 h. Wave heights and periods were adjusted to the sea offshore the Molise coast ("virtual buoy" in Figure 2), according to the relationships originally introduced by Hasselmann [28] in the frame of the JONSWAP project. Assuming the same wind to blow at both the "real" and "virtual" buoy location, it is readily obtained that [29]:

$$\frac{H_{s,\,VIRT}}{H_{s,\,REAL}} = \left(\frac{F_{VIRT}}{F_{REAL}}\right)^{\frac{1}{2}}$$

$$\frac{T_{p,\,VIRT}}{T_{p,\,REAL}} = \left(\frac{F_{VIRT}}{F_{REAL}}\right)^{\frac{1}{3}}$$

(3)

where F indicates the "effective fetch" [30]. Note that, consistently with Hasselmann [28], the mean direction of the waves was assumed to coincide with that of the wind.

The histogram of the wave direction for angular sectors of 22.5° N is displayed in Figure 3, where the offshore directed waves have been removed for the sake of clearness.

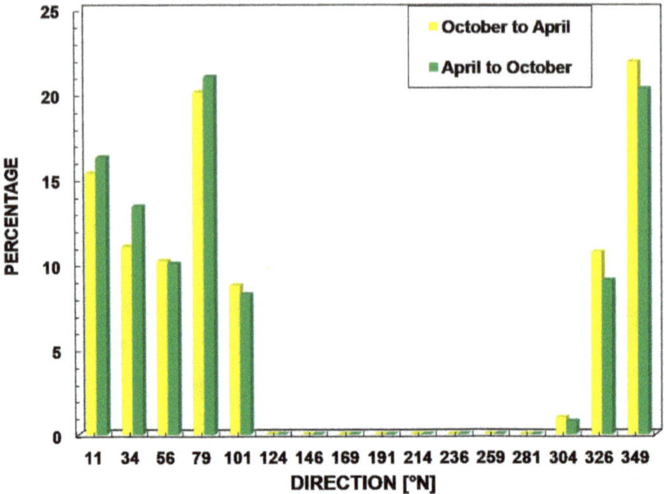

Figure 3. Histogram of the wave directions. Onshore directed angles are removed.

The graph exhibits two well defined modes, which were shown to be seasonally invariant; one is around 350° N and coincides with the NNW dominant direction indicated by Aucelli et al. [24]; the other is close to 80° N, whereas the authors indicated 23° N.

The reason for this apparent inconsistency is because waves from ENE-E have a lower height; hence, they bring forth less energy. This is shown in Figure 4, where along with the mean wave height, the 90th percentile of the directional wave height distribution is also reported. The graph indicates that the highest waves tend to come from 0°–45° N, in accordance with Aucelli et al. [24], with a maximum in the sector 0°–22.5° N.

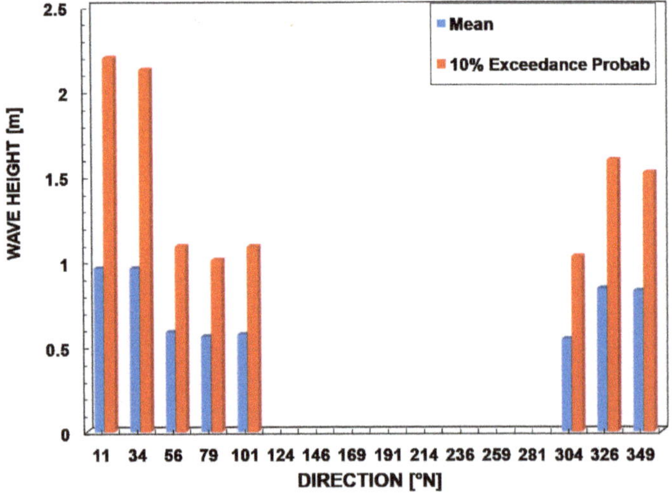

Figure 4. Directional distribution of mean and 90th percentile wave height.

The significant wave height was exceeded by 12 h a year (H_E), and has been estimated for each angular sector after fitting a two-parameter Weibull distribution to the wave data. An average value of

4.08 m was obtained keeping only the onshore directed angles; this value was used for the depth of the closure calculation ($D_c \approx 2\,H_E$) according to Hallermeier [26].

The joint H_s-T_p distribution (e.g., Figure 5) revealed a small wave height (less than 1 m) to have a modal peak period included between 3 and 6 s, whereas for the waves between 1.5 and 4.5 m, the modal T_p progressively moves to the interval 6–9 s. Finally, the period of the largest waves (larger than 4.5 m) is invariably included between 9 and 12 s. The overall mean peak period (direction independent) equals 5.08 s.

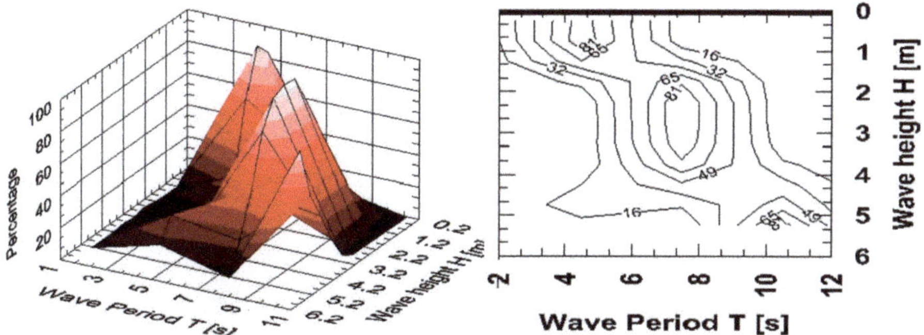

Figure 5. Joint distribution (H_s-T_p) for the angular sector 0–22.5° N. Left panel: 3D distribution. Right panel: contour lines.

6. Analysis of Shoreline Change

A shoreline change analysis was carried out, by comparing the Molise coastline position in 2004, 2011, 2014, and 2016. As mentioned above, according to Crowell et al. [17], the present study can be considered a "medium-term" investigation.

6.1. Approach

As shown in Table 1, data come from the digitalization, in the ArcGis environment, of photograph reliefs from different sources.

Table 1. Summary of the shoreline data.

Date	Source	Reference	Scale	Root Mean Square (RMS) Error (m)
2004	Ortophoto map	Rosskopf et al. [1]	1:2500	2
2011	Ortophoto map	Rosskopf et al. [1]	1:2500	2
2014	Google Earth	Rosskopf et al. [1]	1:500	1
2016	Google Earth	New data	1:500	1

The shoreline location has been assumed to coincide with the instantaneous waterline, which is reasonable in virtue of the low tidal environment [16]. Rough data has been linearly interpolated, and finally re-sampled at a 10 m interval.

Table 1 also reports the uncertainties related to the digitalization procedure (last column, RMS), being the other sources of inaccuracy, such as georeferencing, airphoto, etc., virtually absent; RMS values are consistent with those indicated by Hapke et al. [31] and Crowell et al. [17].

It is worth mentioning that digitalization uncertainties are essentially generated by human factors and measurement routine; accordingly, they can be considered as "chance" or "stochastic" errors, to be treated in the frame of the classical theory of errors [32,33].

Thus, each shoreline measurement $y(x,t)$ can be thought as the sum of the "true" shoreline position, plus a random component, which we will model as a Gaussian random variable with zero mean and a standard deviation equal to RMS (Table 1).

Moreover, since each digitalization has been performed independently of the others, the random components can be considered as mutually independent from a statistical point of view.

The *linear regression rate (LRR)* has been used as an indicator of the rate of shoreline change; it represents the slope of a least-square straight-line, fitted through the shoreline positions at the various available times. As argued in Douglas and Crowell [34] and Maiti and Bhattacharya [35], the use of least square fitting significantly reduces either the effect of random errors or that associated with other cyclical factors, such as tidal fluctuations.

However, differently from previous literature works, here the following procedure has been adopted:

a. The LRR at a given horizontal axis, x, is calculated after gathering all the data falling within a centered window of 40 m width. The window, is then progressively moved forward;

b. The obtained slope, say s_x, is tested for significance at a 95% probability level, according to the well established linear regression theory (see [36]);

c. Whenever the test is not satisfied, *LRR* is finally set to zero.

Note that the point c. implies:

$$LRR \neq 0 \; only \; if : Prob\,[LRR = 0] \leq 0.05 \tag{4}$$

The use of a moving window allows collecting a larger number of points, making the statistical test less unstable. Moreover, the procedure keeps holding its meaning (at least up to certain extent), even when comparing two shorelines only.

Finally, it is worth highlighting that the 95% probability level at the previous point b, has been chosen with the purpose of also including the uncertainties related to the interpolation/resampling procedure.

From the $LRR(x)$ function, accretion and erosion areas have been identified as those segments of coast where the shoreline change rate exceeds a certain limit value, say v_{LIM}, and remains above it for a minimum length, l_{LIM}.

For the scope of the regional analysis here presented, l_{LIM} has been set to 500 m, whereas v_{LIM} has been conveniently related to the uncertainties of the measurement process. The approach here employed is similar to that suggested, among the others, by Hapke et al. [31].

After invoking the gaussianity and independency of the random errors related to each shoreline measurement, it follows, from the theory of probability, that the error related to the difference between the two shoreline measurements at different times, say $y(x, t_2) - y(x, t_1)$, is in turn a Gaussian random variable, with zero mean and variance [36]:

$$VAR[y(t_2) - y(t_1)] = (RMS_2)^2 + (RMS_1)^2 \tag{5}$$

Consequently, we have that at a 95% probability level the error is included in the interval [36]:

$$E_{95} \cong \mp 1.96 \cdot \sqrt{(RMS_2)^2 + (RMS_1)^2} \tag{6}$$

Thus, if the "true" position of the shoreline remains unvaried in the interval $t_2 - t_1$, measurement errors may generate a fictional rate of change, which is included in the interval:

$$v_{E95} \cong \mp 1.96 \cdot \frac{\sqrt{(RMS_2)^2 + (RMS_1)^2}}{t_2 - t_1} \tag{7}$$

with 95% probability.

Hence, it seemed reasonable to assume $v_{LIM} \equiv v_{E95}$, that is to say that a shoreline segment is subject to a significant evolution if its rate of change overcomes that possibly created by measurement errors.

Values of v_{LIM} for the intervals 2004–2016, 2004–2011, 2011–2016 and 2014–2016 are reported in Table 2.

Table 2. Values of v_{LIM} for different time windows.

Period	v_{LIM} (m/Year)
2004–2016	0.37
2004–2011	0.80
2011–2016	0.90
2014–2016	1.41

6.2. Results

Figure 6 shows the 2004–2016 LRR function, whereas Figure 7 displays the corresponding erosion/accretion areas. In both the graphs, the alongshore coordinate, x, is oriented from NW to SE; horizontal dashes indicate detached breakwaters, whereas vertical dashes represent groin fields. From the inspection of Figure 7, it is easily observed that:

1. The foremost erosion areas (E1 and E3) are located in the southern Trigno and Biferno rivers, with a maximum LRR of −8 m/y and −9 m/y, respectively. In both cases, the retreat occurred just southwards a groin field, which in the case of the Biferno river, is further protected by detached breakwaters;
2. The area neighboring the Marina of Santa Cristina harbor has accreted northwards (A4, max LRR +2.2 m/y), and eroded southwards (E4, max LRR −4 m/y); similarly to what was observed above, the erosion zone is located south a groin/breakwater system, which protects the coast for 1.5 km;
3. The area just north of the Saccione stream jetty has accreted by nearly 700 m, at a maximum rate of 1.8 m/y.

All the previous outcomes support the idea of a net NW to SE littoral drift.

As indicated in literature e.g., [16,24], the erosion process at the main river mouths (E1 and E3) has been triggered by a reduction in the sediment delivering caused by dam construction. Rosskopf et al., calculated, that in the period 1954–2014, an "average" annual retreat rate occurred in those areas of −2.69 (Trigno River mouth, segment S1 of the Rosskopf et al., paper) and −2.90 (Biferno river mouth, segment S7), respectively.

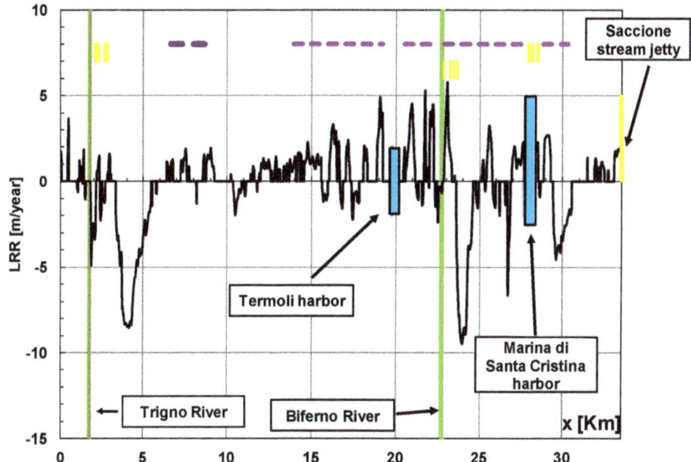

Figure 6. Linear regression rate (LRR) function during the period 2004–2016. Horizontal dashes indicate the detached breakwaters; the vertical dashes indicate groin fields.

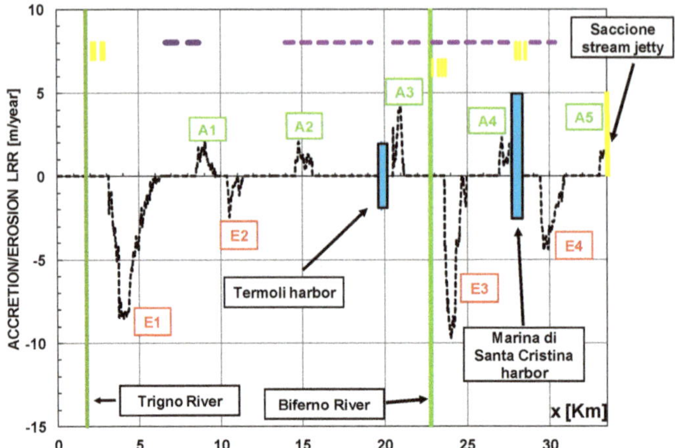

Figure 7. Erosion/accretion areas LRR function during the period 2004–2016. Horizontal dashes indicate the detached breakwaters; the vertical dashes indicate groin fields.

However, it is noteworthy that despite closing the whole waterfront, shore-parallel barriers at the Biferno river mouth did not stop the erosive trend, although mitigating it appreciably. Table 3 indicates that in this area, the average erosion rate has reduced by 50% compared to the period 1986–1998; in contrast to the neighborhood of the Trigno river mouth (S1, not defended by breakwaters) where the rate of retreat increased by 46%.

Table 3. Average LRRs at the Trigno and Biferno river mouths.

Strech	Av. LRR 1986–1998 (m/Year)	Av. LRR 2004–2016 (m/Year)
Trigno (S1)	−1.39	−2.04
Biferno (S7)	−2.31	−1.11

A dominant NW–SE sediment transport would also explain the origin of the accretion zone A3 (+3 m/y). The latter is located within a "densely structured area", where a series of detached breakwaters extends for 2 km on the north side of the Biferno river mouth; moreover, a good deal of coast is further defended by rock revetments.

As shown in Figure 8 (lower panel), the accumulation occurred simultaneously to the construction of a new basin of the Termoli harbor; the sediments, made available by dredging operations, were likely conveyed southwards by littoral currents and finally trapped within the structure system. Figure 8 also displays the north shoreline where the detached breakwaters remained substantially stable through the years, with a small accretion area right at the south of the harbor.

The dynamics from which the areas A2 (+2 m/y), A1 (+1.5 m/y) and E2 (−1.5 m/y) have originated are not readily explained. The accretion areas are located behind long systems of detached breakwaters, which are at a distance from one another of about 5.5 km; however, while A2 is situated at the northern edge of the barriers, A1 is formed at the southern one. Finally, the erosion zone E2 is located nearly 1.5 km southwards of the structures' end.

To have a deeper insight on these areas, the dominant direction of waves (if any) has to be considered, as well as the variability of shoreline orientation. The effects of both these variables are analyzed in the next sections.

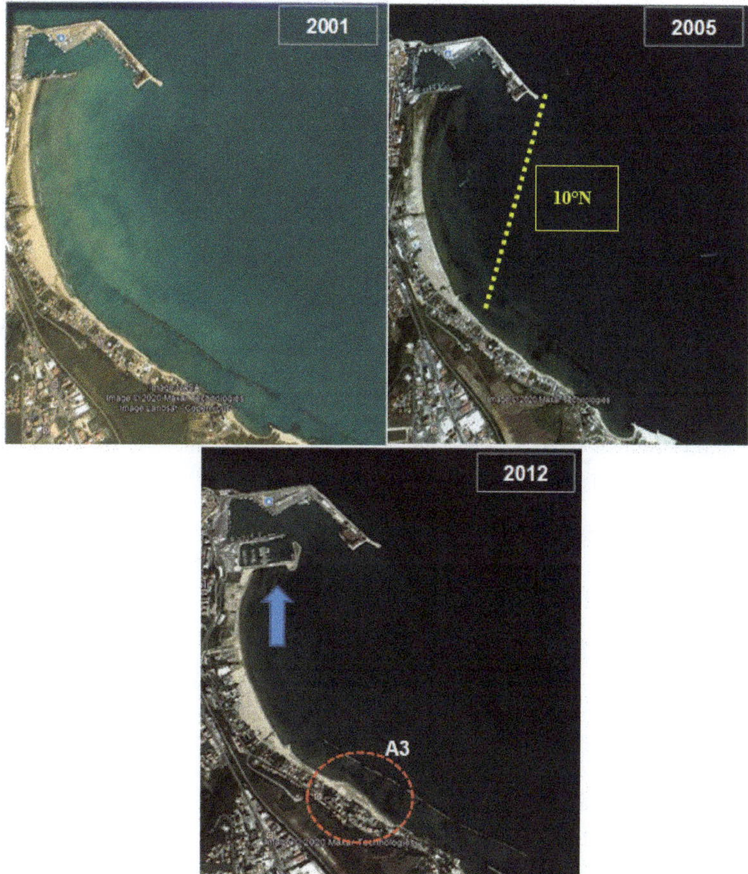

Figure 8. View of the area A3. Yellow dotted straight line represents sheltering by 10° N waves. (Section 5).

7. The Equivalent Wave Concept and Its Applications

According to Rosskopf et al. [16], a primary need for the comprehension of the Molise coast evolution is establishing more stringent relationships between the wave climate and shoreline response; at this first stage of research, this task is accomplished via the so-called "equivalent wave" (EW) concept, i.e., assuming that shoreline changes may be significantly correlated to a single component of the wave climate.

Although well accepted (and employed) in applied coastal engineering, it is noteworthy that scientists do not agree on the existence of the EW; however, it is surprising to observe how well established this idea might be, even among those researchers who tend to negate it.

Silvester [37] reasoned that "it is normal to correlate volumes of accretion taken over a year with some average swell condition for the same period", but concluded that "the selection of some meaningful average, including direction, is in the realms of fantasy". However, the author argued that the annual littoral transport was driven by swells that followed the most intense storms and recognized, in turn, that swells "arrive on a coast from persistent direction". This implicitly supports the idea that a dominant wave attack for shoreline evolution may exist.

On the other hand, Walton and Dean [38], in examining the directional distribution of littoral drift in many coastal areas, explicitly observed "it was surprisingly similar to that which would occur for a single wave component". Lately, the authors provided a mathematical explanation of the above finding [39], by adding up a number of sinusoidal components of littoral transport, according to the CERC formula [27].

Anyhow, even assuming such an EW exists, no universally accepted estimation procedure has been proposed to date.

In this study, the EW direction is inferred empirically from the observation of the shoreline trend, and its height and period are estimated via simple averaging operations. Then, several applications are discussed, which are mainly of qualitative nature.

Differently, in the second paper published in this Special Issue [8], the approach of the Littoral Drift Rose is employed [18], and a quantitative comparison is presented for the case of the Trigno river mouth.

Before discussing the results, it is, however, crucial to point out that the whole analysis performed here assumes that a clear shoreline trend exists, at the considered time scale. This was also explained by Hanson and Kraus [25], who argued that the essence of this hypothesis is that waves producing longshore sediment transport and boundary conditions (such as structures) are the main factors controlling the "steady part" of the beach change signal. This "steady signal" is then superimposed by a "noise component" associated with storms, seasonal changes in waves, tidal fluctuations, and other cyclical and random events, such as, potentially, hurricane and typhoon waves, although they are not present at the latitudes investigated here.

7.1. EW Direction and Parameters

According to Pelnard-Considere [40], shoreline orientation at any "un-bypassed" groin is expected to be in every instant nearly normal to the dominant wave attack; hence, it may be taken as an indicator of the EW direction.

For the case under study, of particular interest are the accretion areas A4 and A5 of Figure 7. However, only little information can be drawn from the first, since the beach is protected by detached breakwaters that diffract the incoming waves (Figure 9a).

Conversely, A5 can freely adjust to the sea and consequently, may represent a more reliable indicator. Panels (b), (c) and (d) of Figure 9 show that this segment of coast has kept a nearly constant orientation in time, which is close to 10° N. Hence, under the hypothesis of straight-parallel bottom contours, the latter can be assumed as the offshore EW direction.

It is noted that 10° N is somehow halfway between the prevailing wave directions indicated by Aucelli et al. [6], i.e., 350° N and 23° N, and corresponds to the angular sector associated with the highest waves in Figure 4. This is consistent with the hypothesis formulated by Silvester [37] that shoreline may be formed by waves correlated to the most intense storms.

Once the EW direction has been established, its height, H_{m0E}, has been estimated as the average wave height in the directional sector 0–22.5° N ($H_{m0E} = 0.96$ m, Figure 3); on the other hand, the peak period, T_{pE}, was simply equal to the mean measured value, 5.08 s.

It is worth noticing that in the applications presented below, the EW characteristics will be assumed constant throughout the coastal area.

7.2. Shoreline Response to Structure Systems

The EW concept has been tentatively used to analyze the shoreline response behind some structure systems. As anticipated, the approach followed here is mainly of qualitative nature and relies on rather crude assumptions; accordingly, the obtained results have to be considered as preliminary.

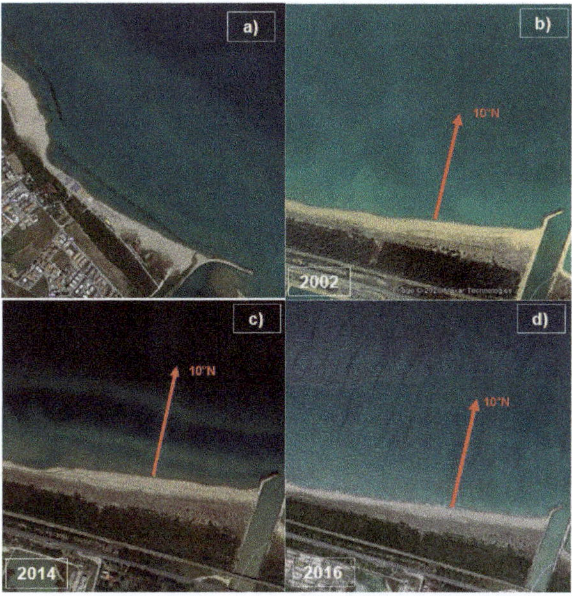

Figure 9. Panel (**a**) shoreline neighboring the area A4; panels (**b–d**), shoreline orientation at the Saccione stream jetty.

As a first point, it could be argued that under a 10° N wave attack, the head of the Termoli harbor breakwater completely shelters the shoreline south of it (upper-right panel of Figure 8), and this would explain the aforementioned stability of that reach of coast; then, the resulting diffraction currents, directed northwards, might be responsible for the accretion observed just below the harbor basin. The effect of diffraction is clearly recognizable from the curvature of the shoreline and is consistent with the inversion in the littoral drift direction observed by Rosskopf et al. [16].

Simplified models implemented in GENESIS were employed for the arrays of detached breakwaters neighboring the accretion area A2 and the erosion area E4 in Figure 7. Hereafter, these structure systems will be referred to as S_{A2} and S_{E4}, respectively. S_{A2} is located nearly 20 km north the Saccione stream jetty and extends for 2.1 km with an average azimuth of 13° N; the latter, is in fact very close to the EW direction. As shown in Figure 10, the shape of the coast behind the system has remained relatively constant through the years.

The breakwaters have variable length and are either emerged or submerged or "partially underwater", with alternating submerged and emerged parts; the typical gap width is 30 m. Additionally, two groins are located behind the barriers, the length of which is equal to 50 and 100 m, respectively.

The GENESIS model, pictured in Figure 11a, is made up on a 30 m wide grid, with the initial shoreline parallel to the barrier system; the distance between the structures and the initial shoreline (180–230 m) has been estimated north the breakwaters, from the first null-point of the LRR(x) function. "Pin points" have been used as lateral conditions at the model boundaries, which are located 2 km away from the breakwater ends. The transmission coefficients K_T, i.e., the ratio between the wave height right behind the barrier and that just in front of it [3–5], have been preliminarily set at 0.4 for emerged barriers and 0.7 for submerged and partially underwater structures. This is basically due to the lack of detailed information on the breakwaters cross section. The equivalent wave has been run for 50 years, to let the beach planform attain a stable shape.

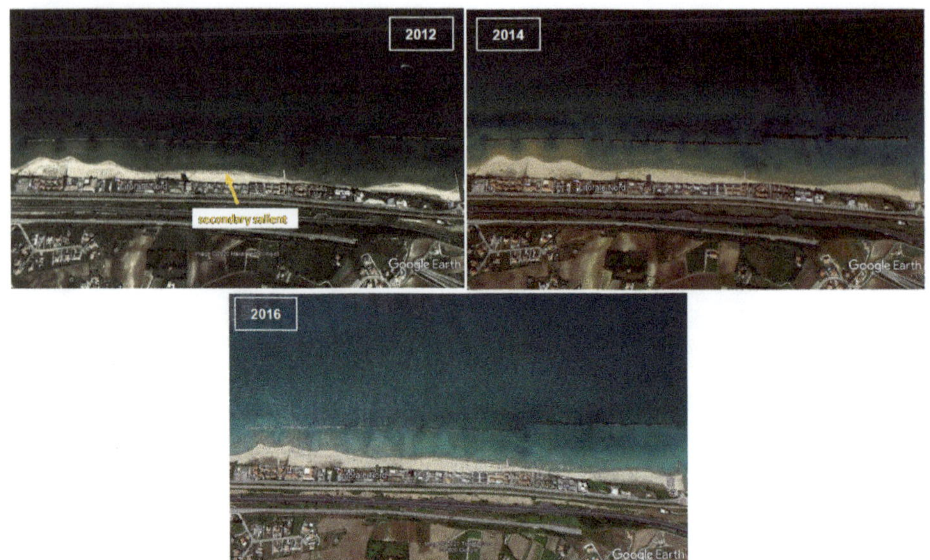

Figure 10. Shoreline planform at S_{A2}.

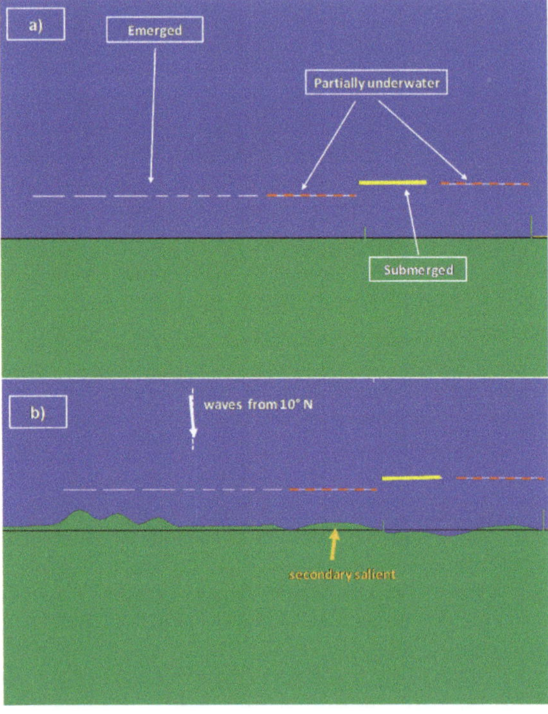

Figure 11. (**a**) S_{A2} as modeled in GENESIS; (**b**) shoreline response under EW ($K_1 = 0.1$, $K_2 = 0.15$).

In general, simulation results correlate quite well with the observed shoreline trend (Figure 11b; $K_1 = 0.1$, $K_2 = 0.15$); the triple humped salient is reasonably reproduced, although slightly tapered, and so is the coast between the groins.

The shoreline trend behind the first group of barriers is instead somewhat flattened; this is either due to the simplified initial shoreline condition or because GENESIS cannot simulate in detail the multiple diffraction emanating from the short breakwater segments. Moreover, since the partially underwater barrier at the right-end side of the first breakwater group is simulated as a continuous structure, the secondary salient (Figures 10 and 11b) is moved nearly 0.4 Km southwards compared to the observed position.

Interestingly, the simulated average rate of shoreline change around A2 is reasonably like the measured one (Figure 12).

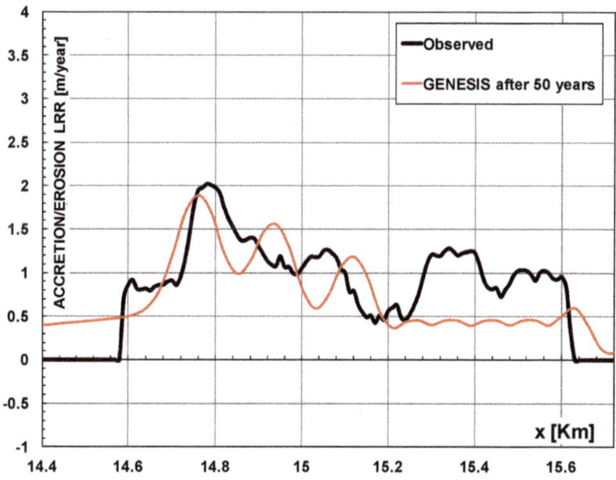

Figure 12. Observed vs. numerical shoreline change rate at A2.

To check the sensitivity of the solution to the wave direction, the EW angle has been shifted by 10° either to the north or to the south. As shown in Figure 13, waves from 0° N lead to a significant skewness of the main salient, whereas waves from 20° N do not capture the shoreline trend. Obviously, this result does not imply that those waves are ineffective to the shoreline evolution, but rather that their "explanatory power" is smaller compared to that of 10° N.

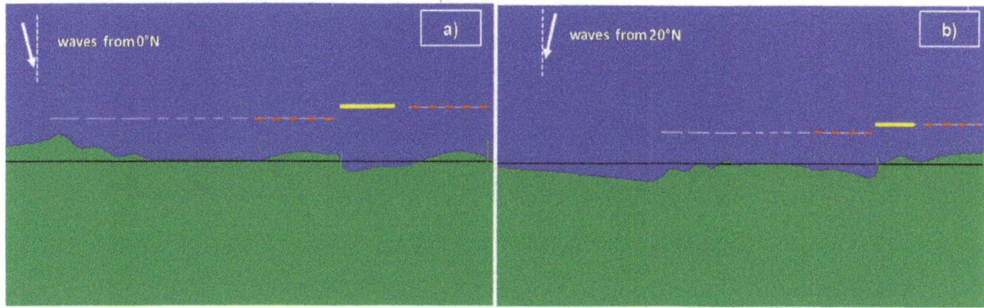

Figure 13. Shoreline response of S_{A2} to waves from 0° N (**a**) and 20° N (**b**). $K_1 = 0.1$, $K_2 = 0.15$.

As already mentioned, S_{E4} follows a field of groins located just southwards of the Marina of Santa Cristina harbor. The system, 3 km north the Saccione stream jetty (Figure 14), is made up of two parts.

The first includes nine short barrier segments (six emerged and three submerged), whereas the second, 450 m southwards, encompasses four submerged breakwaters. The structures are oriented at 25° N and extend for 1.5 km.

Figure 14. Shoreline planform at S_{E4}.

The GENESIS model used to study this area has the same characteristics as those described for S_{A2}; however, a 15 m-wide grid was used to better represent either the breakwater segments (30–75 m length) or the gaps (45 m wide). As far as the lateral boundary conditions are concerned, a "gate" has been imposed northwards to simulate the groin system; on the other hand a "pin" point has been set nearly at 0.7 km south to the breakwaters, at the center of a large area where LRR(x) is uniformly zero.

In this case, the EW seems to lead to a shoreline trend that is reasonably consistent with the observations; as shown in Figure 15 ($K_1 = K_2 = 0.1$), a wide salient is produced behind the first group of structures, followed by a nearly 1 km-long deficit area corresponding to E4. Furthermore, the latter is again realistically simulated in GENESIS (Figure 16).

Figure 17 suggests that a more southern wave attack does not reproduce the shoreline trend properly. However, it stresses that the areas located between the structure groups (breakwater segments or groins) tend to undergo erosion irrespectively of wave direction. For example, the area F1 may erode as "downdrift" effect of either of the breakwater series, depending on whether waves come from the north or the south. The same is valid for the area F2, which is located downdrift from the groin system under wave attacks coming from the north, and downdrift from the first group of breakwaters when the wave climate reverses.

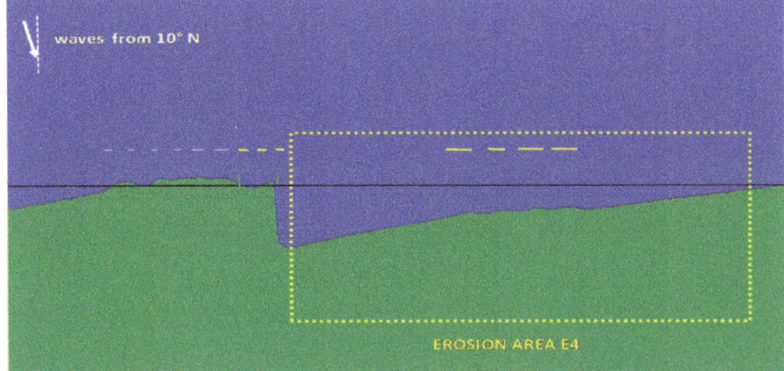

Figure 15. GENESIS model outputs at S_{E4} (K1 = K2 = 0.1).

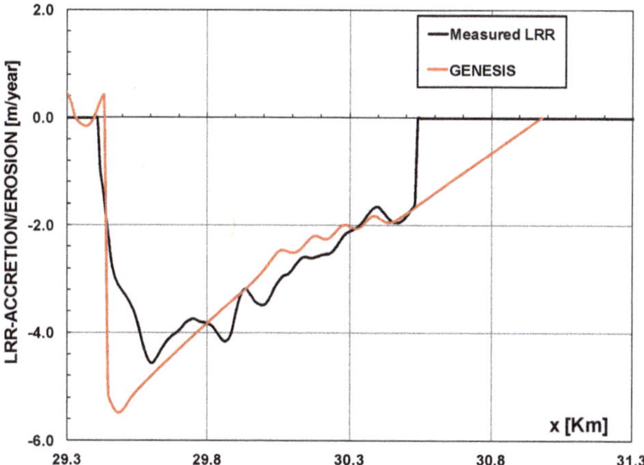

Figure 16. Measured vs. simulated shoreline change rate at E4.

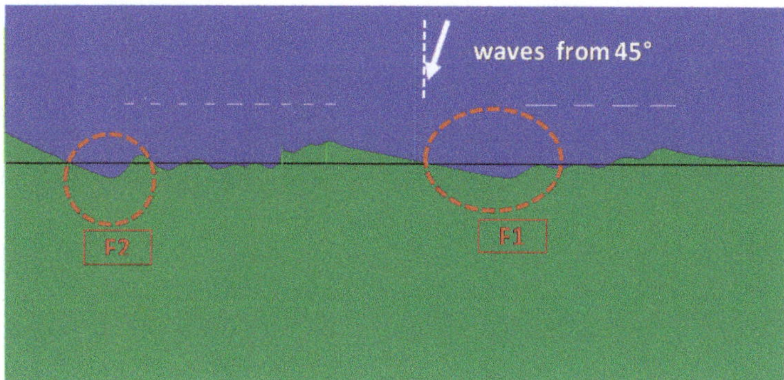

Figure 17. Shoreline response of S_{E4} to a 45° N wave attack.

This concept closely resembles the idea of "negative shoreline diffusivity" introduced by Ashton and Murray [41,42] and Walton and Dean [39]. Ashton and Murray showed that when waves break at a straight coast with large angles (more than 40° relative to the normal coastline), the alongshore sediment transport, rather than smoothing the irregularities in the rectilinear trend of the shoreline, may cause the growth of different types of naturally occurring coastal large-scale landforms, including capes, flying spits, and alongshore sand waves. On the other hand, Walton and Dean, extended the Ashton and Murray reasoning and proved that under a bimodal wave climate where most intense wave attacks come from directions significantly inclined (parallel to the shore in the limit), any natural or man-made perturbations tend to continue growing (cusps will accrete and holes erode).

7.3. Features of Recent Coastline Evolution

Besides the shoreline response at relatively restricted areas, the EW may be used to analyze some more general aspects of the recent evolution of the Molise coast.

To this purpose, it is useful to recall that according to a good number of literature studies (see [43]), the rate of littoral drift at a coast segment, Q, can be represented in the following mathematical form:

$$Q = Q_b \cdot \sin[2 \cdot (\beta - \alpha_b)] \tag{8}$$

in which:

- The module Q_b depends on both the breaking wave height and sediment characteristics;
- β is the azimuth of the normal coastal segment;
- α_b is the azimuth of the breaking wave angle;

For straight parallel bottom contours, Equation (7) can be reformulated in terms of offshore wave parameters, using the conservation of energy and Snell's law. For the particular case of the CERC formula, it has been shown that (e.g., [18]):

$$Q = Q_0 \cdot \sin[2 \cdot (\beta - \alpha_0)] \cdot \left\{ \frac{\cos[2 \cdot (\beta - \alpha_b)]}{\cos[2 \cdot (\beta - \alpha_0)]} \right\}^{0.2} \tag{9}$$

where the subscript "0" denotes deep-water conditions and the module Q_0 is a function of both the offshore wave height and period. Hence, assuming Q_0 to be constant and the cosine term in the braces nearly equal to 1, from Equation (1) it is readily obtained that:

$$\frac{\partial y}{\partial t} = -\frac{2\,Q_0}{D_B + D_C} \cdot \cos[2 \cdot (\beta - \alpha_0)] \cdot \frac{\partial \beta}{\partial x} \tag{10}$$

where x is the alongshore coordinate.

As the first factor on the right hand side of Equation (9) is inherently positive, the sign of the shoreline change (whether accretion or erosion) entirely depends on $\cos[2 \cdot (\beta - \alpha_0)]$ and $\frac{\partial \beta}{\partial x}$. It is, however, worth noticing that as long as Q_0 does not vary with x, no structure effect is being taken into account. The cosine term of Equation (9) is shown in Figure 18 for $\alpha_0 = 10°$ N. In the same graph, the β angle is also reported, which was obtained after passing the 2004 shoreline measurement through a Godin low-pass filter with a 1 km cut-length; this approach allows keeping the general trend of the coast only, removing any local or structure-induced effects.

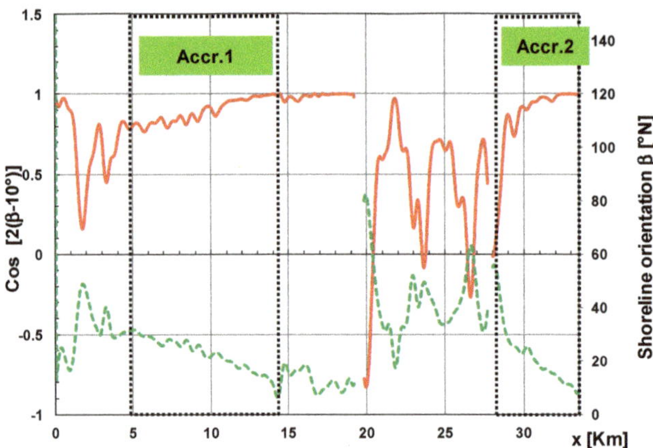

Figure 18. Cosine term of Equation (9) in red and shoreline orientation (in green).

The Figure shows two potential accretion zones, in which β exhibits a clear decreasing trend and $\cos[2 \cdot (\beta - \alpha_0)] > 0$; the former (Accr.1) extends for nearly 10 km south the Trigno river, whereas the latter (Accr.2) is located between the Marina of Santa Cristina harbor and the Saccione stream jetty. On the other hand, no clear trends have been recognized in the remaining part of the coast.

As displayed in Figure 19, the 2004–2011 LRR(x) function is consistent with the indications in Figure 20; shoreline accretion dominates within the "accretion windows", apart from the two reaches labelled as "R1", (corresponding to E1 of Figure 7), and "R2".

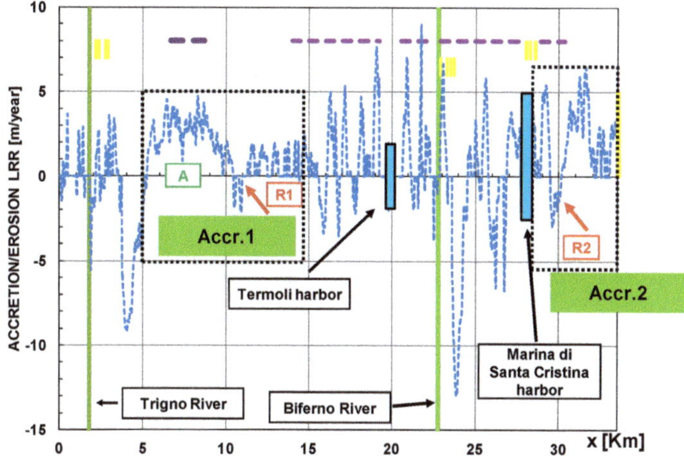

Figure 19. LRR function during the period 2004–2011.

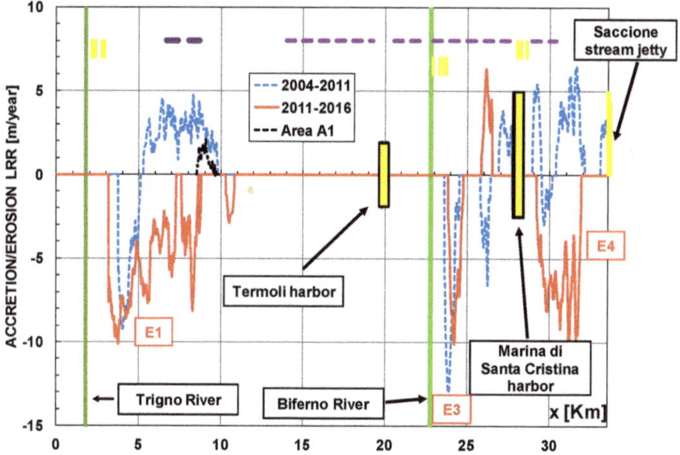

Figure 20. Erosion/accretion areas for the periods 2004–2011 and 2011–2016.

It is realistic to assume the wide accretion zone "A" to be created at a large extent by the natural orientation of the coast with respect to the incoming waves, with a minor effect due the submerged breakwaters protecting the area. This is basically because its length (about 5 km) is more than twice that of the barriers (2.1 km).

As far as the erosion areas are concerned, R2 is located in the main gap of S_{E4}, and is a well predictable "downdrift" effect of the first group of breakwaters in Figure 15. On contrary, R1 can be hardly considered as a breakwater-induced erosion since it is nearly 1.5 km south of the structures. More likely, this area results from a peculiar orientation of the shoreline or is the effect of the wave climate bimodality not considered in the present approach.

The situation depicted in Figure 19 has dramatically changed during the period 2011–2016. As shown in Figure 20, the erosion area just south the Trigno river has widely enlarged and the retreat created by S_{E4} has also significantly extended southwards.

In light of this, it can be concluded that the accretion area A1 of Figure 7 (see also Figure 20) represents the tail of a naturally accreting zone, not yet reached by the erosive front propagating from the Trigno river mouth.

This behavior can be reasonably explained as the downdrift effect of the structure systems (groins protecting the Trigno river and S_{E4}) in response to EW; however, the sudden and violent propagation of the shoreline retreat is somewhat surprising and deserves more research.

In this respect, it is noted that such a fast erosion process has occurred in zones where hard protection measures alternate to "free" coastal reaches, which may suggest it could have been enhanced by the aforementioned "negative shoreline diffusivity effect" associated with the wave climate reversal. However, no clear evidence of this is available at the present stage of knowledge.

8. Conclusions

In this paper, a medium-term shoreline change study of the Molise coast has been presented, with the purpose of:

1. Highlighting the most recent evolution trends and related shoreline change rates;
2. Establishing more stringent relationships between wave climate and shoreline response.

The analysis has been conducted using the LRR as indicator of the coastline evolution; however, a procedure based on the statistical testing of regression results has been adopted, which permits excluding points where the trend is weak or LRR values are influenced by the uncertainties related to data acquisition.

The location and characteristics of the main erosion/accretion areas indicate a dominant NW to SE longshore sediment transport; no hints of possible littoral drift inversion were found.

The present study also confirms the early literature finding that the most intense erosion occurred at the mouth of the main regional rivers Trigno and Biferno. In particular, it has been shown that although the area to the south of the Biferno river is entirely protected by detached breakwaters, shoreline retreat is still continuing at high rates, although the latter are reduced compared to the past.

Section 5 is dedicated to assessing if a dominant wave direction exists (which we call equivalent wave, EW) which could be capable of explaining the main trends of the coastal evolution process. Based on the shoreline development at the Saccione stream jetty, the direction 10° N was selected.

Numerical simulations conducted with the one-line model GENESIS have shown that the use of the "equivalent wave" as a stationary forcing, may lead to a reasonable prediction of shoreline response behind several structure systems, even under crude simplifications of the initial shape of the coast (which has been supposed to be simply straight) and structure characteristics (e.g., uniform transmission coefficient).

Additionally, the EW concept, together with physical reasoning, allowed interpreting the accretion of a significant part of the Molise shoreline between 2004 and 2011.

Finally, in the time interval of 2011–2016, a strong erosive trend was recognized, at both the northern and southern ends of the coast.

Despite that such a strong erosive trend can be reasonably explained as the "downdrift" effect of hard protection measures, the speed at which the erosion areas have propagated southwards is surprising and deserves more research.

In the paper, it is suggested that where rigid structures alternate to undefended shoreline segments, a situation similar to that of *"negative shoreline diffusivity"* may occur, which could lead to the instability of the shore. This idea, which also applies to the case of the area south from the Biferno river mouth should be accurately verified in future research work.

Author Contributions: M.B.: writing, numerical simulations, data processing, data analysis. G.D.P.: data collecting data processing, data analysis, writing. M.C.C.: numerical simulations, data analysis, writing. G.D.G.: writing and data analysis. C.M.R.: data collecting, data analysis, writing. All authors have read and agreed to the published version of the manuscript.

Funding: This research received no external funding.

Conflicts of Interest: The author declares no conflict of interest.

References

1. Dean, R.G.; Chen, R.; Browder, A.E. Full scale monitoring study of a submerged breakwater, Palm Beach, Florida, USA. *Coast. Eng.* **1997**, *29*, 291–315. [CrossRef]
2. Hanson, H.; Kraus, N.C. Shoreline response to a single transmissive detached breakwater. *Coast. Eng. Proc.* **1990**, *22*, 2034–2046.
3. Van der Meer, J.W.; Briganti, R.; Zanuttigh, B.; Wang, B. Wave transmission and reflection at low-crested structures: Design formulae, oblique wave attack and spectral change. *Coast. Eng.* **2005**, *52*, 915–929. [CrossRef]
4. Buccino, M.; Calabrese, M. Conceptual approach for prediction of wave transmission at low-crested breakwaters. *J. Waterway Port. Coast. Ocean. Eng.* **2007**, *133*, 213–224. [CrossRef]
5. Buccino, M.; del Vita, I.; Calabrese, M. Engineering modeling of wave transmission of reef balls. *J. Waterway Port. Coast. Ocean Eng.* **2014**, *140*. [CrossRef]
6. Calabrese, M.; Vicinanza, D.; Buccino, M. 2D Wave setup behind submerged breakwaters. *Ocean Eng.* **2008**, *35*, 1015–1028. [CrossRef]
7. Zanuttigh, B.; Van Der Meer, J.W. Wave reflection from coastal structures in design conditions. *Coast. Eng.* **2008**, *55*, 771–779. [CrossRef]
8. Buccino, M.; D'Anna, M.; Calabrese, M. A study of wave reflection based on the maximum wave momentum flux approach. *Coast. Eng. J.* **2018**, *60*, 1–21. [CrossRef]
9. Bruce, T.; van der Meer, J.W.; Franco, L.; Pearson, J.M. Overtopping performance of different armour units for rubble mound breakwaters. *Coast. Eng.* **2009**, *56*, 166–179. [CrossRef]
10. Van der Meer, J.W.; Allsop, N.W.H.; Bruce, T.; De Rouck, J.; Kortenhaus, A.; Pullen, T.; Schüttrumpf, H.; Troch, P.; Zanuttigh, B. EurOtop, 2018. Manual on Wave Overtopping of Sea Defences and Related Structures. An Overtopping Manual Largely Based on European Research, but for Worldwide Application. 2018. Available online: www.overtopping-manual.com (accessed on 20 October 2018).
11. Vicinanza, D.; Caceres, I.; Buccino, M.; Gironella, X.; Calabrese, M. Wave disturbance behind low crested structures: Diffraction and overtopping effects. *Coast. Eng.* **2009**, *56*, 1173–1185. [CrossRef]
12. Buccino, M.; Daliri, M.; Dentale, F.; Di Leo, A.; Calabrese, M. CFD experiments on a low crested sloping top caisson breakwater. Part 1. Nature of loadings and global stability. *Ocean. Eng.* **2019**, *182*, 259–282. [CrossRef]
13. Buccino, M.; Daliri, M.; Dentale, F.; Calabrese, M. CFD experiments on a low crested sloping top caisson breakwater. Part 2. Analysis of plume impact. *Ocean. Eng.* **2019**, *182*, 345–357. [CrossRef]
14. Masria, A.; Iskander, M.; Negm, A. Coastal protection measures, case study (Mediterranean zone, Egypt). *J. Coast. Conserv.* **2015**, *19*, 281–294. [CrossRef]
15. ARPA Emilia Romagna. *Studio e Proposte di Intervento per Migliorare L'assetto e La Gestione Dei Litorali di Misano e Riccione*; ARPA Emilia Romagna: Bologna, Italy, 2007. (In Italian)
16. Rosskopf, C.M.; Di Paola, G.; Atkinson, D.E.; Rodriguez, G.; Walker, I.J. Recent shoreline evolution and beach erosion along the central Adriatic coast of Italy: The case of Molise region. *J. Coast. Conserv.* **2018**, *22*, 879–895. [CrossRef]
17. Crowell, M.; Leatherman, S.P.; Buckley, M.K. Shoreline change rate analysis: Long term versus short term data. *Shore Beach* **1993**, *61*, 13–20.
18. Aucelli, P.P.C.; Iannantuono, E.; Rosskopf, C.M. Evoluzione recente e rischio di erosione della costa molisana (Italia meridionale). *Ital. J. Geosci.* **2009**, *128*, 759–771. (In Italian) [CrossRef]
19. Aucelli, P.P.C.; Di Paola, G.; Rizzo, A.; Rosskopf, C.M. Present day and future scenarios of coastal erosion and flooding processes along the Italian Adriatic coast: The case of Molise region. *Envrion. Earth Sci.* **2018**, *77*, 371. [CrossRef]
20. Ciccaglione, M.C.; Buccino, M.; Di Paola, G.; Calabrese, M. Trigno river mouth evolution via littoral drift rose application. *Water* **2020**, submitted.
21. De Vincenzo, A.; Molino, A.J.; Molino, B.; Scorpio, V. Reservoir rehabilitation: The new methodological approach of economic environmental defence. *Int. J. Sediment. Res.* **2017**, *32*, 88–94.
22. Scorpio, V.; Aucelli, P.P.C.; Giano, S.I.; Pisano, L.; Robustelli, G.; Rosskopf, C.M.; Schiattarella, M. Riber channel adjustments in Southern Italy over the past 150 years and implications for channel recovery. *Geomorphology* **2015**, *251*, 77–90. [CrossRef]

23. De Vincenzo, A.; Covelli, C.; Molino, A.; Pannone, M.; Ciccaglione, M.C.; Molino, B. Longterm management policies of reservoirs: Possible re-use of dredged sediments for coastal nourishment. *Water* **2019**, *11*, 15. [CrossRef]

24. Aucelli, P.P.C.; De Pippo, T.; Iannantuono, E.; Rosskopf, C.M. Caratterizzazione morfologico-dinamica e meteomarina della costa molisana nel settore compreso tra la foce del torrente Sinarca e Campomarino Lido (Italia meridionale). *Studi Costieri* **2007**, *13*, 75–92. (In Italian)

25. Hanson, H.; Kraus, N.C. GENESIS: A Generalized Shoreline Change Numerical Model for Engineering Use. Ph.D. Thesis, University of Lund, Lund, Sweden, 1989.

26. Hallermeier, W.A. Field data on seaward limit of profile change. *J. Waterw. Port. Coast* **1985**, *111*, 598–602.

27. US Army Corps of Engineers. *Shore Protection Manual*; Coastal Engineering Research Centre, Government Printing Office: Washington, DC, USA, 1984.

28. Hasselmann, K.; Dunckel, M.; Ewing, J.A. Directional wave spectra observed JONSWAP 1973. *J. Phys. Oceanogr.* **1980**, *10*, 1264–1280. [CrossRef]

29. Contini, P.; De Girolamo, P. Impatto morfologico di opere a mare: Casi di studio. In Proceedings of the VIII Conerence Italian Association Offshore and Marina, Forte Santa Teresa, Lerici (la Spezia), Italy, 28–29 May 1998.

30. Saville, T. *The Effect of Fetch Width on Wave Generation*; TM-70; US Army Corps of Engineers, Beach Erosion Board: Washington, DC, USA, 1952.

31. Hapke, C.J.; Himmelstoss, E.A.; Kratzmann, M.; List, J.H.; Thieler, E.R. *National Assessment 528 of Shoreline Change: Historical Shoreline Change along the New England and Mid-Atlantic 529 Coasts*; open-file report 2010-1118; U.S. Geological Survey: Reston, VA, USA, 2010; p. 57.

32. Edwards Demining, W.; Bridge, R.T. On the statistical theory of errors. *Rev. Mod. Phys.* **1934**, *6*, 280–281.

33. Fan, H. *Theory of Errors and Least Square Adjustment*; Royal Institute of Technology (KTH): Stockholm, Sweden, 2010; p. 226.

34. Douglas, B.C.; Crowell, M. Long-term shoreline position prediction and error propagation. *J. Coast. Res.* **2000**, *16*, 145–152.

35. Maiti, S.; Battacharya, A.K. Shoreline change analysis and its application to prediction: A remote sensing and statistics based approach. *Mar. Geol.* **2009**, *257*, 11–23. [CrossRef]

36. Draper, N.R.; Smith, H. *Applied Regression Analysis*; John Wiley and Sons: Hoboken, NJ, USA, 1985.

37. Silvester, R. Fluctuation in littoral drift. In Proceedings of the International Conference on Coastal Engineering, Houston, TX, USA, 3–7 September 1984; ASCE: Reston, VA, USA; pp. 1291–1305, Chapter 88.

38. Walton, T.L.; Dean, R.G. Application of littoral drift roses to coastal engineering problems. In Proceedings of the Conference on Engineering Dynamics in the Surf Zone, Sydney, Australia, 14–17 May 1973; Institution of Engineers: Sydney, VIC, Australia, 1973; pp. 221–227.

39. Walton, T.L.; Dean, R.G. Longshore sediment transport via littoral drift rose. *Ocean. Eng.* **2010**, *37*, 228–235. [CrossRef]

40. Pelnard-Considere, R. Essai de theorie de l'evolution des forms de rivages en plage de sable et de galets. In *4th Journees de l'Hydralique, les Energies de la Mer.*; Question III, Rapport No. 1; Société Hydrotechnique de France: Paris, France, 1956; pp. 289–298. (In French)

41. Ashton, A.D.; Murray, A.B. High-angle wave instability and emergent shoreline shapes: 1. Modeling of sand waves, flying spits, and capes. *J. Geophys. Res.* **2006**, *111*, F04011. [CrossRef]

42. Ashton, A.D.; Murray, A.B. High-angle wave instability and emergent shoreline shapes: 2. Wave climate analysis and comparisons to nature. *J. Geophys. Res.* **2006**, *111*, F04012. [CrossRef]

43. Van Rijn, L.C. A simple general expression for longshore transport of sand, gravel and shingle. *Coast. Eng.* **2014**, *90*, 23–39. [CrossRef]

Publisher's Note: MDPI stays neutral with regard to jurisdictional claims in published maps and institutional affiliations.

Article

Roll Motion of a Water Filled Floating Cylinder—Additional Experimental Verification

Roman Gabl [1,2,*], **Thomas Davey** [1] and **David M. Ingram** [1]

[1] School of Engineering, Institute for Energy Systems, FloWave Ocean Energy Research Facility, The University of Edinburgh, Max Born Crescent, Edinburgh EH9 3BF, UK; tom.davey@flowave.ed.ac.uk (T.D.); David.Ingram@ed.ac.uk (D.M.I.)

[2] Unit of Hydraulic Engineering, University of Innsbruck, Technikerstraße 13, 6020 Innsbruck, Austria

* Correspondence: roman.gabl@ed.ac.uk

Received: 3 July 2020; Accepted: 3 August 2020; Published: 6 August 2020

Abstract: Understanding the behaviour of water filled bodies is important from an applied engineering perspective when understanding the sea-keeping performance of certain floating platforms and vessels. Even by assuming that the deformation is negligible small in relation to the motion of the structure, these fluid-structure-fluid interactions are challenging to model, both physically and numerically, and there is a notable lack of reference data sets and studies to support the validation of this work. Most of the existing information is highly specific to certain hulls forms, or is limited to small motions. A previous study addressed this by modelling a floating cylinder (giving a more generic case) with roll and pitch motions in excess of 20°. The presented experiment expands on that work to further investigate the previously observed switch between pitch and roll in the cylinder under wave action as induced by the sloshing of the internal water volume. An additional experimental investigation, focused on a single draft, was conducted to test open research questions from the previous study. Here we show that the roll response of the water filled cylinder is repeatable, independent of the tank position and wave amplitude, provided the observation time is long enough to capture the fully developed motion response of the floating object. The mooring system used comprised four soft lines connected on two points on the cylinder. This arrangement resulted in slightly different restoring forces in different wave directions. A relative change of the wave direction by 90° led to a larger wave frequency band in which the roll motion occurred. These cases were, again, also conducted with the solid ballast. Both sets of data provide an interesting validation case for future work on water ballast inside a floating object.

Keywords: floating cylinder; water filled; motion capturing; wave tank; wave gauges; free surface; sloshing; validation experiment

---

1. Introduction

Fluid-Structure-Fluid-Interactions (i.e., the behaviour of fluids within a floating body) influence the behaviour of a number of floating platforms, vessels and devices (e.g., certain wave energy converters) under the assumption of a negligible small deformation in relation to the large motions. Motivated by the paucity of studies and data relating to generic vessel shapes with large motions, which in turn limited numerical model validation, a previous study by Gabl et al. [1] examined a water filled floating cylinder. This experimental investigation explored four different drafts and two different ballast options, namely water and solid, under regular waves with a variable wave frequency, with the full dataset available for modelling validation purposes [2]. Large motions in the range of up to 20° were observed and allowed the investigation of the influence of the sloshing of the inside water on the motion of the structure. It could be shown that the water filled cylinder showed a switch from pitch to roll (rotation around the axis in the wave direction) for a particular frequency band. This could not be observed in the solid ballast option and is consequently concluded to be caused, or at least enhanced, by the motion of the inner water body. The work presented here extends the work from the previous investigation, but with a narrower scope. The open research questions (RQ) focus on the position of the apparatus in the test tank, the importance of the observation time, sensitivity to wave amplitude, and sensitivity to wave direction, as well as the influence of the mooring system. These RQ are further discussed in Section 3.1. The observations from the previous experimental programme are used to narrow down the experimental scope in terms of cylinder draft and ballast variables, allowing this study to concentrate to the interesting behaviours observed in pitch/roll response switching and its connection to outlined research questions.

Fossen and Nijmeimer [3] define parametric resonance as a system under external excitation including a time-varying parameter. This can lead to oscillations larger than caused by resonance, under which the acting force varies at the same frequency as the natural frequency of the system. Such a behaviour can be observed in ships, for which it is called parametric rolling. It can also occur for ships sailing against waves with a length close to the total length of the vessel, which can lead to a wave excitation frequency near twice the natural roll frequency. A change of the direction as well as the speed can reduce this effect significantly as well as adequate roll damping. A number of particularly severe incidents involving container ships and roll motions of up to 40° intensified the research efforts in this particular field [3,4], but smaller roll angles will also have a significant influence on vessel usability and their detailed design [5,6]. A wide range of numerical models are suppose to predict the roll motion correctly [7–10] along with experimental investigations which provide case-specific validation data [11,12]. Zhou et al. [13] and Zhou [14] showed good agreement between experimental investigations and a hybrid model, which connects a 3D-numerical calculation of the roll damping coefficient and in a second step the solution of the 3 Degree of Freedom (DoF) motion of the ship. Specific investigations comparing liquid and frozen ballast showed that there is no significant influence of the roll response [15,16].

The cylindrical geometry used in this investigation is most akin to a cylindrical platform [17,18], but it not intended to model any specific prototype. It is instead intended to provide a generic geometry which removes the influence of specific hull forms and ensures the data is more usable in numerical modelling development and validation (at which point hull form influences can be reintroduced). The hull form of ships has a significant influence on the behaviour but it is far more complex than the presented geometry and often impractical for early stage modelling implementation. Tarrant and Meskell [19] investigated parametric rolling for point absorbers deployed as wave energy converter (WEC) with two relative oscillating bodies (Wavebob). They describe this phenomena as a coupling of two DoF when wave frequency is twice of the natural frequency. The fluid structure interaction in the time domain was supported by WAMIT calculations and validated with a 1:17 scaled model. Radhakrishnan et al. [20] presented a study of a spherical buoy moored with a single line in three different depth. The sway motion caused by the instability was depending on the relative position to the still water position and could be observed in the frequency range of 1.5 to 2.25 with a peak around double the natural period.

As observed in the previous floating cylinder, this roll motion only occurs with the water ballast option [1] (i.e., not when a solid internal ballast was utilised). Hence, it can be assumed that this instability is caused, or at least intensified, by the sloshing of the inner water body. Sloshing effects are widely studied using numerical and experimental approaches and have many different applications. For example, Jiang et al. [21] studies the free surface inside a rectangular tank introduced by a big jet. More common is the usage of an oscillation machine [22] or 6-DoF sloshing platforms [23]. Those experimental investigations are used as a validation experiment for numerical simulation and, in particular, the meshless Smoothed Particle Hydrodynamics (SPH) method which has the potential to combine big motions of floating structures with sloshing [24–29]. The geometries include cylindrical and rectangular forms [30] but also more complex geometries such as fuel tanks [31] and other large liquid tanks [32]. A successful method to reduce the run-up as well as pressure peaks at the walls is the addition of additional structures inside the tank. This can be fixed inner structures [33–37] as well as potentially flexible elements [38]. Such a reduction of the impact of the sloshing is a next step planned for the floating cylinder.

This paper expands the previous experimental investigation and focuses on the mooring system to identify the potential influences on modelling assumptions. Therefore, different pre-tensions, wave tank deployment locations, as well as wave directions are investigated and discussed. An additional data set, which includes the wave direction of 180°, is provided via Edinburgh DataShare [39].

2. Methodology

2.1. Experimental Set-up

The experiment investigated a floating cylinder filled with water, and a similar solid ballast option, in the wave tank. It was identical to the system used for Gabl et al. [1,2,40] and extends the previously published results to provide a deeper understanding of the influences (e.g., mooring) on the interesting pitch/roll behaviour and other responses. The initial investigations compared the two ballast options with four different drafts under regular waves. The main goal was to identify the influence of the sloshing of the inner water body under large motions.

This experiment uses a single draft d (the second lowest from the previous study) to further investigate the switch between the pitch and roll response, which only occurred for the water ballast configuration. A draft of d equal 227.9 mm was chosen as it provides the clearest motions in the previous studies. The corresponding inner water level h was 170.7 mm and the filled cylinder with an outer diameter D of 500 mm had a total weight including ballast of 44.65 kg. The cylinder is open at the top but even the most extreme motion did not result in spilling or overtopping of the cylinder walls. Figure 1 presents an overview of the experimental set-up, which was investigated in the FloWave Ocean Energy Research Facility [41,42]. FloWave is a unique wave and current testing facility providing a full 360° capability for flow and waves. 168 wave makers are located in a circle (diameter of 25 m) and are used to generate as well as absorb waves. The upper volume has a constant water depth of 2 m and is separated by a 1 m thick floor (a centre section can be brought up to allow dry installation) with a similar lower volume. Both volumes are connected at the outside with 28 flow-drive units, which area also arranged in a circle. Currents of up to 1.6 m/s can be generated in any direction independently of the waves. Typical applications are the investigation of tidal turbines [43], novel velocity measurement approaches [44,45] as well as remotely operated vehicles [46]. For the presented investigation no current was used and the different wave directions were investigated without changing or rotating the experimental set-up in the tank.

Station keeping of the floating cylinder was achieved by four soft mooring lines. Two of them are connected to one point on the cylinder at the water line, so that the connection point is always at the height of the still water surface. This allows symmetry along the main axes in the tank but rotations around the y-axis were minimally influenced due to the low mooring stiffness in this degree of freedom (Figure 2). The previous experimental runs primarily investigated waves in the positive x-direction (90° in the tank definition; Figure 1) and this set-up influences the dominant pitch rotation to a lesser extent. The comparison of the solid and water ballast showed that the sloshing of the inner water led to a significant roll response of the water filled body. This paper aims to provide further insight into this effect and, when combined with the two available data sets [2,39] from the experiments in Gabl et al. [1], can be used as a validation experiments for numerical simulations focusing on Fluid-Structure-Fluid interaction with negligibly small deformations under big motions.

Figure 1. Experimental set-up in the tank on the raised tank floor (**a**) including the global coordinate system and the labelled mooring lines—(**b**) solid ballast case under waves coming from 180°—(**c**) side view of the experiments including the wave gauge (WG) array in the back—(**d**) overview of the full experimental setting in the tank.

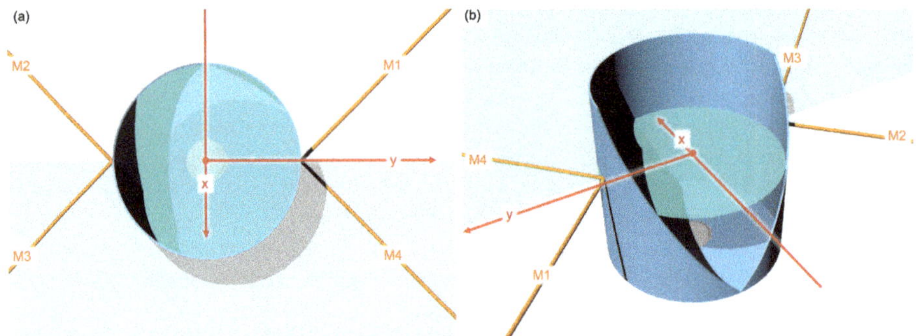

Figure 2. Rendered view of the cylinder from top (**a**) and from the side (**b**)—coordinate system moved due to better visibility.

2.2. Instrumentation

Two measurement instrumentation systems were used: (a) motion capturing system (MoCAP) and (b) wave gauges (WG). Synchronising these systems allowed the six Degree of Freedom (DoF) motion response of the floating cylinder as well as the free surface elevation to be documented. This system was similar to the previous experimental investigation [1,2].

The origin of the global coordinate system used by the MoCAP was located in the centre of the circular tank. A right-handed coordinate system was defined with a vertical z-axis. The x-axis was aligned with the main wave direction. As shown in Figure 1, this wave direction is defined as 90° in the tank definition. Therefore, the tank is split into two halves with one side covering 0 to 180° and the other one with a negative sign. Both boundaries are identical. The eight Qualisys cameras (Göteborg, Sweden) of the upper water MoCAP were located at different heights on the tank side in an arc arrangement over approximately ±60°. Each camera captured the motion of the six markers mounted on the floating cylinder and the tracking is conducted with the software Qualisys Track

Manager (QTM, version 2019.3, Qualisys, Göteborg, Sweden). A rigid body was defined based on those markers and the 6-DoF motion calculated. The local body coordinate system used the same orientation as the global one and it was ensured that the origin is in the height of the still water surface of the tank. Regular refinement calibrations ensured a very high accuracy of the system, which was typical smaller 1 mm.

In the previous investigation, seven wave gauges (WG) were installed to cover a wide range in the x-direction of the tank. This provided an overview of the waves in front as well as behind the floating cylinder but no simultaneous values. For the current experiments, five WG were installed on the movable gantry in a reflection array with a constant offset in y-direction. The spacings between the WG were defined by a Golomb ruler with an order of 5 (marks (11, 9, 4, 1, 0); base length of 1 m; Table 1). Figure 1c shows the highlighted WG in the tank and the numbering in the main wave direction. The WG4 is located at $x = 0$ m, which is the initial position of the cylinder, unless otherwise stated. A regular 5 point calibration covering ±100 mm was conducted to ensure the high accuracy of this instrumentation, which is typically smaller than 1 mm [40,47].

Table 1. Location of the wave gauges (WG) in relation to the global coordinate system. The origin is in the centre of the tank and the spacing is defined by a Golomb ruler—outside diameter of the cylinder D equal 0.5 m.

WGNr	WG1	WG2	WG3	WG4	WG5
x (m)	−0.82	−0.73	−0.45	0.00	0.18
y (m)			$-1.50 = 3{\cdot}D$		

The MoCAP used a sampling frequency of 128 Hz and the WG used 32 Hz. A digital pulse provided by the tank was used to synchronise both systems. In the majority of tests and if not otherwise stated, the capture time was 180 s and the periodic repeat time 128 s. The first 52 s covers the ramp-up phase, which was not included in the following presented results.

2.3. Experimental Conditions

All experiments were conducted with regular waves having a single wave direction. FloWave can provide a full 360° capacity for waves as well as for current but for this investigation, only wave directions between ±90 and 180° were used. Waves coming from the other half of the tank would have moved the floating cylinder in the direction of the WG, which could have led to a collision. It is assumed that the experimental setup is symmetric along the x-axis (Figure 2).

A constant requested wave amplitude a_W (input value for the wave makers) of 50 mm was used (unless otherwise stated) and the requested wave frequency f_W varied between 0.3 to 1.0 Hz. A Fast Fourier Transform (FFT) analysis was conducted for each WG and the maximum value of the amplitude spectrum was identified. The motion data were similarly analysed. Figure 3 presents a summary of the comparison of the requested and measured WG values for the investigated cases. The differences Δa and Δf were calculated by subtracting the requested value from the measured amplitude. Additional auxiliary lines are included to show representative percentages of the requested wave amplitude and frequency. Beside the five individual results for the WG, mean values over all WG are presented in Figure 3. This mean value is later used to calculate the RAO.

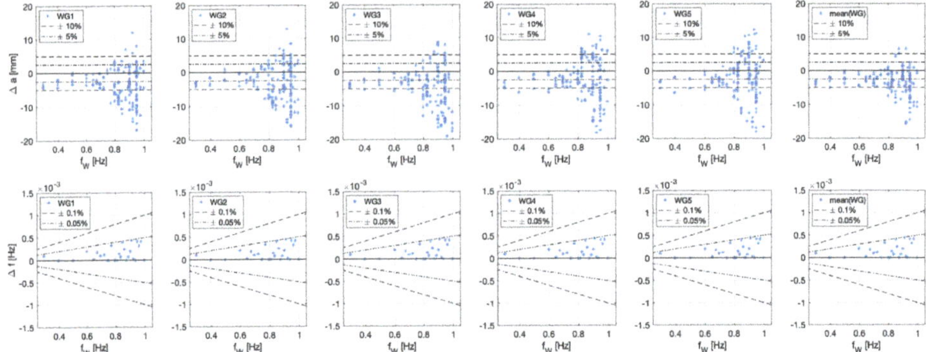

Figure 3. Analysis of the five individual wave gauge (WG) data, as well as the mean value conducted for the wave amplitude (**upper row**) and frequency (**lower row**)—difference $\Delta a = a_R - a_W$ (measured amplitude minus the requested amplitude); similar for the frequency.

The analysis of the wave amplitudes show that for the higher frequencies the spreading of the results increased. FloWave typically delivers reduced wave amplitudes [2,46], which are corrected as part of open tank calibration tests. The observed higher values are likely caused by reflections coming from the cylinder. This effect is smoothed by averaging over the five WG and these averaged values are used for the normalisation. The discrepancies in the wave frequency are very small ($<5 \times 10^{-4}$ Hz) and in the following sections the requested wave frequency f_W is used.

3. Results

3.1. Overview

The presented research addresses four research questions (RQ): position of the cylinder in the tank (RQ1), length of the observation time (RQ2), sensitivity of the wave amplitude (RQ3) and wave direction (RQ4). In the first step, the sensitivity of the position of the floating object under wave conditions was investigated to ensure that the roll response was not caused by a very unlikely but potential local hotspot of reflection in the tank. This first RQ is investigated in Section 3.2, which also includes a change of the pre-tension of the mooring lines. Section 3.3 presents the results of the test investigating the influence of the length of acquisition time (RQ2). A sensitivity study was conducted for the wave amplitude and the results of RQ3 can be found in Section 3.4. The fourth RQ utilises the 360° wave capacity of FloWave and different wave directions are investigated (Section 3.5). The main focus was the 180° wave direction case, which was orthogonal to the previous experiments. For this research question, the solid ballast option was also investigated to provide a direct comparison between the two options for this additional wave directions.

All the presented investigations are limited to one draft of 227.9 mm (determined from the previous experimental programme [1,2]) with an inner water level h of 170.7 mm and a total weight of 44.65 kg. Unless otherwise mentioned, the presented experiments were conducted with the water filled ballast option, a requested wave amplitude a_W of 50 mm and under a wave direction of 90° based on the tank definition (Figure 1).

3.2. Position in the Tank

The first research questions focuses on the location of the floating cylinder in the wave tank, which has a diameter of 25 m. In the previous investigation the initial position in still water was always the tank centre and the regular waves pushed the cylinder in the positive x-direction. Large motions in surge were observed caused by the very soft mooring system. As a first step of the presented investigation the frequency sweep was repeated to test how repeatable the results are. The main focus was on the occurrence of the roll motion. In the following figures the graphs labelled with *reference* uses the published data based on Gabl et al. [1,2] and *repetition* marks the similar experimental set-up conducted as part of this additional investigation.

The four individual mooring lines consisted of a hollow elastic of 3 m long (diameter 3 mm) with a very high stretch factor connected directly to the cylinder at two points, which could be adjusted so that it was always at the height of the still water surface of the tank. Each elastic was expanded with standard non-stretch rope to the tank side, which had a fixed length of 6.5 m. A small pre-tension was added to avoid the possibility of slack lines. For the *pre-tension* cases the rope was reduced by approximately 3 m for all mooring lines. Hence, the initial position stayed the same but the pre-tension in the mooring lines was increased. By releasing two mooring lines, namely M3 and M4 (Figures 1 and 2), the initial position of the cylinder was moved into the negative x-direction. Similarly, the mooring lines M1 and M2 was brought back to the original condition to investigate the initial motion in the positive x-direction. For these tests the wave direction was changed to $-90°$, which brought the cylinder approximately to the same relative location under wave loads. The definition of the DoF are not changed for those specific cases with a different wave direction.

Figure 4 presents the mean values of the repeat time for the previously described cases in relation to the wave frequency f_W. The difference of the modified set-ps can be seen in the surge analysis. The *reference* and *repetition* data as well as the increased *pre-tension* show similar response starting from 0 for the lower frequencies and up to around 1.5 m around 0.8 Hz. A nearly constant offset can be observed for this analysis due to the change of the initial position the floating cylinder in the tank. For frequencies around 0.8 Hz the floating cylinder reaches the centre of the tank and the WG array covers this range (Table 1, Figure 1). The limited cases conducted with an initial motion in positive x-direction show a symmetrical behaviour around $x = 0$ m. The differences for all other DoF are relatively small. Nevertheless, the mean value of pitch reaches approximately up to $5°$, which is not detected by the corresponding FFT when examining the maximum value of the amplitude spectrum. The *moved* x^+ cases show a negative mean pitch value indicating that the wave direction for those cases is in the opposing direction and the definition of the DoF was not changed.

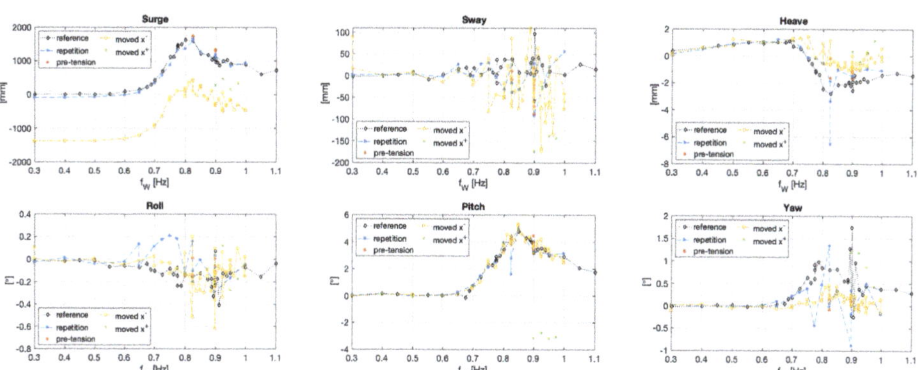

Figure 4. Average value of the repeat time for all DoF—the reference measurement uses data from Gabl et al. [1] and is compared to the repetition of those experiments as well as changes of the pretension and different initial position in the tank.

The maximum amplitude response based on the FFT analysis is presented in Figure 5 and the corresponding response frequencies in Figure 6. Both analyses are presented in relation to the wave frequency f_W. The surge and sway responses are larger for the *moved* x^- but the response frequencies for this frequency band are small. It is highly likely that this was caused by the asymmetric pre-tension in the mooring lines (M1 and M2 was larger than M3 and M4, Figure 1). The other DoF do not show significant differences considering the unequal distribution of the pre-tension in all four mooring lines. The heave motion shows nearly no difference between the investigated cases. Considering the rotations, the *repetition* generally resulted in a slightly smaller response in comparison to the previous experimental result.

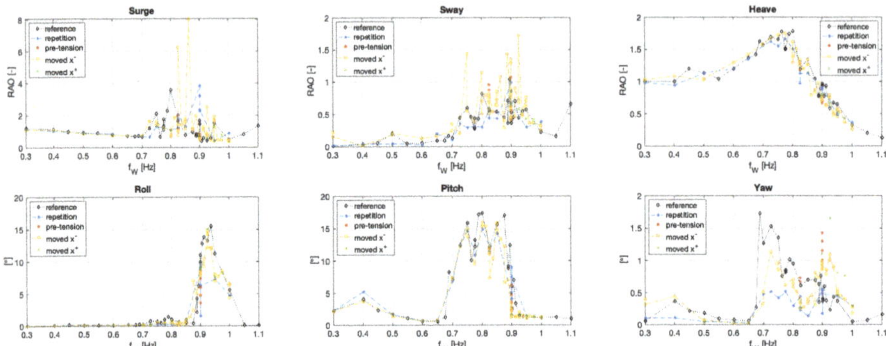

Figure 5. Amplitude response in the six DoF in relation to the requested wave frequency f_W.

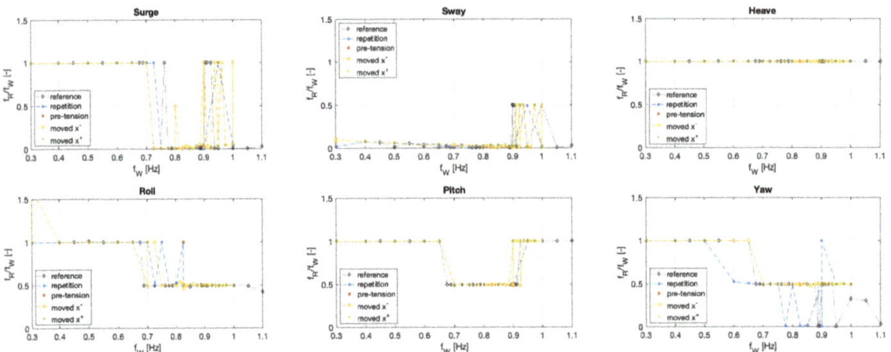

Figure 6. Frequency response f_R normalised by and presented in relation to the requested wave frequency f_W for the amplitude results in Figure 5.

In addition to the overview in Figure 5, the responses in roll and pitch are analysed in Figure 7 for the individual cases with adapted scaling of the y-axis (wave frequency f_W). The first set of graphs show the comparison between the results of the *reference* and *repetition* measurements. The switch between roll and pitch happens at a frequency of approximately 0.9 Hz and therefore eight repeat tests were conducted for this case. In general a relatively good agreement with the previous investigation could be found, which is in the range of the differences between different repetitions (back to back testing). The largest differences were observed for the wave frequency band with a significant response in pitch and roll, which is discussed in Section 3.3. The higher frequency band could be further investigated to provide more points in this comparison. Nevertheless, the very similar pitch response shows a good agreement and this frequency band was further investigated as part of the *moved x⁻* cases. This comparison is shown on the right side of the Figure 7. The changed *pre-tension* as well as the *moved x⁺* cases result are very similar to the other results.

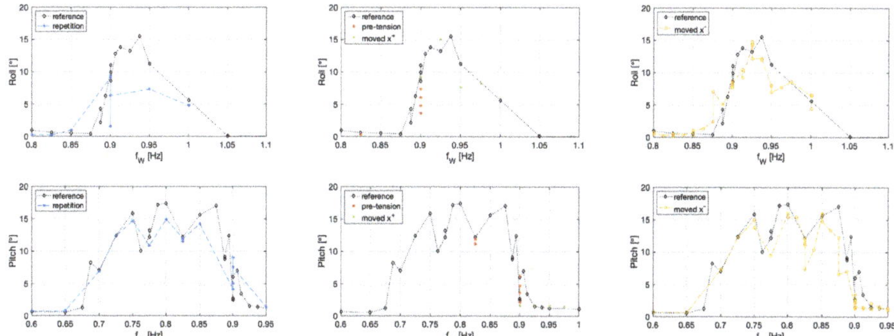

Figure 7. Detailed view of roll (**upper row**) and pitch (**lower row**) presented in Figure 5—the used boundaries for wave frequency f_W are limited to focus on a specific frequency band for each DoF.

The analyses presented in Figures 5–7 investigate the pre-tension and location of the floating cylinder in the tank. All cases showed a repeatable occurrence of the roll motion response and consequently the RQ1 can be answered in that the roll response is not depending on the specific tank position. Furthermore, it can be summarised that the results for the majority of wave frequencies show no significant influence of these variables. For the transient frequency band the results can vary significantly between similar test and this instability is further investigated in Section 3.3. The largest differences are caused by the different pre-tension in the mooring lines or occur with at a small response frequency.

3.3. Length of the Test

As mentioned in Section 3.2, the largest differences between the *reference* data and the *repetition* could be found in the wave frequency f_W band around 0.9 Hz. The switch between roll and pitch is observed around this frequency for the wave direction of 90° in the tank definition. As shown in Section 3.5, this value is significantly reduced by changing the wave direction to 180°.

The second research question focuses on the length of the capture time. A typical total length of 180 s was used. To ensure that no effects are missed, extended experimental runs were conducted. Therefore, the length is tripled from 180 s (3 min) to 540 s (9 min). Considering a similar ramp-up time of 52 s, the extended observation time contained 488 s. Figure 8 presents the time series of roll and pitch for six examples. The ramp-up time is marked with a red vertical line and the horizontal lines provide the results of the FFT analysis as a response amplitude around the mean value of the repeat time.

The first three time series show a wave frequency f_W of 0.9 Hz with one normal and two repetitions of the extended run. Comparing the chosen ramp-up time for these cases show that the floating cylinder first responds with a pitch rotation. Roll starts later and when it occurs pitch is reduced. The short run shows that an extended ramp-up time could potentially improve the analysis. But the comparison at the same wave frequency with an extended observation time (second and third set in Figure 8) show that such a switch between roll and pitch can also occur later in the experiment. The repetition show that only the basic behaviour can be reproduced. A splitting of the complete time series in sub-section could improve the analysis but the main finding is that the results in this transition zone show a large variance of responses. Consequently, this frequency band should be used primarily for a qualitative and not quantitative comparison of numerical results. A similar comparison for the f_W equal 0.925 Hz is also presented in Figure 8. This frequency results in a clear roll response starting with a very small initial pitch response. An extension of the observation time proves that the stable behaviour is not changing. Similar results are shown with the further increased wave frequency of 0.9375 Hz (last set in Figure 8).

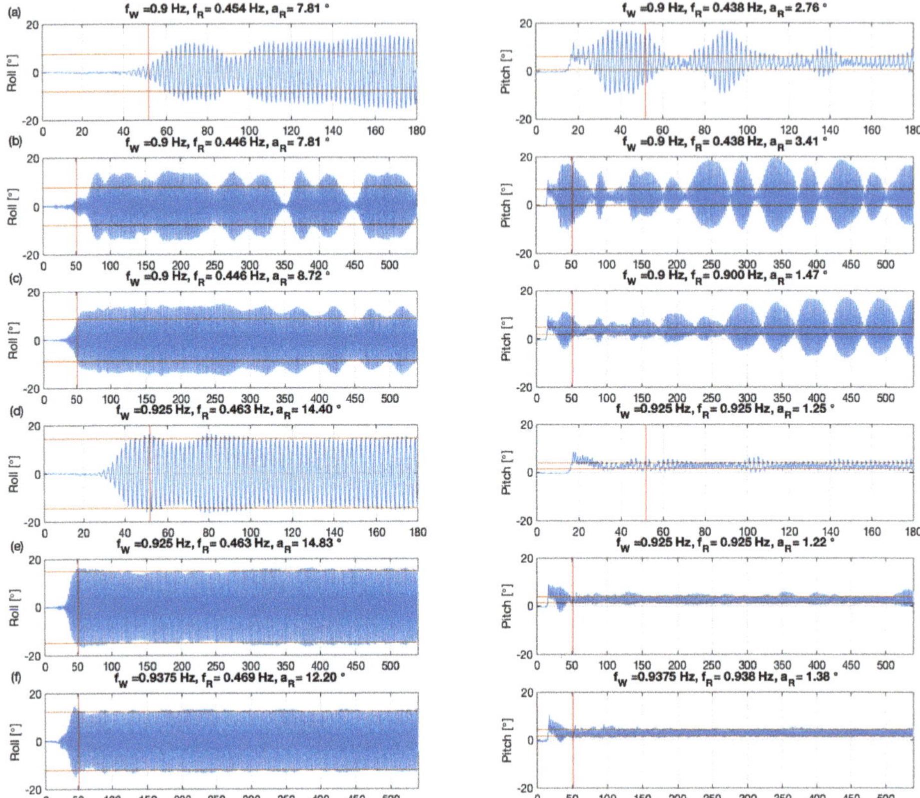

Figure 8. Sample time series of roll and pitch for six test runs with different requested wave frequency f_W and expanded capture time—(**a**) $f_W = 0.90$ Hz, 180 s, (**b,c**) 0.90 Hz, 512 s, (**d**) 0.925 Hz, 180 s, (**e**) 0.925 Hz, 512 s, (**f**) 0.9375 Hz, 512 s—vertical red line indicates the 52 s rump-up time. The horizontal red lines mark the response amplitude $\pm a_R$ and the connected response frequency f_R is provided in the title of the individual graph.

In addition to this individual time series, Figure 9 presents additional analysis for those two rotation responses, while the first 52 s are coloured in red. Therefore, the pitch response is plotted against the measured roll values in the right column of Figure 9. The oscillation shows a positive average pitch value, which can also be seen in Figure 4. The results for a wave frequency of 0.9 Hz show a butterfly like response covering both rotation directions. Outside of this transient zone a clear response can be observed, which for the additional three wave frequencies takes the form of a significant roll response. In the other two columns of Figure 9 the specific angle is compared to the corresponding velocity. A simple harmonic oscillator would show a circular behaviour with the maximum speed at an angle equal to zero. Only for one repetition of the wave frequency 0.9 Hz could such a comparable circular behaviour be found and the response of both DoF are in the same magnitude. A clear pattern of circular behaviour can be found for cases with a higher frequency and a dominant roll response. The response in pitch is similar and it has to be considered that this would be obvious response direction based on the incident waves.

Based on the conducted comparisons and the RQ2, it can be stated that the 180 s capture time is sufficient. In the transient zone a non-stable switch between pitch and roll could be observed, which complicates the quantification of the separate rotations and reduces the RAO results to a very rough assumption. Consequently, it is advisable to use a frequency band with a clear response for future validation of numerical simulations.

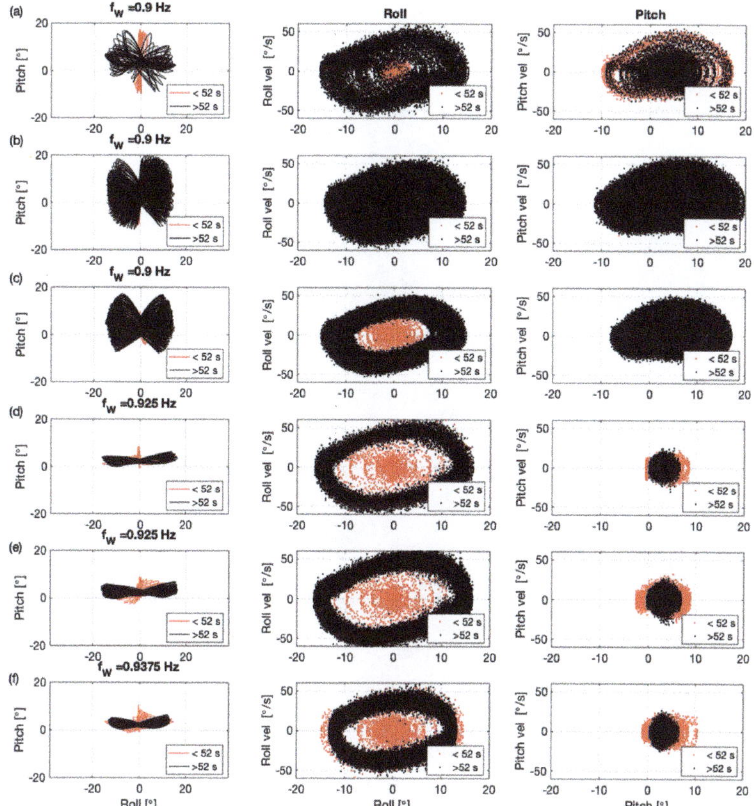

Figure 9. Direct comparison of pitch and roll as well as the comparison of pitch and roll angle with the corresponding velocity for the time series presented in Figure 8—(**a**) f_W = 0.90 Hz, 180 s, (**b,c**) 0.90 Hz, 512 s, (**d**) 0.925 Hz, 180 s, (**e**) 0.925 Hz, 512 s, (**f**) 0.9375 Hz, 512 s—the first 52 s are presented in red.

3.4. Wave Amplitudes

Early in the research project and based on unpublished preliminary experiments the decision was taken to run the experiments with a constant requested wave height of 50 mm. The intention is that the amplitudes should be large enough to produce a measurable response without being so large as to negate any assumptions of linear behaviour required for numerical modelling validation. The main comparison of water and solid ballast options is based on this assumption [1,2]. As part of this additional experiments, this decision was again tested. It has to be mentioned that this was not a comprehensive investigation and the main aim is to understand how far the results are independent of the wave amplitude. This is important if numerical codes simulating in the frequency domain are be used to reproduce the experiments.

Figure 10 presents the response amplitudes for all six DoF for the conducted experiments with different requested wave amplitudes and Figure 11 the corresponding response frequencies f_R normalised by the wave frequency f_W. Both figures use the measured wave amplitude a_R for the x-axis of the graph. The results are presented as Response Amplitude Operators (RAO) and grouped by f_W and wave direction.

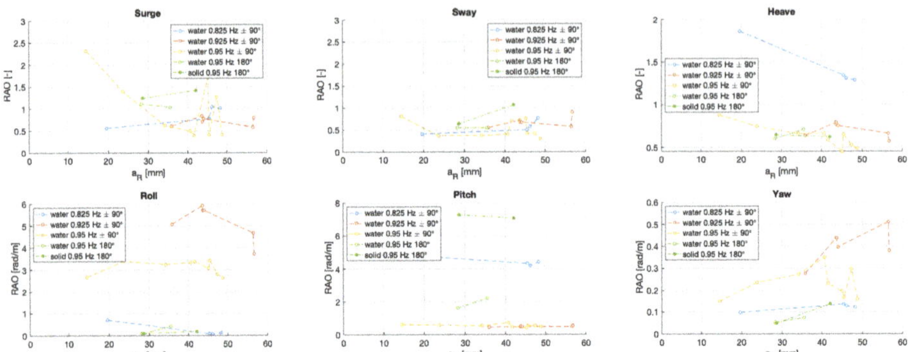

Figure 10. Amplitude response in the six DoF in relation to the measured wave amplitude a_R—individual graphs with a fixed requested wave frequency f_W and wave direction

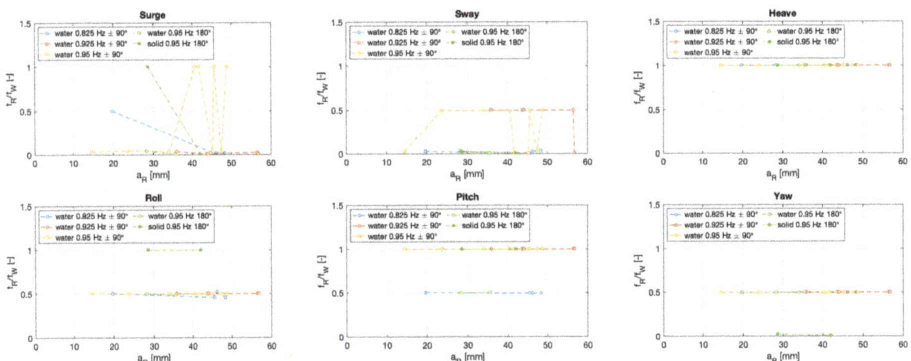

Figure 11. Frequency response f_R normalised by the requested wave frequency f_W and presented in relation to the measured wave amplitude a_R in addition to Figure 10.

The motion responses show a relatively constant response. The bigger jumps in surge are connected with changes in the response frequency. Lower wave amplitudes show larger differences to the other investigated amplitudes. This is especially noticeable for the 0.825 Hz wave case, which is very close to the peak response in heave. Further experimental runs are needed to fully investigate this.

The research question RQ3 focuses on the sensitivity of the wave amplitude. The limited cases show that the differences for the motion RAO are mainly in the range of 0.5. This range expands to typically 1 rad/m for the rotations. A relatively constant response for roll and pitch can be observed and the yaw components are relatively small. Further experimental investigations including a wide range of wave amplitudes are needed to fully assess this.

3.5. Direction Including Comparison with the Solid Case

The main focus of this investigation lies in the change of wave direction. Hence the mooring system is symmetrical around the x and y-axis, waves coming from the orthogonal direction are particularly interesting. Previously, the main wave direction was 90° in the tank direction and along the positive global x-axis. As shown in Figure 1, two mooring lines were connected in one point, which allowed a smaller influence for the rotation around the y-axis (pitch for 90°). Rotations around the x-axis (roll for 90°) were more influenced by the mooring system. Consequently, it was assumed that the observed switch from pitch to roll for the water ballast case was suppressed rather than caused by the mooring. This was tested as part of the current experimental investigation by rotating the wave direction to 180° (Figure 1). If not otherwise stated, the local body coordinate system was aligned with the wave direction. Hence, roll is defined as rotation around an axis in the direction of the waves and pitch orthogonal to it. The following presented investigations were also conducted with the solid ballast option to provide the direct comparison for the additional wave direction. The plotted cases with water ballast are indicated with an empty marker and a solid one for the experimental runs with a solid ballast.

Figure 12 shows the summary of the additional comparison and focuses on the key three DoFs, namely roll, pitch and heave. The corresponding frequency response can be found in Figure 13. Both figures also include the reference values for the solid ballast option. Those experimental results were measured under a wave direction of 90° and good reproducibility of the occurrence of the roll response is shown in Section 3.2. A full frequency sweep was conducted with the 180° wave direction as part of the additional experimental investigation as well as additional wave runs in ±135°. For the latter, it has to be highlighted that the investigated mooring system was symmetrical along both main axes, but not identical—two mooring lines are connected in one point on both sides of the cylinder. For waves along one mooring line the system is not balanced and the mooring system influences the motion response significantly. Those results were not ideal, nevertheless they are included in the paper to show why only the main directions are considered in this study for this mooring configuration.

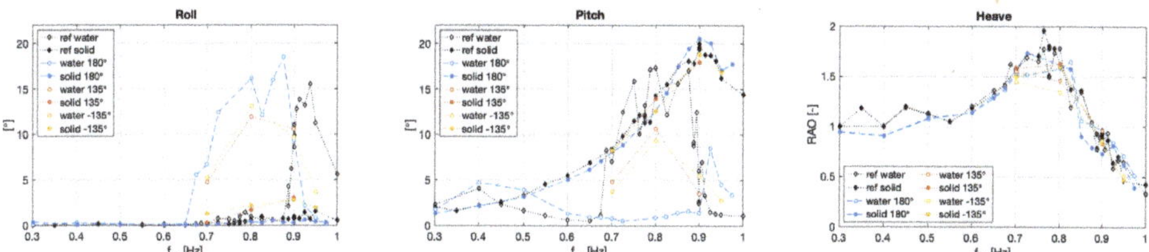

Figure 12. Roll, pitch and heave response for the wave directions 180° and ±135° compared for the references data (90°; [1])—water ○ and solid ● ballast option.

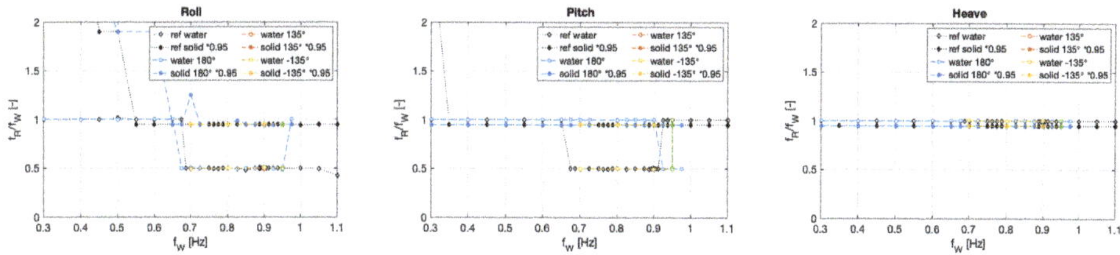

Figure 13. Frequency response f_R normalised by the wave frequency f_W in addition to Figure 12—water ○ and solid ● ballast option, which is multiplied by 0.95 to allow a better visibility of the results.

The response in the vertical direction (heave, Figure 12) is very similar for all investigated cases, which was already shown in Gabl et al. [1]. As presented in Figure 13, the response frequency for heave is identical for all wave frequency f_W (the values of the solid cases are multiplied by a factor of 0.95 so that the overlay of the curves is reduced). Changing the wave direction consequently has no influence on the heave motion.

The obvious difference in the motion response between the two investigated wave directions, namely 90 and 180°, can be found in the roll response (rotation around the axis in wave direction) and for the water filled case. In the *reference* investigations, the floating cylinder responded in the lower frequencies with a pitch response. Crossing a transition frequency band (Figure 8), a further increase of the frequency results in a roll oscillation. This can only be observed for the water filled cylinder and consequently is caused by the sloshing of the inner water body. By changing the wave direction to 180°, the principal main rotation stays very similar. Hence, for this wave direction the cylinder responds for a wide frequency band with a roll motion instead of pitch. In the higher frequency, where roll previously occurred, rotations are reduced and a small response in pitch can be observed. The response frequencies in roll (Figure 13, left graph) are very similar for both wave directions. A reduction to 0.5 for f_R/f_W can be seen over a f_W of 0.65 Hz. This indicates that the potential for the roll response is present in both wave direction but the weaker response of the mooring system is around the global y-axis, which leads to a pitch motion in the 90° case. The pitch response of the 180° remains relatively small in the lower frequency band and is mainly in the same frequency as f_W. This is in contrast to the 90° case, which response at half f_W with relatively large pitch values.

Figure 14 presents three pair of time series for waves with the similar wave frequency f_W of 0.85, 0.9 and 0.95 Hz for both wave directions (90° and 180°), and Figure 15 shows the detailed analysis of the two rotation around the x and y-axis, which is comparable to Figure 9. In this particular case the measured rotations around the global axis are presented. Hence, the left column in Figure 14 includes the results for the rotation around the x-axis and the right column the one around y-axis (definition presented in Figure 1). Based on the actual wave direction, roll and pitch are mentioned in the label of the axis. The first row shows the results of the 0.85 Hz wave along the global x-direction (90°). After the waves reach the floating cylinder an initial increase in pitch is observed and a stable rotation around y builds up. Only a very small roll motion is visible. The same wave train is used for the 180° direction presented in the second row of Figure 14. An identical initial motion is observed for the rotation around the x-axis (the different signs are caused by the definition of the positive direction; in both cases the top of the cylinder moved first in the wave direction). This initialised oscillation in the wave direction (around the x-axis) switches to a response around y.

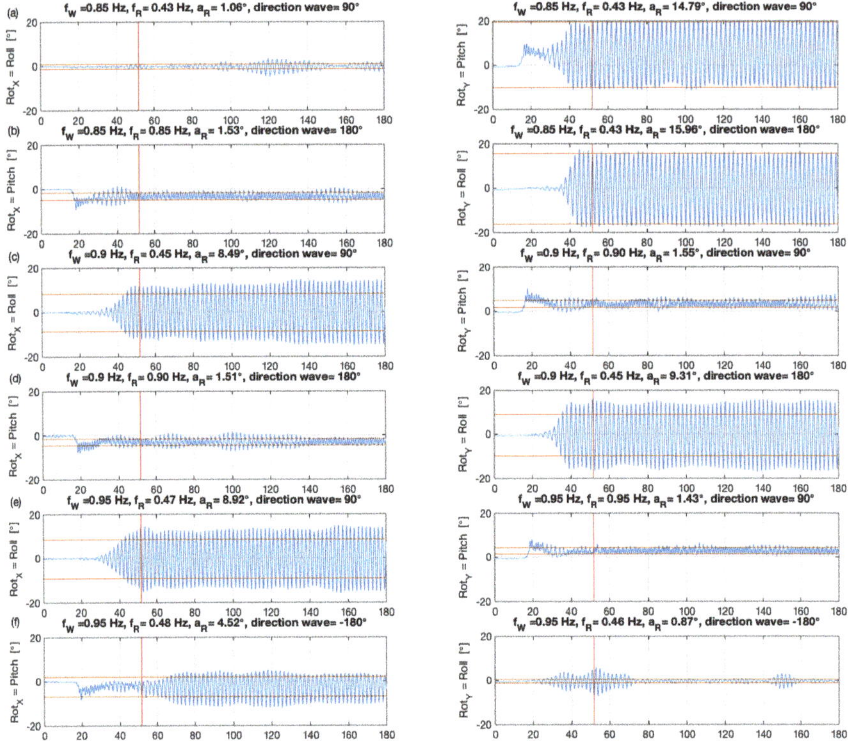

Figure 14. Sample time series of roll and pitch for six test runs with different requested wave frequency f_W and comparison of the wave direction of 90° and 180°— (**a**) f_W = 0.85 Hz, 90°, (**b**) 0.85 Hz, 180°, (**c**) 0.90 Hz, 90°, (**d**) 0.90 Hz, 180°, (**e**) 0.95 Hz, 90°, (**f**) 0.95 Hz, −180°—vertical red line indicates the 52 s ramp-up time. The horizontal red lines mark the response amplitude $\pm a_R$ and the connected response frequency f_R is provided in the title of the individual graph.

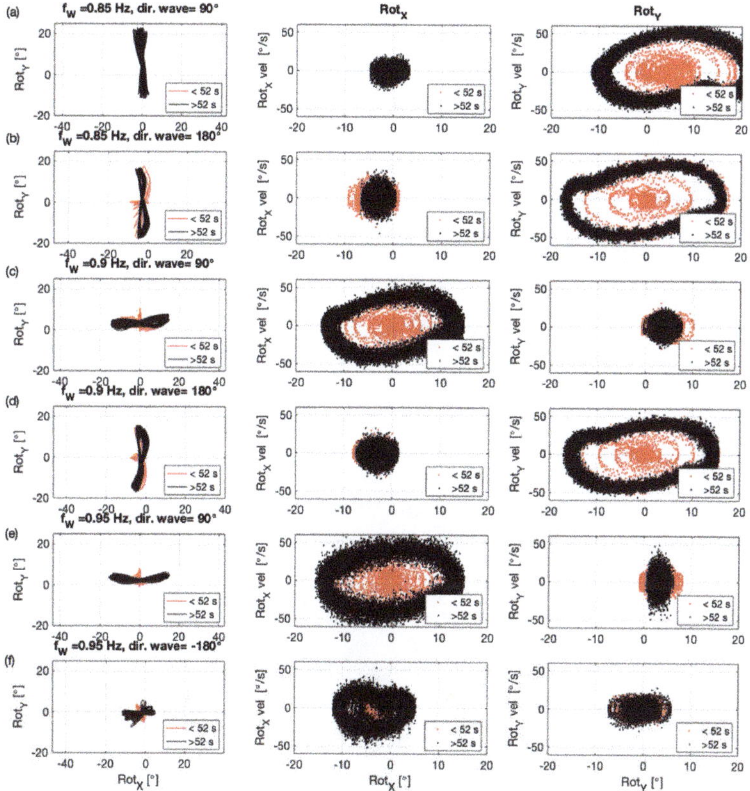

Figure 15. Direct comparison of pitch and roll for the time series presented in Figure 14—(**a**) f_W = 0.85 Hz, 90°, (**b**) 0.85 Hz, 180°, (**c**) 0.90 Hz, 90°, (**d**) 0.90 Hz, 180°, (**e**) 0.95 Hz, 90°, (**f**) 0.95 Hz, −180°—the first 52 s are presented in red.

A full change of the dominant rotation with the wave direction can be found for the wave frequency of 0.9 Hz. By further increasing the frequency to 0.95 Hz, a comparable behaviour to 0.85 Hz could be observed. The roll response for the 90° cases remains in the same direction for the 180°, which is now pitch.

As mentioned previously, only the main direction along the global horizontal axis provides an unbiased response of the floating cylinder, hence the mooring system does not possess a rotational symmetry (Figure 1). Figures 12 and 13 include results for some sample frequencies with wave directions of ±135°. The solid ballast option response is very close to the other wave directions but also a small additional rotation in roll can be observed. This influence of the mooring lines are larger for the water filled cylinder and makes it hard to extract an exact response. Nevertheless, a further variation of the wave direction was conducted for a constant wave frequency of 0.95 Hz for both ballast options. Figure 16 summarises the amplitude responses in relation to the wave direction and the associated response frequency f_R is provided in Figure 17. The influence of the mooring system can be seen in the small increase of the roll response for the solid version between −180° (equal to +180°) and −90°. Overall similar range, the response amplitude for roll decreases approaching −180°, which was also observed for the last pair of time series with a wave frequency of 0.95 Hz presented in Figure 14. The pitch response is relatively large in relation to the other wave directions. This can also be clearly observed in the response frequency f_R in pitch at a ratio of 0.5 in relation to the wave frequency f_W (Figure 17). The change in heave is small.

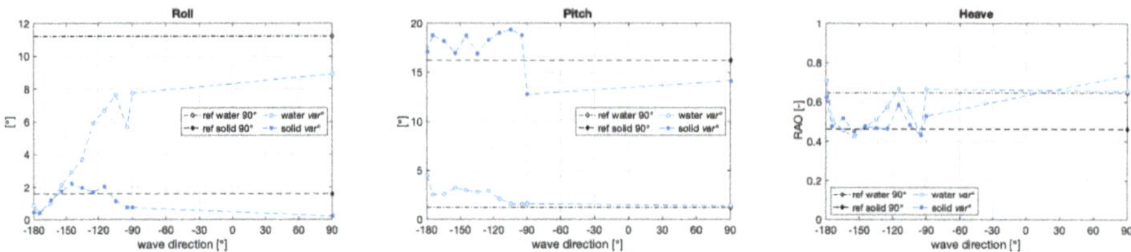

Figure 16. Roll, pitch and heave response for different wave directions with a fixed wave frequency of 0.95 Hz—water ○ and solid ● ballast option.

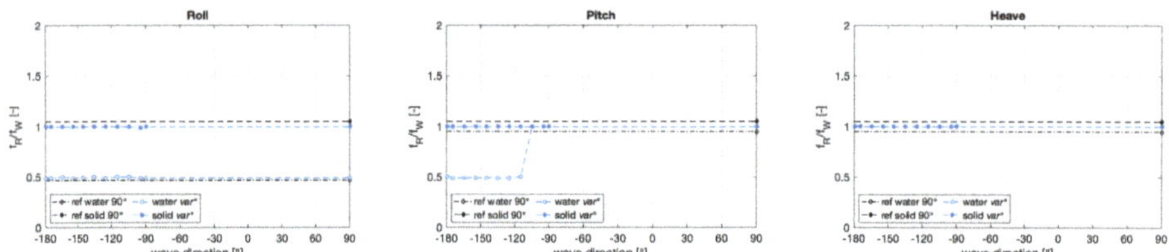

Figure 17. Frequency response f_R normalised by the wave frequency f_W in addition to Figure 16—the reference values (either 0.5 or 1[-]) are multiplied by a factor to make them better visible—water ○ and solid ● ballast option.

The key finding of the variation of the wave direction and the research question RQ4 is that the chosen mooring system is not the cause of the roll response and more likely prevented it. A change of the wave direction to 180° results in a significant increase of the frequency band associated with a roll response, while the solid ballast option showed almost no dependence on the changed wave direction. Consequently, those results are a vital expansion of the previously published data set and is available as an addition via Edinburgh DataShare [39].

4. Discussion

The chosen mooring system and potential alternatives are discussed below. Furthermore, the changes in the position of the wave gauges (WG) from the initial layout, presented in Gabl et al. [1,2], to the current one are highlighted and a brief overview of future work provided.

The mooring system was introduced as part of the previous experimental investigation exploring four different drafts of the floating cylinder with water and solid ballast. The station keeping mooring configuration deliberately provided a smaller resistance for the rotations around the y-axis to ensure a minimum of influence for this degree of freedom (Figure 2). This asymmetry was intended to provide a maximum freedom for rotations in the wave direction and a slightly higher resistance in the orthogonal one. This lower resistance occurred for the roll response by turning the wave direction by 90°, which lead to a significant increase of the frequency band with a roll response. Consequently, the rolling of the floating cylinder is more likely to be restricted by the mooring configuration, rather than caused by it. A higher degree of symmetry could be achieved by connecting each mooring line to one point on the cylinder. Such a system would be symmetrical for all four quadrants but also lead to two different main directions, namely along a mooring line as well as between two mooring lines. Ideally a fully rotationally symmetric mooring system would be desirable. One option to achieve this could be a single mooring line connection at the bottom of the cylinder. A disadvantage of this approach is, that the introduced mooring force might have an increased influence on the motion response of the cylinder. It could also introduce a greater self-aligning torque and prohibit the oscillation. The chosen system was a good compromise between station keeping and allowing the motion response to develop freely. Nevertheless, a complete change of the mooring system could be a further step forward including a variation of the mooring line stiffness. Alternative approaches for mooring systems, which might restrict different DoF will also be considered.

For the previous experimental runs the wave gauges (WG) were distributed over a wide range in front as well as behind the floating cylinder. This allowed the provision of a very good measurement of the incoming waves and potentially changes caused by the cylinder. A direct comparison of the motion response with the WG was not possible. For the new experiments, the WG were arranged with a y-offset to the floating cylinder (Figure 1). Especially in cases with a moved initial position (Section 3.2), a very good side by side alignment could be reached. The distance between the floating cylinder and the WG array was chosen to be 1.5 m, which is 3 times the diameter of the cylinder D. The results of the measured free surface elevation show bigger influences of reflections caused by the floating body, which has to be considered in future usage of the data as a validation experiment. A further translation of the WG in the y-direction would have reduced this influence. This would also allow the use of the full 360° capacity of FloWave without risking a collision of the cylinder with the WG. Furthermore, the WG remained at the same position for the different wave directions. Future experiments will include a adjustable WG array mounted on a rotatable arm. The alignment can be guaranteed with the MoCAP.

As previously mentioned in Section 3 an expansion of the investigated wave amplitudes as well as a further refinement of the wave frequencies is desirable. A further aim of the research project is to measure the changes of the free surface of the inner water body [40] as well as the velocities of the sloshing water body inside of the cylinder.

5. Conclusions

This work extended the work of Gabl et al. [1] in exploring the influence of the water ballast in a floating cylinder. In those original tests a pronounced roll response was observed towards the higher end of explored wave frequency range. This result was reproduced in these tests and is confirmed as a real effect.

It could be shown that the tests are reproducible and that the pre-tension of the soft mooring lines as well as the position of the floating cylinder in the tank have only a negligible influence on the response of the floating cylinder. An extension of the capture time proves that the 180 s minus the 52 s ramp-up time are sufficient to ensure that the motion response is fully developed and stable. In a transition frequency band the response between pitch and roll switches constantly and similar wave conditions can lead to varying results. Consequently, those frequencies should be avoided for validations experiments which target the investigation of the roll motion of the water filled structure, or at least this instability should be kept in mind. The variation of the requested wave amplitude from the standard value of 50 mm were limited to some exploratory cases but resulted in relatively stable responses. Nevertheless, for a direct comparison in the time domain the actual measured wave amplitude should be used.

A significant change could be observed for the changed wave direction. In contrast to the solid ballast option, the roll motion of the water filled cylinder started with a far smaller wave frequency and was the dominant rotation response for this case. This is caused by the mooring design, which allowed a smaller resistance for the roll rotation in case of the wave direction in 180°. Based on this result, it can be stated that the initial roll response of the water filled cylinder was not caused by the mooring system. It quite contrary prohibited it for the 90° direction. Consequently, the 180° wave direction is especially interesting for validation experiments aiming to reproduce the roll response of the water filled structure.

Author Contributions: R.G., T.D. and D.M.I. are responsible for the conceptualisation of the experimental investigation. R.G. measured the data and analysed the data. R.G. and T.D. wrote the initial draft and D.M.I. reviewed and edited the paper. All authors have read and agreed to the published version of the manuscript.

Funding: This work was supported by the Austrian Science Fund (FWF) under Grant J3918.

Conflicts of Interest: The authors declare no conflict of interest.

Notation

a_R	amplitude waves (mm) measured
a_W	amplitude waves (mm) requested from the wave makers
d	cylinder draft (m)
D	cylinder diameter (m)
f_R	response frequency (Hz) measured
f_W	frequency wave (Hz) requested from the wave makers
h	water depth inside the cylinder (m)
H	height of the cylinder (m)
x	distance (m) in the main wave direction defined as 90°
X	motion in x-direction (mm)
y	distance orthogonal to the main wave direction (m)
Y	motion in y-direction (mm)
z	distance vertical direction (m)
Z	motion in z-direction (mm)
DoF	degree of freedom
MoCAP	motion capturing system
RAO	response amplitude operator
RQ	research question
WG	wave gauge

References

1. Gabl, R.; Davey, T.; Nixon, E.; Steynor, J.; Ingram, D.M. Comparison of a Floating Cylinder with Solid and Water Ballast. *Water* **2019**, *11*, 2487. [CrossRef]
2. Gabl, R.; Davey, T.; Nixon, E.; Steynor, J.; Ingram, D.M. Experimental Data of a Floating Cylinder in a Wave Tank: Comparison Solid and Water Ballast. *Data* **2019**, *4*, 146. [CrossRef]
3. Fossen, T.; Nijmeijer, H. *Parametric Resonance in Dynamical Systems*; Springer: New York, NY, USA, 2012.
4. France, W.; Levadou, M.; Treakle, T.; Paulling, J.; Michel, R.; Moore, C. *An Investigation of Head-Sea Parametric Rolling and Its Influence on Container Lashing Systems*; The Society of Naval Architects and Marine Engineers: Alexandria, VA, USA, 2001; pp. 1–24.
5. Cheng, P.; Liang, N.; Li, R.; Lan, H.; Cheng, Q. Analysis of Influence of Ship Roll on Ship Power System with Renewable Energy. *Energies* **2020**, *13*, 1. [CrossRef]
6. Li, C.T.; Wang, D.Y.; Cai, Z.H. Experimental and numerical investigation on the scaled model of lashing bridge coupled with hull structure and container stack. *Ships Offshore Struct.* **2020**. [CrossRef]
7. Neves, M.A.S.; Rodríguez, C.A. On unstable ship motions resulting from strong non-linear coupling. *Ocean Eng.* **2016**, *33*, 1853–1883. [CrossRef]
8. Wang, L.; Tang, Y.; Li, Y.; Zhang, J.C.; Lui, LQ. Studies on Stochastic Parametric Roll of Ship with Stochastic Averaging Method. *China Ocean Eng.* **2020**, *34*, 289–298. [CrossRef]
9. Guze, S.; Wawrzynski, W.; Wilczynski, P. Determination of Parameters Describing the Risk Areas of Ships Chaotic Rolling on the Example of LNG Carrier and OSV Vessel. *J. Mar. Sci. Eng.* **2020**, *8*, 91. [CrossRef]
10. Kianejad, S.S.; Enshaei, H.; Duffy, J.; Ansarifard, N. Prediction of a ship roll added mass moment of inertia using numerical simulation. *Ocean Eng.* **2019**, *173*, 77–89. [CrossRef]
11. Olivieri, A.; Francescutto, A.; Campana, E.F.; Stern, F. Parametric Roll: Highly Controlled Experiments for an Innovative Ship Design. In Proceedings of the ASME 2008 27th International Conference on Offshore Mechanics and Arctic Engineering, Estoril, Portugal, 15–20 June 2008; Volume 4, pp. 291–299.
12. Huang, S.; Duan, W.; Han, X.; Nicoll, R.; You, Y.G.; Sheng, S.W. Nonlinear analysis of sloshing and floating body coupled motion in the time-domain. *Ocean Eng.* **2018**, *164*, 350–366. [CrossRef]
13. Zhou, Y.; Ma, N.; Lu, J.; Gu, M. A study of hybrid prediction method for ship parametric rolling. *J. Hydrodyn.* **2016**, *28*, 617–628. [CrossRef]
14. Zhou, Y. Further validation study of hybrid prediction method of parametric roll. *Ocean Eng.* **2019**, *186*, 106103. [CrossRef]

15. Zhao, W.; Taylor, P.; Wolgamot, H.; Taylor, R.E. Identifying linear and nonlinear coupling between fluid sloshing in tanks, roll of a barge and external free-surface waves. *J. Fluid Mech.* **2018**, *844*, 403–434. [CrossRef]

16. Zhao, W.H.; McPhail, F. Roll response of an LNG carrier considering the liquid cargo flow. *Ocean Eng.* **2017**, *129*, 83–91. [CrossRef]

17. Zhu, X.; Yoo, W. Numerical modeling of a spar platform tethered by a mooring cable. *Chin. J. Mech. Eng.* **2015**, *28*, 785–792. [CrossRef]

18. Orszaghova, J.; Wolgamot, H.; Draper, S.; Taylor, PH.; Rafiee, A. Onset and limiting amplitude of yaw instability of a submerged three-tethered buoy. *Proc. R. Soc. A* **2020**, *476*, 20190762. [CrossRef]

19. Tarrant, K.; Meskel, C. Investigation on parametrically excited motions of point absorbers in regular waves. *Ocean Eng.* **2016**, *111*, 67–81. [CrossRef]

20. Radhakrishnan, S.; Datla, R.; Hires, R.I. Theoretical and experimental analysis of tethered buoy instability in gravity waves. *Ocean Eng.* **2007**, *34*, 261–274. [CrossRef]

21. Jiang, H.; You, Y.; Hu, Z.; Zheng, X.; Ma, Q. Comparative Study on Violent Sloshing with Water Jet Flows by Using the ISPH Method. *Water* **2019**, *11*, 2590. [CrossRef]

22. Trimulyono, A.; Hashimoto, H.; Matsuda, A. Experimental Validation of Single- and Two-Phase Smoothed Particle Hydrodynamics on Sloshing in a Prismatic Tank. *J. Mar. Sci. Eng.* **2019**, *7*, 247. [CrossRef]

23. Kim, D.H.; Kim, E.S.; Shin, S.-C.; Kwon, S.H. Sources of the Measurement Error of the Impact Pressure in Sloshing Experiments. *J. Mar. Sci. Eng.* **2019**, *7*, 207. [CrossRef]

24. Delorme, L.; Colagrossi, A.; Souto-Iglesias, A.; Zamora-Rodríguez, R.; Botía-Vera, E. A set of canonical problems in sloshing, Part I: Pressure field in forced roll-comparison between experimental results and SPH. *Ocean Eng.* **2009**, *36*, 168–178. [CrossRef]

25. Amicarelli, A.; Manenti, S.; Albano, R.; Agate, G.; Paggi, M.; Longoni, L.; Mirauda, D.; Ziane, L.; Viccione, G.; Todeschini, S.; et al. 2020 SPHERA v.9.0.0: A Computational Fluid Dynamics research code, based on the Smoothed Particle Hydrodynamics mesh-less method. *Comput. Phys. Commun.* **2009**, *250*, 107157. [CrossRef]

26. Gunn, D.F.; Rudman, M.; Cohen, R.C.Z. Wave interaction with a tethered buoy: SPH simulation and experimental validation. *Ocean Eng.* **2018**, *156*, 306–317. [CrossRef]

27. Guo, K.; Sun, P.-N.; Cao, X.-Y.; Huang, X. A 3D SPH model for simulating water flooding of a damaged floating structure. *J. Hydrodyn.* **2017**, *29*, 831–844. [CrossRef]

28. Ming, F.R.; Zhang, A.M.; Cheng, H.; Sun, P.N. Numerical simulation of a damaged ship cabin flooding in transversal waves T with Smoothed Particle Hydrodynamics method. *Ocean Eng.* **2018**, *165*, 336–352. [CrossRef]

29. Cheng, H.; Zhang, A.M.; Ming, F.R. Study on coupled dynamics of ship and flooding water based on experimental and SPH methods. *Phys. Fluids* **2017**, *29*, 107101. [CrossRef]

30. Royon-Lebeaud, A.; Hopfinger, E.J.; Cartellier, A. Liquid Sloshing and Wave Breaking in Circular and Square-Base Cylindrical Containers. *J. Fluid Mech.* **2007**, *577*, 467–494. [CrossRef]

31. Frosina, E.; Senatore, A.; Andreozzi, A.; Fortunato, F.; Giliberti, P. Experimental and Numerical Analyses of the Sloshing in a Fuel Tank. *Energies* **2018**, *11*, 682. [CrossRef]

32. Chong, W.; Hongde, Q.; Ge, L.; Ting, G. Study on Sloshing of Liquid Tank in Large LNG-FSRU Based on CLSVOF Method. *Int. J. Heat Technol.* **2016**, *34*, 616–622.

33. Liu, J.; Zang, Q.S.; Ye, W.B.; Lin, G. High performance of sloshing problem in cylindrical tank with various barrels by isogeometric boundary element method. *Eng. Anal. Bound. Elem.* **2020**, *114*, 148–165. [CrossRef]

34. Wei, Z.; Feng, J.; Ghalandari, M.; Maleki, A.; Abdelmalek, Z. Numerical Modeling of Sloshing Frequencies in Tanks with Structure Using New Presented DQM-BEM Technique. *Symmetry* **2020**, *12*, 655. [CrossRef]

35. Sanapala, V.S.; Sajish, S.D.; Velusamy, K.; Ravisankar, A.; Patnaik, B.S.V. An experimental investigation on the dynamics of liquid sloshing in a rectangular tank and its interaction with an internal vertical pole. *J. Sound Vib.* **2019**, *449*, 43–63. [CrossRef]

36. Ye, W.B.; Liu, J.; Lin, G.; Zhou, Y.; Yu, L. High performance analysis of lateral sloshing response in vertical cylinders with dual circular or arc-shaped porous structures. *Appl. Ocean Res.* **2018**, *81*, 47–71. [CrossRef]

37. Demirel, E.; Aral, M.M. Liquid Sloshing Damping in an Accelerated Tank Using a Novel Slot-Baffle Design. *Water* **2018**, *10*, 1565. [CrossRef]

38. Dinçer, A.E. Investigation of the sloshing behavior due to seismic excitations considering two-way coupling of the fluid and the structure. *Water* **2019**, *11*, 2664. [CrossRef]

39. Gabl, R.; Davey, T.; Ingram, D.M. *Additional Experimental Data of a Floating Cylinder in a Wave Tank—Verification Experiments*; DataShare Edinburgh [Dataset]; University of Edinburgh: Edinburgh, UK, 2020. [CrossRef]

40. Gabl, R.; Steynor, J.; Forehand, D.I.M.; Davey, T.; Bruce, T.; Ingram, D.M. Capturing the Motion of the Free Surface of a Fluid Stored within a Floating Structure. *Water* **2019**, *11*, 50. [CrossRef]

41. Ingram, D.; Wallace, R.; Robinson, A.; Bryden, I. The Design and Commissioning of the First, Circular, Combined Current and Wave Test Basin. In Proceedings of Oceans 2014 MTS/IEEE, Taipei, Taiwan, 7–10 April 2014.

42. Draycott, S.; Sellar, B.; Davey, T.; Noble, D.R.; Venugopal, V.; Ingram, D. Capture and Simulation of the Ocean Environment for Offshore Renewable Energy. *Renew. Sustain. Energy Rev.* **2019**, *104*, 15–29. [CrossRef]

43. Noble, D.R.; Draycott, S.; Nambiar, A.; Sellar, B.G.; Steynor, J.; Kiprakis, A. Experimental Assessment of Flow, Performance, and Loads for Tidal Turbines in a Closely-Spaced Array. *Energies* **2020**, *13*, 1977. [CrossRef]

44. Jourdain de Thieulloy, M.; Dorward, M.; Old, C.; Gabl, R.; Davey, T.; Ingram, D.M.; Sellar, B.G. On the Use of a Single Beam Acoustic Current Profiler for Multi-Point Velocity Measurement in a Wave and Current Basin. *Sensors* **2020**, *20*, 3881. [CrossRef]

45. Jourdain de Thieulloy, M.; Dorward, M.; Old, C.; Gabl, R.; Davey, T.; Ingram, D.M.; Sellar, B.G. Single-Beam Acoustic Doppler Profiler and Co-Located Acoustic Doppler Velocimeter Flow Velocity Data. *Data* **2020**, *5*, 61. [CrossRef]

46. Gabl, R.; Davey, T.; Cao, Y.; Li, Q.; Li, B.; Walker, K.L.; Giorgio-Serchi, F.; Aracri, S.; Kiprakis, A.; Stokes, A.; et al. Experimental Force Data of a hung up ROV under Wave and Current. *Data* **2020**, *5*, 57. [CrossRef]

47. MARINET (2012) Work Package 2: Standards and Best Practice—D2.1 Wave Instrumentation Database. Revision: 05. Available online: http://www.marinet2.eu/wp-content/uploads/2017/04/D2.01-Wave-Instrumentation-Database.pdf (accessed on 30 May 2020).

Article

Experimental Evidence of the Influence of Recurves on Wave Loads at Vertical Seawalls

Dimitris Stagonas [1,*, Rajendran Ravindar [2], Venkatachalam Sriram [2] and Stefan Schimmels [3]**

[1] School of Water, Energy and Environment, Centre for Thermal Energy Systems and Materials, Cranfield University, College Rd, Wharley End, Bedford MK43 0AL, UK
[2] Department of Ocean Engineering, Indian Institute of Technology, Madras, Chennai 600036, India; itsravindar@gmail.com (R.R.); vsriram@iitm.ac.in (V.S.)
[3] Forschungszentrum Kuste (FZK), Leibniz University Hannover & Technische Universität Braunschweig, Merkurstraße 11, D-30419 Hannover, Germany; schimmels@fzk.uni-hannover.de
* Correspondence: D.Stagonas@Cranfield.ac.uk

Received: 21 December 2019; Accepted: 13 March 2020; Published: 21 March 2020

Abstract: The role of recurves on top of seawalls in reducing overtopping has been previously shown but their influence in the distribution and magnitude of wave-induced pressures and forces on the seawall remains largely unexplored. This paper deals with the effects of different recurve geometries on the loads acting on the vertical wall. Three geometries with different arc lengths, or extremity angles (α_e), were investigated in large-scale physical model tests with regular waves, resulting in a range of pulsating (non-breaking waves) to impulsive (breaking waves) conditions at the structure. As the waves hit the seawall, the up-rushing flow is deflected seawards by the recurve and eventually, re-enters the underlying water column and interacts with the next incoming wave. The re-entering water mass is, intuitively, expected to alter the incident waves but it was found that the recurve shape does not affect wave heights significantly. For purely pulsating conditions, the influence of α_e on peak pressures and forces was also negligible. In marked contrast, the mean of the maximum impulsive pressure and force peaks increased, even by a factor of more than two, with the extremity angle. While there is no clear relation between the shape of the recurve and the mean peak pressures and forces, interestingly the mean of the 10% highest forces increases gradually with α_e and this effect becomes more pronounced with increasing impact intensity.

Keywords: recurves; recurve geometry; vertical seawalls; wave loads and pressures; pulsating and impulsive conditions

1. Introduction

Wave recurves and parapets are used to reduce overtopping without considerably increasing the seawall height. The primary purpose of a recurve is to deflect the wave rushing up the wall seawards, thereby reducing overtopping. Compared with parapets, recurves form a smoother angle with the vertical wall and deflect the flow gradually. In contrast, chamfered parapets form a sharp angle with the seawall and rapidly alter the flow trajectory. Figure 1 shows the functional principle of a recurve and two examples for a sea wall equipped with a recurve and a chamfered parapet, respectively.

In Figure 2, the working principle of a recurve is further illustrated by some snapshots taken during the present experiments. It is seen how a wave is (a) approaching the sea wall and (b) hits the sea wall, producing an up-rushing water jet. The recurve (c) alters the trajectory of the up-rushing water, which is deflected seawards and (d) eventually re-enters the underlying water column and interacts with the next incoming wave (the latter interaction is not visible in Figure 2).

The positive effect of recurves and parapets in reducing overtopping has been illustrated for a range of coastal defences. [1,2] provided results showing overtopping reduction when parapets are installed on sea-dikes. After analysing small-scale test data with recurves and parapets on a vertical seawall, [3,4] found the performance of such elements in reducing overtopping depends on the freeboard (R_c) to significant wave height (H_s) ratio. Specifically, overtopping becomes negligible for non-dimensional freeboard (R_c / H_s) values of 1.5 and higher, while for ratios less than 1.2, the positive effect of the parapet vanishes. It is also noteworthy that several seawall shapes were proposed, where the vertical wall is completely replaced by a curved wall (recurve walls) to mimic the action of recurves and reduce overtopping, e.g., [5–8].

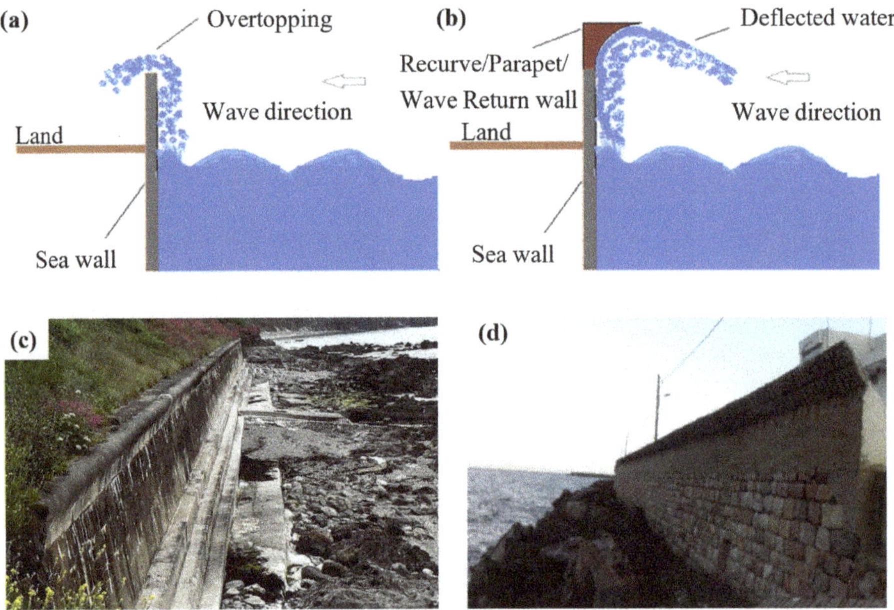

Figure 1. Schematic of the operation principle of a seawall equipped with a recurve, (**a**,**b**) and photographs of a seawall equipped with a recurve (**c**) and a chamfered parapet (**d**).

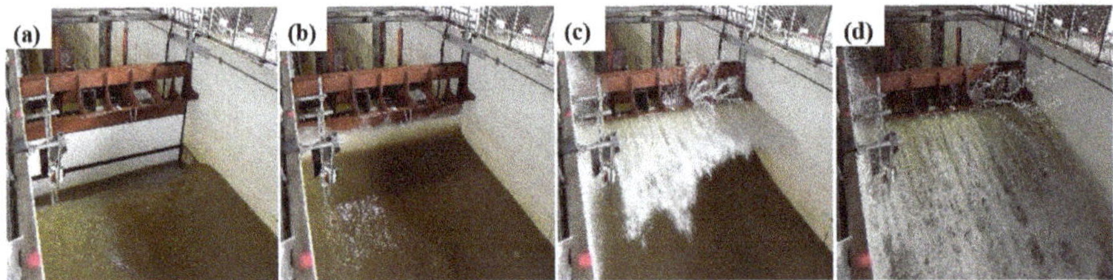

Figure 2. Sequence of snap-shots (from **a**–**d**) from the present experiments showing a wave approaching and interacting with a vertical seawall with a recurve on top.

Compared with the good understanding of the effect of recurves and parapets on the reduction of overtopping rates at vertical seawalls, much less is known about the loads on these structures, which, therefore, are often estimated by experience. Based on large-scale experiments with different recurves on a vertical seawall, [9] describe increasing wave-induced loads on the super-structure with increasing seawards protruding length, i.e., increasing extremity angle, of the recurve. Later, [10] conducted small-scale experiments for measuring the loads of non-breaking waves on recurves at the top of vertical breakwaters and also found a protruding length—wave load effect, similar to that of [9].

In this context, it is important to distinguish between pulsating loads caused by non-breaking waves where the water just goes up and down the wall and impulsive loads caused by breaking or broken waves when a more or less vertical wall of water or a mixture of air and water hits the wall, producing large pressure and force peaks and a water jet rushing vertically upwards. While the experiments of [9] covered both cases [10] were mainly focused on pulsating conditions even if [11] showed by high-fidelity numerical simulations that impulsive conditions on the recurve can be induced even by non-breaking waves.

Earlier, [3] observed that in the presence of a chamfered parapet, the wave loads acting on a model seawall increased by a factor of 1.7 and 2.0 for impulsive and pulsating conditions, respectively. These observations, however, contrasted the cases of a seawall with and without a parapet at its top, and did not consider the influence of different geometries. In addition, [3] considered the forces developed on the seawall-parapet system and not on the seawall only. At the same time, [9], [10] and [11] focused in wave pressures and loads acting on recurves with different geometries installed at the top of the same vertical wall. Therefore, and to the best of the authors' knowledge, the influence of the recurve's geometry on the pressures and loads acting on the seawall and not on the seawall-super-structure system or on the super-structure alone has seldom been considered. Given that recurves are often retrofitted on pre-existing walls, the a-priori knowledge of any influence at the loading regime on the seawall will feed into the decision-making process.

Thus, the present paper compares experimental measurements of wave loads acting on a seawall equipped with three different recurves. For the same incoming wave conditions, the shape of the recurve is altered by increasing the length of its arc, which is expressed here through the extremity angle (α_e), see Figure 3. As the extremity angle (and hence the length of the arc) increases, the protruding seaward length (B_r) of the recurve increases as well, leading to a gradual rise of the freeboard, Figure 3. Following [4], increasing the freeboard improves the overtopping performance of the recurve. In the present work, the freeboard for all three α_e considered ensures optimum overtopping performance for all recurves, thereby enabling the comparison of the wave-induced loads on the seawall without the need to consider overtopping measurements.

In the remainder, the function of the different recurves and in particular, the influence of the extremity angle to the trajectory of the seawards deflected water is described first. Then, the effect of α_e to the incoming wave conditions is considered. Finally, and still with respect to α_e, measurements of the pressure distribution and the horizontal force at the seawall are presented and discussed. At this point, it should be noted that due to the time restrictions, the case of a recurve-free vertical wall was not considered in the experiments, and this might be considered as a limitation of the present work. However, as the present study is focused on the inter-comparison of different recurve shapes and their general effect on the wave loads at a vertical wall, we believe that it is justified not to consider the pure wall case in this context.

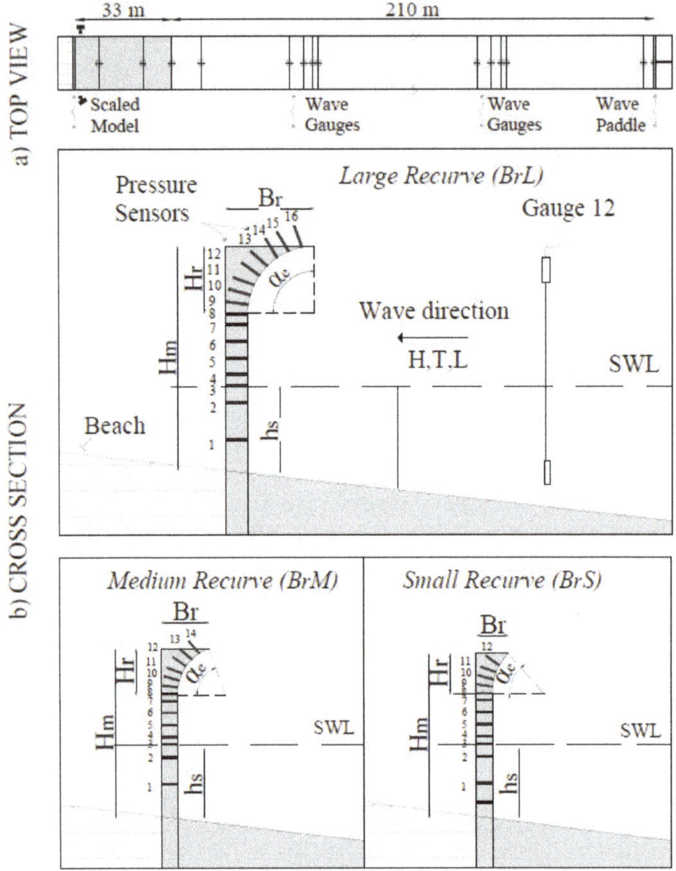

Figure 3. Experimental setup and instrumentation of the seawall and the recurves; (**a**) top view and (**b**) cross section.

2. Experimental Setup

The experiments were carried out in the Large Wave Flume (Großer Wellenkanal, GWK), Hannover, Germany. The flume is about 300 m long, 5 m wide, and 7 m deep and waves are generated by a piston-type wavemaker equipped with active wave absorption. A model seawall was installed at a distance of 243 m from the wave maker, at the end of a 33 m long 1:10 approaching slope. Twelve capacitance type, wave gauges were used to measure surface elevation in the flume, with a sampling rate of 100 Hz. Figure 3 illustrates the experimental setup.

On top of the steel-made sea wall, three different recurves with varying extremity angle (α_e), i.e., different protruding lengths B_r, were installed, giving special attention to eliminate any discontinuities at the interface between the wall and the super-structure. The three different geometries considered were as follows:

Small recurve (B_rS): $\alpha e = 48°$, $B_r = 0.20$ m, Hr = 0.45 m, Hm = 1.84 m

Medium recurve (B_rM): $\alpha e = 70°$, $B_r = 0.40$ m, Hr = 0.57 m, Hm = 1.96 m

Large recurve (B_rL): $\alpha e = 90°$, $B_r = 0.61$ m, Hr = 0.61 m, Hm = 2.00 m

Experiments were initially conducted with the small recurve B_rS, followed by tests with the medium B_rM and the large recurve B_rL. In total, 16 pressure transducers sampled at 5 kHz were used to measure pressures on the seawall and the recurves. Seven of these transducers were located on the seawall, and 5, 7 and 9 transducers were installed in the small, medium, and large recurve, respectively.

The total horizontal force on the seawall is computed from the pressure measurements as follows:

$$F_h = \sum_j P_j \times \Delta z_j \tag{1}$$

P_j: Pressure recorded by the transducer j, with $j = 1 \ldots 7$.

Δz_j: Distance between two successive transducers on the seawall. For the lowest transducer ($j = 1$), Δz_1 is the distance between the toe of the wall and the transducer.

It is reminded that for the calculation of the horizontal force, only pressure measurements from transducers 1 to 7 were used. Therefore, F_h is the (shoreward) force acting solely on the vertical seawall and not on the whole seawall-recurve system.

Finally, two video cameras were used to record the interaction of the incoming waves with the wall and the recurves. Camera 1 was positioned inside the flume, facing the seawall at an angle, while camera 2 was placed outside and over the flume, facing its sidewall. The first camera (Camera 1) recorded videos with 300 fps and the second (Camera 2) with 30 fps.

3. Testing Conditions

Experiments were carried out at a water depth of d = 4.1 m, i.e., a water depth of h_s = 0.8 m at the toe of the wall. Six regular wave cases are considered for the present study with incident wave heights (H_i) and periods (T_i) ranging between 0.5 m < H_i < 0.8 m and 4 s < T_i < 8 s. These conditions were selected to yield non-dimensional freeboard to wave height ratios falling within the optimum overtopping performance range according to Kortenhaus et al. (2002) and resulting in both, pulsating and impulsive conditions at the vertical wall.

Table 1 summarises the wave conditions for the six cases and outlines observations made during the tests and later through the analysis of the video footage. It can be seen that the testing conditions vary from pulsating to impulsive cases, i.e., from non-breaking to plunging with small and large air pockets.

Table 1. Summary of the wave conditions. The wave height and period correspond to the target values, while the wavelength is calculated at the deep section of the flume (d = 4.1 m).

Test Case	Wave Height (m)	Wave Period (s)	Wave Length (m)	Wave Steepness	Load Condition	Observations
H07T4	0.7	4	21.02	0.033	Pulsating	Non-breaking waves running up and down the vertical wall.
H05T8	0.5	8	48.55	0.01	Pulsating	Non-breaking waves running up and down the vertical wall
H06T6	0.6	6	35.13	0.017	Pulsating (transition to impulsive)	Waves slightly breaking on the vertical wall, i.e., breaking cannot be clearly observed in the flume, but the pressure signals show an initial peak, which is higher than the following quasi-static peak. This case can be considered as transition from pulsating to impulsive conditions.
H06T8	0.6	8	48.55	0.012	Impulsive	Waves breaking on the slope, about 15 m in front of the wall. When the post-breaking wave reached the vertical wall, it overturned again and broke on the structure, forming a large air pocket between the plunging crest and the vertical wall.

Table 1. *Cont.*

Test Case	Wave Height (m)	Wave Period (s)	Wave Length (m)	Wave Steepness	Load Condition	Observations
H07T8	0.7	8	48.55	0.014	Impulsive	Waves breaking on the slope as above, but with more intense breaking of the secondary wave on the seawall.
H07T6	0.7	6	35.13	0.02	Impulsive	The wave crest overturned directly at the wall, forming a large air pocket between the wave and the structure and leading to considerable impacts with loud noise and vibrations transmitted through the structure and the flume. This case also showed the highest velocities of the up-rushing aerated water jet.

In each test, about 100 waves were generated, but for the following analysis of the surface elevation and pressure measurements, the first and last parts of the time histories were omitted, i.e., only measurements acquired after the establishment of quasi-steady conditions in the flume were considered. The minimum number of omitted waves was 15 on each side of the time history and was varied depending on the incoming wave conditions and the conditions at the wall (pulsating or impulsive).

4. Influence of the Recurve on the Incoming Wave Conditions

In all cases, pulsating and impulsive, the water mass that runs up the wall is deflected by the recurve and re-enters the underlying water surface at a certain distance in front of the wall. The angle of deflection corresponds to the angle of the recurve and during the experiments, the distance of re-entry was physically observed to vary between less than 10 m and 23 m for the cases considered here. The distance is related to the deflection angle and the speed of the up-rushing water mass at the wall, where the latter depends on the incoming wave conditions and is naturally higher for impulsive conditions than for pulsating conditions. The largest distance of about 23 m, therefore, occurred for case H07T6 and the shortest recurve B_rS, i.e., most intense breaking on the wall and an extremity/deflection angle of about 48°.

The deflected water mass surely disturbs the incoming waves, but a more detailed analysis of this interaction was beyond the scope of the present study, not at least as in most of the cases the point of re-entry was out of the field of view of the video cameras. However, in order to assess if the incoming wave conditions are differently influenced by the shape of the recurve (α_e) and by the point of re-entry relative to the phase of the incoming waves, the surface elevation records at 33 m (toe of the approaching slope) and 9 m (closest wave probe to the seawall) in front of the structure were analyzed.

In terms of the elevation record analysis, a zero down-crossing approach was used to calculate the wave period and height of each wave in every record. Then, the statistical properties of each file—e.g., the mean wave height and the standard deviation—were calculated for the part of the record corresponding to quasi-stable conditions in the flume, as explained previously. The ratio of the mean wave height nearest to the wall (H_{234}) over the mean wave height at the beginning of the slope (H_{210}) is plotted over α_e in Figure 4.

At first sight, a clear difference between the wave height ratios can be observed, which reflects the combined effects of shoaling and breaking on the slope (cases H06T8 and H07T8 only) and in particular, the re-entering water mass deflected from the structure. More importantly, the results in Figure 4 do not indicate any significant influence of α_e on the incoming wave heights, with some cases being slightly more influenced than others. However, this might also be attributed to the breaking of the approaching wave on the slope in front of the last wave probe (H_{234}), at least for H06T8 and H07T8. Overall, the height of the incoming waves does not seem to strongly depend on the shape of the recurve, i.e., the re-entry point.

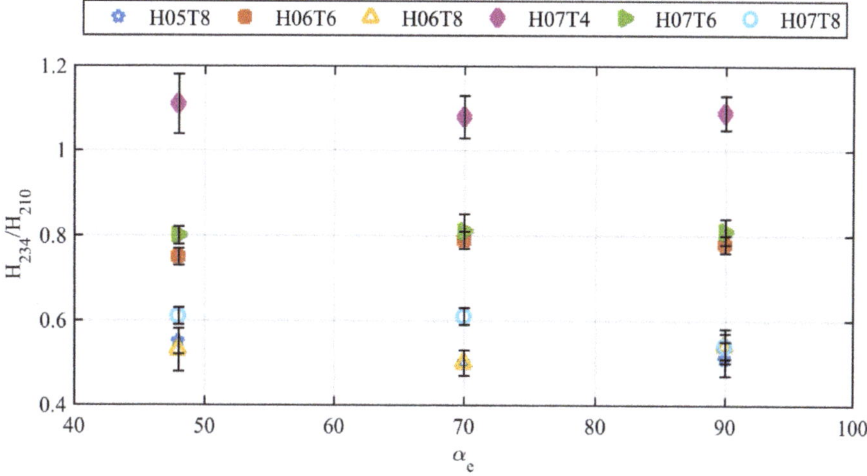

Figure 4. Ratio of mean wave heights close to the wall (H_{234}) and at the toe of the slope (H_{210}) against the extremity angle of the recurve for all test cases.

Nonetheless, the rather small differences in the incoming wave heights for the three recurve geometries do not automatically entail similar small differences for the pressures and forces on the wall. In particular, for impulsive conditions, a small change of the hydrodynamic conditions close to the wall may lead to slightly different breaking conditions, which in turn, may have a considerable impact on the wave-induced pressure and force distributions and magnitudes. This shall be further explored in the following sections considering pulsating and impulsive conditions separately.

5. Pulsating Conditions

The three test cases H07T4, H05T8 and H06T6 were considered as pulsating, while H07T4, H05T8, corresponded to waves that did not break on the seawall and H06T6 resulted in slightly breaking waves at the structure and can, therefore, be considered as transitional to impulsive conditions (cf. Table 1). According to [12], waves breaking slightly on a vertical wall induce a short first peak in the pressure time series, which is a few times larger than the following quasi-static peak. An example of such pressure records for H06T6 is shown in Figure 5.

The distribution of peak pressures along the vertical wall for all three wave conditions and all three recurves is shown in Figure 6. The colours indicate the wave conditions; green: H07T4, red: H05T8, blue H06T6, and the markers indicate the different recurve shapes; diamond: B_rS, cross: B_rM, circle: B_rL. Additionally, the pressure distribution curve proposed by [13] is also plotted (original: dashed-dotted black line; with factor 3: dashed-dotted dark grey line). While Figure 6 shows the peak pressures for each single wave, Figure 7 shows the mean values and a separate plot for each wave condition.

The highest pressure peaks occur above the still water level for all cases, which is in qualitative agreement with the observations of [14] and an indication of pulsating conditions. While the distribution of peak pressures is quite similar for all test cases, the magnitude of the pressure peaks differs considerably with the incoming wave conditions. In particular, the steeper but purely pulsating waves (H07T4) yield the smallest pressures on the wall, while less steep waves (H06T6) breaking slightly on the wall result in the highest pressure peaks, as could be expected from Figure 6. For the non-breaking wave cases (H07T4 and H05T8), some events were also significantly higher than the quasi-static pressure (Figure 6), indicating slightly breaking and deviation from purely pulsating conditions for those particular waves. However, these events occurred rather rarely for H07T4 and slightly more often for H05T8, confirming the quite good agreement of the mean pressure peaks with the empirical pressure distribution curve proposed by [13] for H07T4 and the slightly higher values for H05T8 (Figure 7).

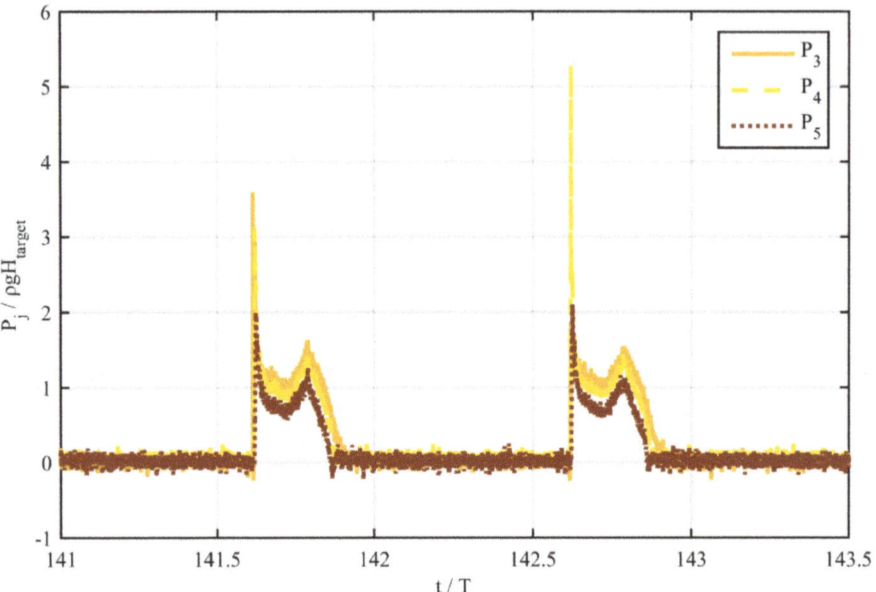

Figure 5. Example pressure time histories at three different locations on the seawall (see also Figure 3) for test case H06T6. Pressure is normalized with hydrostatic pressure of the target wave height (0.6 m), and time with wave period (6 s).

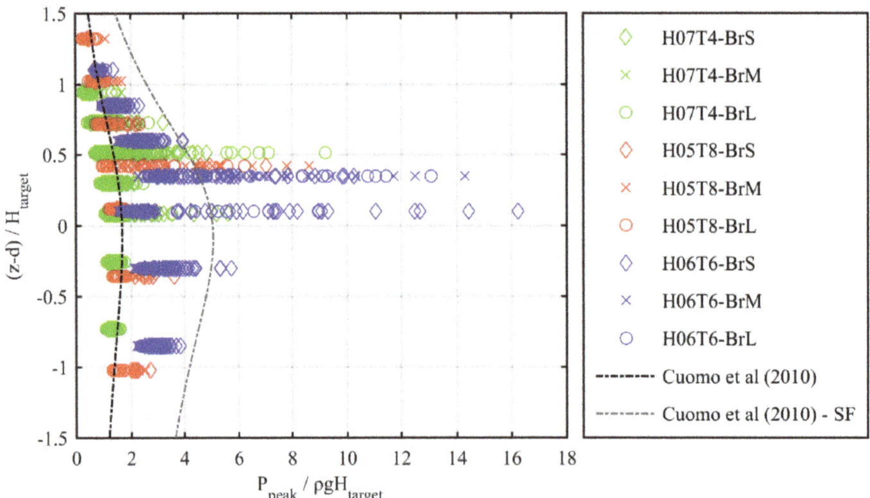

Figure 6. Distribution of pressure peaks over relative location of pressure transducers along the seawall for pulsating conditions.

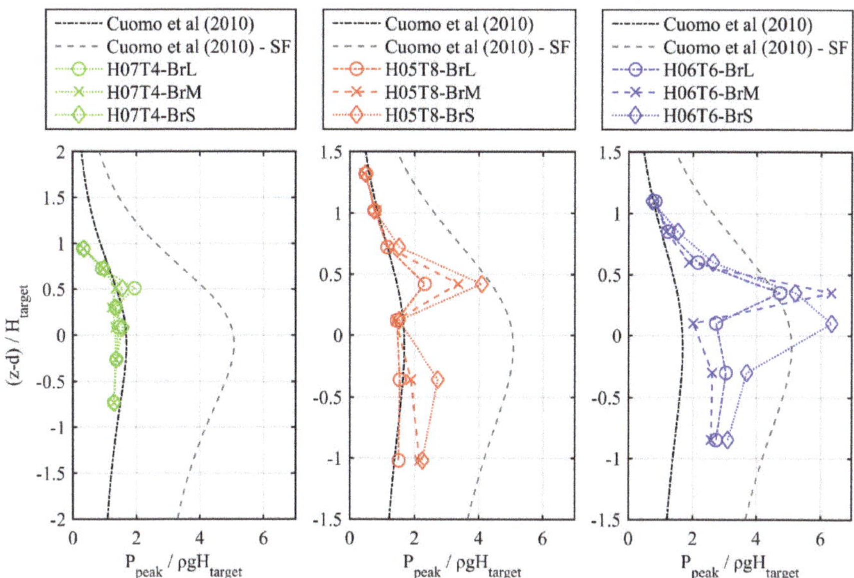

Figure 7. Mean peak pressures at each measuring location for pulsating conditions. Left: H07T4; middle: H05T8; right: H06T6.

In general, as the conditions at the wall diverge gradually from the purely pulsating regime (H07T4 → H05T8 → H06T6), the agreement with the empirical curve reduces as well. With increasing occurrence of slightly breaking waves, the highest mean peak pressures increase as expected, but also the other pressures above and below the still water level increase and the pressure distribution tends towards the empirical curve with a safety factor of 3. In particular, pressures below the still water level increase as there is also a tendency for the location of the highest peak pressures towards still water level. It should be noted here that the empirical curve of [13] was derived from experiments with irregular waves considering the mean of the highest four pressure peaks at each location and normalizing with the significant wave height H_{m0}. The safety factor of 3 was suggested as due to the random nature of irregular waves, single events were observed to be about three times larger than the values provided by the design formula. The present experiments with regular waves confirm good agreement with the design formula for purely pulsating conditions when taking the average of all pressure peaks and normalizing with the incident wave height. Furthermore, it can be noted that the re-entering of the water mass deflected by the recurve introduces a similar kind of randomness in the pressure peaks, which may exceed the mean value also by a factor of about 3.

The influence of the recurve shape can also be clearly identified in Figure 7 by the differences of the mean peak pressures, particularly at the location of the highest pressures and below. This effect increases with increasing divergence from purely pulsating conditions and can be attributed to small changes of the hydrodynamic conditions in front of the wall (cf. Section 5). Changes, which appear to depend on the location of re-entry of the deflected water mass, i.e., at which phase the incident wave is hit by the previous wave deflected from the recurve. The more severe impulsive conditions become, i.e. for more intense wave breaking, the higher and the more sensitive to local wave hydrodynamics the loads on the structure become. This is further illustrated by the mean force peaks shown in Figure 8. While basically no dependence of the forces on the extremity angle can be observed for H07T4, slight deviations can be seen for H05T8 and large differences are obvious for H06T6.

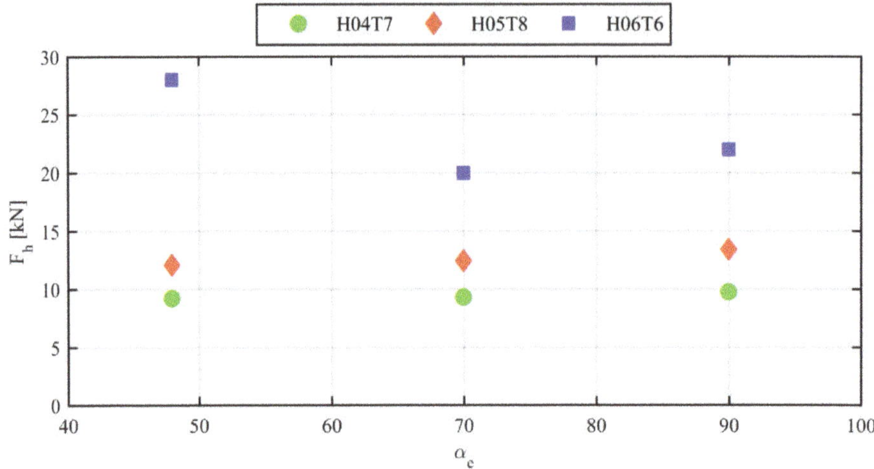

Figure 8. Mean peak force over the extremity angle of the recurve for pulsating conditions on the wall.

6. Impulsive Conditions

The location and magnitude of pressure peaks on seawalls are known to vary with the breaker type. [14], reported numerous experimental observations showing the location of maximum pressures to occur just above the still water for nearly breaking waves, at still water level for waves breaking on the seawall and forming air pockets, and below still water level when the wave crest overturns at a distance from the wall, resulting in the interaction of a plunging bore with the structure.

For the tests presented in the current work, waves for H06T8 and H07T8 plunged on the slope approximately 15 m from the seawall. The residual wave was then observed to propagate and plunge on the wall but bellow the nominal still water level line. Hence, these cases differ from the traditionally broken wave cases where an aeriated bore interacts with the wall. On the contrary, waves for H07T6 were directly breaking on the wall, forming a large air pocket between the wave and the structure (cf. Table 1).

In analogy to the discussion on pulsating conditions above, Figure 9 shows the distribution of peak pressures along the vertical wall for all three wave conditions and all three recurves, and Figure 10 shows the mean values with separate plots for each wave condition. Colours indicate the wave conditions; green: H06T8, red: H07T8, blue H07T6, and the markers indicate the different recurve shapes; diamond: B_rS, cross: B_rM, circle: B_rL.

In agreement with [13] and [14], the highest pressure peaks were recorded at and around the still water level. Even if H06T8 and H07T8 do not represent the classical broken wave cases, the highest peak pressures are found to be below still water level with H07T8 showing significantly stronger impacts than H06T8 as the residual waves were steeper and plunged with higher intensity on the structure. Waves plunging directly on the wall (H07T6) expectedly resulted in the highest pressures, with mean values up to 50 times larger than the quasi static pressures and extreme events, almost 150 times larger.

Just as for the pulsating conditions discussed above, the effect of the recurve shape on the magnitude and distribution of peak pressures along the wall can also be clearly observed for the impulsive conditions in Figure 10. It is again basically restricted to the pressures below the location of maximum impact and it is expectedly even more remarkable than for the pulsating conditions. This confirms the sensitivity of the peak pressures to only slight differences in the local (breaking) wave hydrodynamics and explains the increasing differences with increasing impact magnitude (H06TT8 → H07T8 → H07T6). Even if case H07T6 suggests that impacts also become larger with increasing extremity angle of the recurve, this might just be a coincidence as for the other two cases this relation cannot be observed and also, the mean force peaks do not show this dependence.

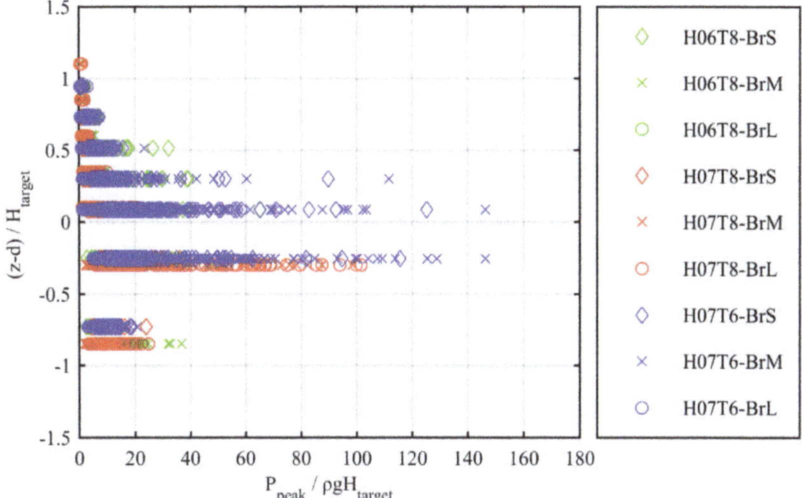

Figure 9. Distribution of pressure peaks over relative location of pressure transducers along the seawall for impulsive conditions.

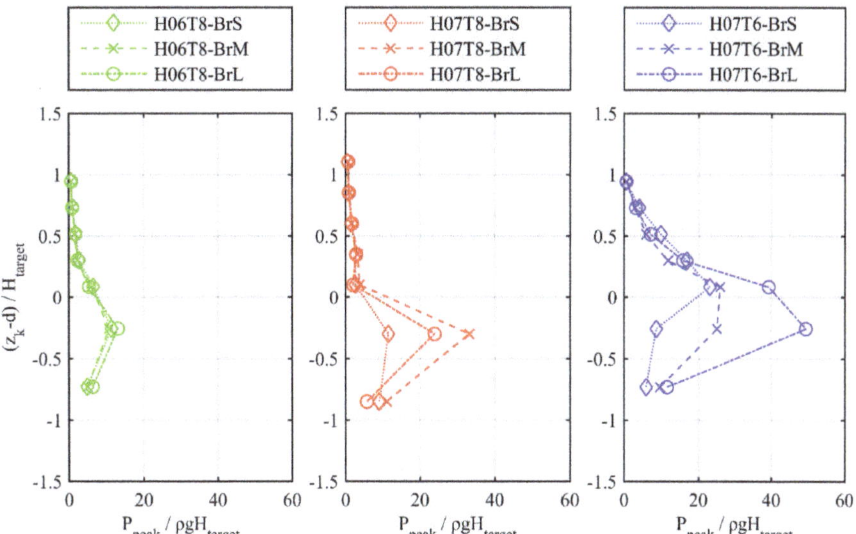

Figure 10. Mean peak pressures at each measuring location for pulsating conditions. Left: H07T4; middle: H05T8; right: H06T6.

Interestingly, this picture changes when only the largest impact events are considered. This is illustrated in Figure 11 where the mean values of the 10% highest force peaks measured on the seawall are plotted over the extremity angle (α_e). The linear trend lines have just been shown for reasons of better illustration and should in no case been interpreted as design formulas not at least as they are only valid for the particular cases considered here and they all go through the origin, implying that no forces act on the wall without recurve ($\alpha_e = 0°$), which is surely unphysical. However, it can be clearly seen that the mean of the 10% highest force peaks increases with increasing extremity angle of the recurve for all three considered wave conditions and that this trend becomes more pronounced with increasing impact intensity. Similar results were also found when the mean of the 33% highest force peaks was considered, but it is beyond the scope of the present study to analyse this in more detail, not at least as the available data does not allow for that.

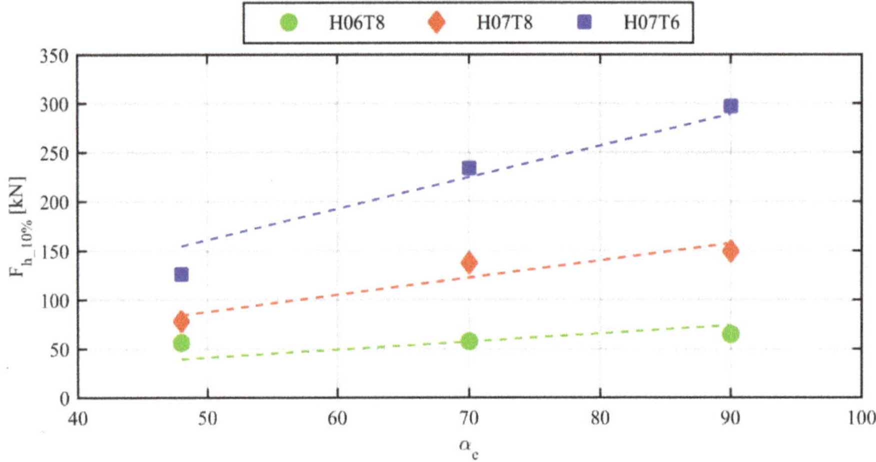

Figure 11. Mean of the 10% highest force peaks recorded for each test case over the extremity angle.

7. Conclusions

The influence of a recurve on wave-induced pressures and loads on a vertical seawall has been examined in large-scale physical model tests. Six different regular wave conditions ranging from pulsating (non-breaking) to impulsive (breaking) wave loads on the structure have been considered and three different recurves with extremity angles of 48° (B_rS), 70° (B_rM) and 90° (B_rL) were tested under the same wave conditions. The water mass running up the vertical wall is deflected by the recurve and re-enters the water in front of the structure at different distances depending on the wave conditions and the extremity angle of the recurve. Although the re-entering water mass may indeed alter the incident waves, the surface elevation measurements presented indicate that the effect of the recurve shape on the incoming wave heights is insignificant. On the other hand, pressures and forces on the vertical wall may change considerably with the recurve shape. While for purely pulsating conditions almost no influence of the recurve extremity angle can be considered, α_e becomes increasingly significant for impulsive conditions. There is no clear relation between the extremity angle of the recurve and the mean peak pressures and forces, but it was found that the mean of the largest force peaks increases with increasing α_e. Characteristically, for the same wave conditions, the mean of the 10% highest force peaks (e.g., $F_{10\%}$) may differ by a factor of more than two when impulsive conditions occur at the seawall; although not presented here, the same behaviour was also found for the mean of the 33% highest force peaks ($F_{33\%}$). Nevertheless, this effect must be further investigated and verified in future studies.

Author Contributions: D.S.: authorship; funding; design; set-up and delivery of experiments; data analysis and presentation; editing and proof read. S.S.: co-authorship, design; set-up and delivery of experiments; data analysis and presentation. R.R.: co-authorship; data analysis and presentation; editing. V.S. co-authorship; data analysis. All authors have read and agreed to the published version of the manuscript.

Funding: Funded from HYDRALAB IV transnational access scheme, contract number 261520.

Acknowledgments: This work has been carried out in the scope of project 'Large-scale measurements of wave loads and mapping of impact pressure distribution at the underside of parapets (HyIV-FZK-06)' under the funding of HYDRALAB IV, contract number 261520. The authors would like to thank the whole team at GWK for the hospitality, technical and scientific support.

Conflicts of Interest: The authors declare that to the best of their knowledge there is no conflict of interest.

References

1. Van Doorslaer, K.; De Rouck, J. Reduction on wave overtopping on a smooth dike by means of a parapet. *Coast. Eng. Proc.* **2011**, *1*, 6. [CrossRef]

2. Veale, W.; Suzuki, T.; Verwaest, T.; Trouw, K.; Mertens, T. Integrated design of coastal protection works for Wenduine, Belgium. *Coast. Eng. Proc.* **2012**, *1*, 70. [CrossRef]

3. Kortenhaus, A.; Haupt, R.; Oumeraci, H. Design aspects of vertical walls with steep foreland slopes. In *Breakwaters, Coastal Structures and Coastlines: Proceedings of the International Conference Organized by the Institution of Civil Engineers and Held in London, UK on 26–28 September 2001*; Thomas Telford: London, UK, 2001; pp. 221–232.

4. Kortenhaus, A.; Pearson, J.; Bruce, T.; Allsop, N.W.H.; van der Meer, J.W. Influence of parapets and recurves on wave overtopping and wave loading of complex vertical walls. *Coast. Struct.* **2003**, 369–381. [CrossRef]

5. Owen, M.W.; Steele, A.A.J. *Effectiveness of Recurved wave Return Walls*; HR Wallingford: Wallingford, UK, 1993.

6. Murakami, K.; Irie, I.; Kamikubo, Y. Experiments on a non-wave overtopping type of seawall. *Coast. Eng. Proc.* **1996**, *1*, 1840–1851.

7. Kamikubo, Y.; Murakami, K.; Irie, I.; Hamasaki, Y. Transportation of water spray on non-wave overtopping type seawall. In Proceedings of the Twelfth International Offshore and Polar Engineering Conference, Kitakyushu, Japan, 26–31 May 2002; Volume 3, pp. 821–826.

8. Allsop, N.W.H.; Alderson, J.S.; Chapman, A. Defending buildings and people against wave overtopping. In Proceedings of the Conference on Coastal Structures, Venice, Italy, 2–4 July 2007.

9. Stagonas, D.; Lara, J.L.; Losada, I.J.; Higuera, P.; Jaime, F.F.; Muller, G. Large scale measurements of wave loads and mapping of impact pressure distribution at the underside of wave recurves. In Proceedings of the HYDRALAB IV Joint User Meeting, Lisbon, Portugal, 2–4 July 2014.

10. Martinelli, L.; Ruol, P.; Volpato, M.; Favaretto, C.; Castellino, M.; De Girolamo, P.; Franco, L.; Romano, A.; Sammarco, P. Experimental investigation on non-breaking wave forces and overtopping at the recurved parapets of vertical breakwaters. *Coast. Eng.* **2018**, *141*, 52–67. [CrossRef]

11. Castellino, M.; Sammarco, P.; Romano, A.; Martinelli, L.; Ruol, P.; Franco, L.; De Girolamo, P. Large impulsive forces on recurved parapets under non-breaking waves. A numerical study. *Coast. Eng.* **2018**, *136*, 1–15. [CrossRef]

12. Kortenhaus, A.; Oumeraci, H. Classification of wave loading on monolithic coastal structures. *Coast. Eng. Proc.* **1998**, *1*. [CrossRef]

13. Cuomo, G.; Allsop, W.; Bruce, T.; Pearson, J. Breaking wave loads at vertical seawalls and breakwaters. *Coast. Eng.* **2010**, *57*, 424–439. [CrossRef]

14. Hull, P.; Müller, G. An investigation of breaker heights, shapes and pressures. *Ocean Eng.* **2002**, *29*, 59–79. [CrossRef]

Article

Two-Dimensional Free-Surface Flow Modeling for Wave-Structure Interactions and Induced Motions of Floating Bodies

Der-Chang Lo [1,*], **Keh-Han Wang** [2] and **Tai-Wen Hsu** [3]

[1] Department of Maritime Information and Technology, National Kaohsiung University of Science and Technology, Kaohsiung City 80543, Taiwan

[2] Department of Civil and Environmental Engineering, University of Houston, Houston, TX 77004, USA; khwang@uh.edu

[3] Department of Harbor and River Engineering, National Taiwan Ocean University, Keelung City 20224, Taiwan; twhsu@mail.ntou.edu.tw

[*] Correspondence: loderg@nkust.edu.tw; Tel.: +886-7810-0888 (ext. 25320)

Received: 31 December 2019; Accepted: 12 February 2020; Published: 15 February 2020

Abstract: In this study, the level set (LS) and immersed boundary (IB) methods were integrated into a Navier–Stokes equation two-phase flow solver, to investigate wave-structure interactions and induced motions of floating bodies in two dimensions. The movement of an interfacial boundary of two fluids, even with severe free-surface deformation, is tracked by using the level set method, while an immersed object inside a fluid domain is treated by the IB method. Both approaches can be implemented by solving the Navier–Stokes equations for viscous laminar flows with embedded objects in fluids. For accurate treatment of the solid–liquid phase, appropriate source terms as forcing functions to take into account the hydrodynamic effects on the body boundaries are added into the governing equations. The integrated compact interfacial tracking techniques between the interfaces of gas–liquid phase and the solid–liquid phase allow the use of a combined Eulerian Cartesian and Lagrangian grid system. Problems related to the fluid-structure interactions and induced motions of a floating body, such as (a) a dam-break wave over a dry bed; (b) a dam-break wave over either a submerged semicircular or rectangular cylinder; (c) wave decomposition process over a trapezoid breakwater; (d) a free-falling wedge into a water body; and (e) wave packet interacting with a floating body are selected to test the model performance. For all cases, the computed results are found to agree reasonably well with published experimental data and numerical solutions. For the case of modeling wave decomposition process, improved solutions are obtained. The detailed features of flow phenomena described by the physical variables of velocity, pressure and vorticity are presented and discussed. The present two-phase flow model is proved to have the advantage of simulating the cases with induced severe interfacial oscillations and coupled gas (or air) motions where the single-phase model may miss the contributions of the air motions on the interfaces. Additionally, the proposed method with uses of the LS and IB methods is demonstrated to be able to achieve the reliable predictions of complex flow fields.

Keywords: two-phase flows; fluid-structure interactions; wave decomposition; floating body

---

1. Introduction

Obtaining solutions of a hydrodynamic model for the studies of complex two-phase flows, such as the propagation of surge fronts, movement of free-surface by a moving body, or waves pass over either completely or partially immersed structures, is an important subject in coastal and ocean engineering applications [1,2]. The key issue of modeling free-surface flows is the treatment of moving interfaces because the physical variables describing the free-surface boundary conditions are not known a priori. The aspect of solving the free-surface flow problem is mainly the description of the mesh movement, using one of the following methods: Eulerian [3–6], Lagrangian [7–10], or arbitrary Lagrangian–Eulerian [11–14].

In the Eulerian method, the coordinates of the grid points are fixed during computation. The main advantage of this method is that the two-phase flow can undergo large deformations. Notably, the significant studies of the Eulerian method are to track the fluid interfaces as the fluid moves through the meshes. The most popular method under this category is the volume of fluid (VOF) method [4], which was originally developed for finite volume method, and the marker and cell (MAC) method [4,15]. In the VOF method, the conservative principle of the volume fraction that moves in the fluid field is applied. The other methods used for tracking the free-surfaces in two-phase flow modeling include the level set method [16–18] and the front-tracking method [19–22]. In the Lagrangian method, it qualitatively gives accurate free-surface shapes that moves with a fluid particle. However, the numerical errors may be introduced due to the overly twisted of re-mesh. Notably, the implementation of this scheme in three dimensions would be very difficult indeed. The advantages of both the Eulerian and the Lagrangian methods are combined to yield the arbitrary Lagrangian–Eulerian (ALE) technique, in which the grid points can be moved independently of the fluid motion. As far as the ALE method is concerned, the nodes in the interior of the domain are displaced in an arbitrarily prescribed way to obtain a mesh of proper shape and thus to avoid mesh crossing. The ALE method can be considered as a moving deformable control volume problem in which the node velocity is different from the fluid velocity. However, the drawback of the ALE method is that the mesh regeneration is not an efficient technique.

While computational fluid dynamics (CFD) used for solving incompressible Navier–Stokes equations remains an important subject, the development of new and effective numerical algorithms for tackling challenging CFD problems have been observed, together with the advancement of the computer technology for simulation efficiency. Solutions for flow-structure interactions with irregular geometries is one of such significant subjects. The simplicity in generating the structured Cartesian grids finds a niche for implementing numerical formulations in comparison with other grid systems, such as body-fitted or unstructured meshes. The present study has thus selected the coordinates in the fluid domain, with the grid points to be fixed in space and time throughout the computation. The fluid can still undergo large deformations without a loss of required accuracy while using the fixed grid system. The immersed boundary (IB) method, first proposed by Peskin [23,24], involves a combined Eulerian and Lagrangian grid system approach, while all the target regions are simulated by using a Cartesian coordinate system. The effect of an immersed object can be modeled through the introduced forcing terms on the right-hand side of the Navier–Stokes equations. Notably, this method enables the application of simple orthogonal grid lines to irregular objects. This approach is particularly effective for analyzing moving objects. The IB method has been applied by various scholars [25–31]. This continuous forcing IB method is also implemented in the present study to solve the Navier–Stokes equations involving a fixed or moving obstacle interacting with fluids. Evidently, the usage of Cartesian grid system in the imposition of boundary conditions at an immersed boundary is complicated. As the immersed boundary can cut through the underlying Cartesian meshes in an arbitrary manner, the main challenge is to formulate a boundary treatment which does not adversely affect the accuracy and the conservation properties of the numerical solvers.

One of the contributions of the present study made to the LS and IB methods is to use a combined Eulerian Cartesian and Lagrangian grid system, to avoid the re-meshing procedure for two-phase flow model involving coupled fluid-structure interaction. A fourth-order Runge–Kutta integration in time and WENO scheme [32] in space are adopted to discretize the LS equation to track the free-surface positions with severe deformations, including wave breaking. Geometries featuring the solid obstacles in a flow domain are embedded in the Cartesian grids, with special treatments with discrete surface markers (Lagrangian grid points) to determine the force density vector at the Lagrangian points. It is a straightforward to compute the force density vector at the Eulerian grid points by using a spreading scheme involving the Lagrangian marker points. The solution with velocity and pressure variables can be easily implemented based on the finite difference method on a Cartesian grid system using a fractional-step projection method. Carrica et al. [33] used a dynamic overset grids for the analysis of hydrodynamics of ship motion. The dynamic overset mesh has a good numerical stability over ALE, as constant re-meshing of the solution domain is not necessary. The accuracy of the method lies in establishing the connections between the overset grids and the underlying grid points in the overlapping region. Later, Yang et al. [34] adopted a sharp interface method instead of the overset grids, to handle the coupling computation for solid–fluid interactions, where the interfacial flow was treated with a two-phase method. More recently, Calderer et al. [35] presented a hybrid curvilinear immersed boundary and level set approach for the simulation of air/water interaction with complex floating structures. Additionally, the present study adopts a special treatment based on the normal momentum equation with the pressure projection boundary condition method to calculate the forces on a body. Bihs and Kamath [36] used an open-source CFD code REEF3D for floating body simulations based on a six degrees of freedom (6DOF) algorithm. That model is represented with a combination of the level set method and ghost cell immersed boundary method. Notably, we present a compact and effect two-phase flow code in modeling fluid-structure interactions involving gas, liquid, and solid phases, with the use of a combined Eulerian Cartesian and Lagrangian grid system. In this study, the model is used extensively in simulating complex wave hydrodynamics and fluid-structure interactions, such as the dam-breaking flow, waves over submerged structures, and wave breaking. With a novel approach, the present model is also applied to investigate the dynamic response of a 2D floating body that can move in the three degrees of freedom (3DOF). The proposed numerical model can be comfortably extended to study three-dimensional two-phase flow problems by including a 6DOF algorithm. In this study, a numerically efficient two-phase flow model is developed to simulate two-dimensional nonlinear viscous flows including those initialized by fluid-structure interactions and the motion of floating body. The governing equations with the use of a combined LS and IB approach are solved to investigate the complex viscous free-surface flow problems. The LS method is applied to track the free-surface positions with severe deformations. Numerically, the flow equations with velocity and pressure as the physical variables are discretized over a Cartesian grid system and solved using the projection method and Adams–Bashforth time-stepping finite difference scheme. Featuring the solid bodies in a flow domain, the geometries with Lagrangian-based boundary marker points are embedded in the Cartesian grids with the imposed interpolation treatments, to ensure the accuracy of the solutions in the cut cells. Selected cases, including a uniform flow passing around a cylinder, two cylinders moving against each other in a viscous fluid, dam-break-induced surge waves, a periodic wave over a submerged trapezoidal bar, a free-falling wedge, and a wave packet interacting with a floating body, were simulated to examine the phenomena of interfacial flows involving the computations in gas–liquid and liquid–solid phases. Comparisons between the present modeling results with other published solutions are also carried out to demonstrate the robust and accuracy of the developed two-phase flow model.

2. Research Methods

The two-phase free-surface modeling is briefly described in this section. The level set (LS) and immersed boundary (IB) methods are involved to investigate wave-structure interactions and induced motions of floating bodies.

2.1. Two-Phase Flows Using Level Set Formulation

A two-phase flow model in which the interface between two fluids is captured using a level set (LS) method is developed in this study. Additionally, an immersed boundary (IB) method is adopted to obtain the hydrodynamic forces on structures after the interaction by moving fluids. The movement of the interfacial boundary between two fluids and the calculation of surface tension are based on the fixed grids assigned in the domain. Figure 1 depicts a schematic diagram showing a fixed Cartesian grid system that is used for solving the Navier–Stokes equations. Under the assumption of incompressible fluid medium and considering the effects of surface tension between two fluids and an immersed body, the governing equations for two-phase flows can be formulated as follows:

$$\nabla \cdot \mathbf{u} = 0 \tag{1}$$

$$\frac{\partial \mathbf{u}}{\partial t} + \nabla \mathbf{u}\mathbf{u} = -\frac{\nabla p}{\rho(\phi)} + \frac{\mu(\phi)}{\rho(\phi)} \nabla \cdot \left(\nabla \mathbf{u} + \nabla^T \mathbf{u}\right) - \frac{\sigma k(\phi)\nabla\phi\delta(\phi)}{\rho(\phi)} + \mathbf{g} - A_b \mathbf{u} + \mathbf{f} \tag{2}$$

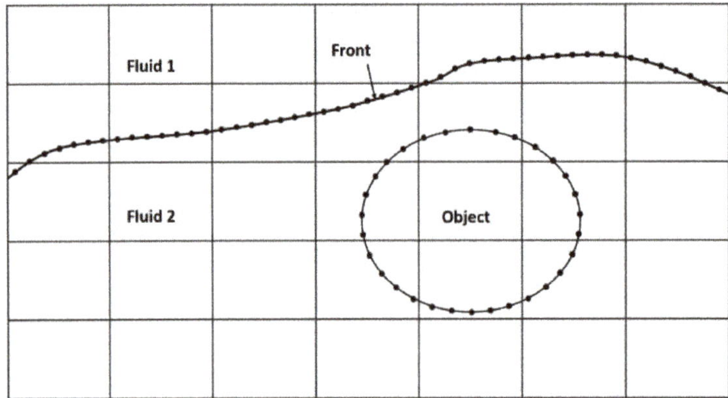

Figure 1. Computations of flow containing two-phase, including immersed boundary. The governing equations are solved on a fixed grid, but the interface between two fluids is represented by a moving front in which it is consisting of connected marker points, and the immersed object is included in the governing equation by the immersed boundary method.

Here, the computational domain contains two different fluids, and δ is the Dirac delta function. σ is the coefficient of surface tension, k is the curvature of an interface, $\mathbf{f}$ denotes hydrodynamic forces induced by an immersed body, A_b as defined later represents the wave absorbing coefficient, $\mathbf{g}$ is the gravitational acceleration vector, t is time, and ϕ denotes the level set function. The 2D velocity vector, $\mathbf{u}$, describes the velocity components in x and z directions, and p is the pressure. More detailed descriptions of the physical variables given in Equation (2) can be found in Sethian and Smereka [37].

For the level set function, $\phi(\mathbf{x}, t)$, typically, when $\phi < 0$, it denotes the gas region, while $\phi > 0$ is for the liquid region. The fluid properties, such as the density (ρ) and viscosity (μ) can be interpolated according to the following:

$$\rho(\phi) = \rho_l H(\phi) + \rho_g (1 - H(\phi)) \text{ and } \mu(\phi) = \mu_l H(\phi) + \mu_g (1 - H(\phi)) \tag{3}$$

where $H(\phi)$ is the Heaviside function [38].

$$H(\phi) = \begin{cases} 0 & \phi < -\varepsilon \\ \frac{1}{2}\left(1 + \frac{\phi}{\varepsilon} + \frac{1}{\pi}\sin\left(\frac{\pi\phi}{\varepsilon}\right)\right) & |\phi| \le \varepsilon \\ 1 & \phi > \varepsilon \end{cases} \tag{4}$$

The interfacial thickness of the Heaviside function is approximately 2ε, where ε is a small parameter that is related to the order of the size of the mesh cells close to the interface and can be selected through the numerical tests. The density (ρ) and viscosity (μ) with the subscripts l and g represent, respectively, the corresponding properties of liquid and gas.

The level set function (ϕ) describing the interface that moves with the fluid particles is governed by the transport equation:

$$\frac{\partial \phi}{\partial t} + \mathbf{u} \cdot \nabla \phi = 0 \tag{5}$$

For the numerical accuracy, a fourth-order Runge–Kutta integration in time and WENO [32] scheme in space are adopted to discretize Equation (5).

In Equation (2), the smoothed delta function can be written as follows:

$$\delta(\phi) = \frac{dH}{d\phi} = \begin{cases} \frac{1}{2}\left(1 + \cos\left(\frac{\pi\phi}{\varepsilon}\right)\right)/\varepsilon & |\phi| \le \varepsilon \\ 0 & \text{otherwise} \end{cases} \tag{6}$$

Moreover, the curvature $k(\phi)$ can be computed as follows:

$$k(\phi) = \nabla \cdot \mathbf{n} = \nabla \cdot \left(\frac{\nabla \phi}{|\nabla \phi|}\right)_{\phi=0} \tag{7}$$

With the consideration of a wave-absorbing region, the proposed absorbing coefficient (A_b) by Lin and Liu [39] is adopted:

$$A_b = C_\alpha \frac{\exp\left[\left(\frac{|x-x_{xt}|}{x_{ab}}\right)^{n_c}\right] - 1}{\exp(1) - 1}, \quad x_{xt} < x < x_{xt} + x_{ab} \tag{8}$$

where A_b denotes the absorbing coefficient, and x_{xt} and x_{ab} are the starting position and length of the absorbing region. C_α and n_c are the empirical damping related coefficients. Here, the coefficients of $C_\alpha = 200$ and $n_c = 10$, as recommended by [39], are used in numerical simulations.

2.2. Numerical Formulations

The governing equations (Equations (1) and (2)) are solved by using the projection method [40], where the predicted intermediate velocity field with the known values obtained at the n (previous) time level are firstly calculated. According to the two-step calculation procedure and Adams–Bashforth time-stepping approach, the solutions of velocity vector ($\mathbf{u}^*$) at each intermediate (between n and n + 1) time level are determined by solving Equation (2) with the expressions of temporal difference as follows:

$$\frac{\hat{\mathbf{u}} - \mathbf{u}^n}{\Delta t} = \mathbf{RHS} \tag{9}$$

$$\frac{\mathbf{u}^* - \mathbf{u}^n}{\Delta t} = \mathbf{RHS} + \mathbf{f}\left(\hat{\mathbf{u}}\right) \tag{10}$$

where Δt is the time step, and $\mathbf{u}^n$ is the velocity vector at the n time level. The terms with the effects of convection, diffusion, surface tension, wave absorbing, and gravity are combined with a notation of $\mathbf{RHS} = 1.5\mathbf{A}^n - 0.5\mathbf{A}^{n-1}$:

$$\mathbf{A} = -\nabla \mathbf{u}\mathbf{u} + \frac{\mu(\Phi)}{\rho(\Phi)}\nabla \cdot \left(\nabla \mathbf{u} + \nabla^T \mathbf{u}\right) - \frac{\sigma k(\Phi)\nabla\Phi\delta(\Phi)}{\rho(\Phi)} + \mathbf{g} - A_b\mathbf{u} \tag{11}$$

In Equation (10), $\mathbf{f}$ is an added term describing force density induced by the immersed bodies. Its calculation procedure, based on the IB method, is provided in the following section. With a flow passing over a rigid body, the calculation domain is separated by two phases, where one is the ordinary fluid flow surrounding the rigid body, and the other is the body phase with an inclusion of a body forcing. As can be seen in Equation (10), the velocity vectors, $\hat{\mathbf{u}}$, are first computed without considering the effect of immersed boundary. Then, the forcing term, $\mathbf{f}$, is computed by a continuous forcing IB method with the computed velocity vectors, $\hat{\mathbf{u}}$. By imposing a body force, $\mathbf{f}$, in those cells, which has the property of a signed distance function near the immersed boundary, the intermediate velocity field is updated from Equation (10).

Taking the divergence of Equations (1) and (2), the pressure field can be obtained by solving the Poisson equation, which is derived by satisfying the continuity equation:

$$\nabla \cdot \left(\frac{1}{\rho(\phi)} \nabla p \right) = \frac{1}{\Delta t} (\nabla \cdot \mathbf{u}^*) \tag{12}$$

The pressure Neumann boundary conditions should be satisfied. We have the following:

$$\frac{1}{\rho(\phi)} \nabla p \cdot \mathbf{n} = - \left(\frac{\mathbf{u}^{n+1} - \mathbf{u}^*}{\Delta t} \right) \tag{13}$$

Once the velocity vectors at the intermediate time level are determined, Equation (12) can be solved, using a spline alternating direction implicit method (SADI) with a standard tri-diagonal matrix solver applied for the calculations of the pressure field. Here, the SADI method is used in the z direction. The iteration consists of a sweep through the i index for each (i, k) pair, solving implicitly for the pressures in the k-index direction, where i and k represent, respectively, the grid indices along x and z directions. Then, the velocity vectors at the new (n + 1) time level can be obtained by solving the following equation:

$$\frac{\mathbf{u}^{n+1} - \mathbf{u}^*}{\Delta t} = - \frac{1}{\rho(\phi)} \nabla p \tag{14}$$

In terms of the evaluation of the level set function (ϕ) for capturing the positions of the two-phase interfaces, Equation (5), when expressed in Lagrangian description, is given as follows:

$$\frac{d\phi}{dt} = L(\phi) \tag{15}$$

where $L(\phi)$ in Equation (15) is an approximation (e.g., WENO approximation in these lecture notes). Using a uniform grid in one space dimension as an example, Equation (15) results in an ordinary differential equation (ODE) with a discrete conservative finite difference scheme.

Then, the fourth-order Runge–Kutta scheme, as shown below, is used to advance the interfaces:

$$\phi = \phi^0 + \left(\frac{k_1}{6} + \frac{k_2}{3} + \frac{k_3}{3} + \frac{k_4}{6} \right) \tag{16}$$

where ϕ^0 is the coordinates of the interfacial points from the previous time step, and k_1, k_2, k_3, and k_4 are obtained as follows:

$$k_1 = L\left(\phi^0, t \right) \Delta t \tag{17}$$

$$k_2 = L\left(\phi^0 + \frac{k_1}{2}, t + \frac{\Delta t}{2} \right) \Delta t, \tag{18}$$

$$k_3 = L\left(\phi^0 + \frac{k_2}{2}, t + \frac{\Delta t}{2} \right) \Delta t \tag{19}$$

$$k_4 = L\left(\phi^0 + k_3, t + \Delta t \right) \Delta \tag{20}$$

Here, the computed $\mathbf{u}^{n+1}$ are used as the advection needed velocity field for the calculation of the level set function.

2.3. Immersed Boundary (IB) Method

The IB method basically follows the concept that the effect of an immersed structure on its surrounding flow is modeled through a force density vector (f), which is included in the Navier–Stokes equations. Figure 2 depicts an example plot showing the arrangement of the Eulerian and Lagrangian grid points used for transferring the values of variables between them through an interpolation procedure with a weighted sum of regularized Dirac delta function, δ_h. Here, the body boundary (fluid–solid interface) is distributed with discrete surface markers (Lagrangian grid points). The force density vector (f) on the right-hand side of Equation (2) can be formulated as follows:

$$\mathbf{f}(\mathbf{x},t) = \int \mathbf{F}(\mathbf{X}_l,t)\delta_h(\mathbf{x}-\mathbf{X}_l)d\mathbf{x} \tag{21}$$

where $\mathbf{F}(\mathbf{X}_l,t)$ is the force density vector at the Lagrangian points with coordinates $\mathbf{X}_l$ along the body boundary. The $\mathbf{F}(\mathbf{X}_l,t)$ values at Lagrangian points are computed beforehand and then transferred into the Eulerian meshes for $\mathbf{f}(\mathbf{x},t)$, using Equation (21). The discrete Dirac delta function δ_h developed by Peskin [23] is adopted in this study. The expression of δ_h for a 2D case is given as follows:

$$\delta_h(\mathbf{x}-\mathbf{X}) = \delta_h^{1D}(x_1 - X_1)\delta_h^{1D}(x_2 - X_2) \tag{22}$$

where $\delta_h^{1D}(x_1 - X_1) = \frac{1}{h}\varphi(r_1)$ is the regularized one-dimensional delta function. The formulations of $\varphi(r)$, proposed both by Roma et al. [25] and Yang et al. [29], are as follows, respectively:

$$\varphi(r) = \begin{cases} \frac{1}{6}\left(5 - 3|r| - \sqrt{-3(1-|r|)^2 + 1}\right) & 0.5 \le |r| \le 1.5 \\ \frac{1}{3}\left(1 + \sqrt{-3|r|^2 + 1}\right) & |r| \le 0.5 \\ 0 & \text{otherwise} \end{cases} \tag{23}$$

$$\varphi(r) =$$
$$\begin{cases} \frac{55}{48} - \frac{\sqrt{3}\pi}{108} - \frac{13|r|}{12} + \frac{r^2}{4} + \frac{2|r|-3}{48}\sqrt{-12r^2 + 36|r| - 23} + \frac{\sqrt{3}}{36}\arcsin\left(\frac{\sqrt{3}}{2}(2|r|-3)\right) & 1.0 \le |r| \le 2.0 \\ \frac{17}{48} + \frac{\sqrt{3}\pi}{108} + \frac{|r|}{4} - \frac{r^2}{4} + \frac{1-2|r|}{16}\sqrt{-12r^2 + 12|r| + 1} - \frac{\sqrt{3}}{12}\arcsin\left(\frac{\sqrt{3}}{2}(2|r|-1)\right) & |r| \le 1.0 \\ 0 & \text{otherwise} \end{cases} \tag{24}$$

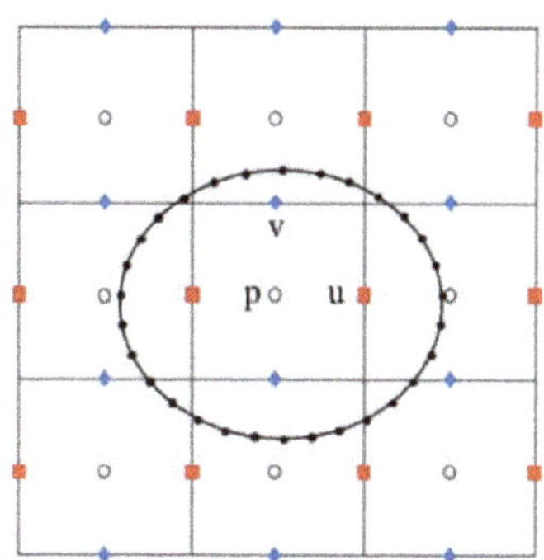

Figure 2. Arrangements of staggered Cartesian grid and embedded interface with marker points in a two-dimensional case.

These equations are used together with the selection of the influential radius, r, to be either $1.5\Delta h$ (Equation (23)) or $2\Delta h$ (Equation (24)) for the numerical sensitivity study. The force density vector, $\mathbf{F}(\mathbf{X}_l,t)$, at the interfacial Lagrangian marker points is calculated based on the time rate change of velocity between the body velocity ($\mathbf{U}_b$) and the Lagrangian point velocity ($\mathbf{U}^\Gamma$):

$$\mathbf{U}^\Gamma(\mathbf{X}_l,t) = \sum\sum \hat{\mathbf{u}}(\mathbf{x},t)\delta_h(\mathbf{x}-\mathbf{X}_l)h^2 \tag{25}$$

$$F(\mathbf{X}_l, t) = \frac{\mathbf{U}_b(\mathbf{X}_l, t) - \mathbf{U}^{\Gamma}(\mathbf{X}_l, t)}{\Delta t} \tag{26}$$

where the velocity vector ($\mathbf{U}_b$) of an immersed body is determined from the equations of motion. Accordingly, the force density vector (f) at the Eulerian grid points can be calculated by using a spreading scheme involving the Lagrangian marker points as follows:

$$f(\mathbf{x}, t) = \sum \sum F(\mathbf{X}_l, t)\delta_h(\mathbf{x} - \mathbf{X}_l)\Delta V_l \tag{27}$$

where ΔV_l is the control volume defined about the m-th Lagrangian marker. The resulting Eulerian force (f) from Equation (27) is then substituted into Equation (10) for solution computation.

2.4. Rigid Body Dynamics

The movements of a body subject to the external forces and moments are governed by the equations of motion for linear and angular momentum of a rigid body. In this vertical two-dimensional modeling study, the equations of motion for a body subject to external forcing can be formulated by considering the effects of inertial motion and internal mass. According to [41], the equations describing the translational and rotational motions of a moving body are expressed as follows:

$$\rho_b V_b \frac{d\mathbf{U}_b}{dt} = (\rho_b - \rho)V_b \mathbf{g} - \rho \int f dV + \rho \frac{d}{dt} \int \mathbf{u} dV \tag{28}$$

$$I_b \frac{d\omega_b}{dt} = -\rho \int \mathbf{r} \times f dV + \rho \frac{d}{dt} \int (\mathbf{r} \times \mathbf{u})dV \tag{29}$$

Here, V_b, I_b, and ρ_b are volume, moment of inertia, and density of the body, respectively; ρ is the fluid density; $\mathbf{r}$ represents the position vector relative to the body centroid, $\mathbf{g}$ is the vector of gravitational acceleration, and the symbol $\times$ denotes the cross product. $\mathbf{U}_b$ and ω_b are respectively the translational and angular velocity vectors of the rigid body. In the present study, the Euler forward difference method is used to solve Equations (28) and (29) for $\mathbf{U}_b$ and ω_b. The velocity vector at the surface of the rigid body can be calculated as follows:

$$\mathbf{U}(\mathbf{X}_l) = \mathbf{U}_b + \omega_b \times \mathbf{r} \tag{30}$$

3. Results and Discussions

Using the developed 2D two-phase flow model that integrates the approaches of level set (LS), immersed boundary (IB), and structural responses, selected fluid-structure interaction problems are investigated. The study cases include the following: (a) a uniform flow passing through a circular cylinder to verify the adopted IB methodology, (b) two cylinders moving against each other to test the moving boundary effect, (c) dam-break flows to examine the free-surface tracking capability, (d) dam-break wave flowing over a submerged structure, (e) wave decomposition process over a submerged trapezoid breakwater, (f) a free-falling wedge, and (g) wave packet interacting with a floating body. It should be noted the simulations carried out involve the interfacial conditions specified at the gas–liquid phase and the liquid–solid phase.

3.1. Flow Passing through a Cylinder

In order to verify the present IB method, cases with steady uniform flows passing through a circular cylinder are simulated for validation. The flow results are obtained at Re = 20, 40, 80, 100, and 200, where Re is the Reynolds number defined as Re = UD/υ. Here, υ denotes the kinematics viscosity of a fluid, U is the uniform flow velocity, and D denotes the diameter of the cylinder. In the test cases, a cylinder with a diameter of one unit is embedded in a 32 × 16 computational domain. Upstream flow velocity of one unit is assumed to be uniformly distributed in vertical direction. The accuracies of the predicted drag and lift coefficients, and other physical variables for the selected cases, are evaluated. For the flow conditions of Re = 20 and 40, the computed reattachment length, Lw, and drag coefficient, C_D, as given in Table 1, are compared with the published experimental [42] and numerical [43–47] results. Additionally, to ensure the grid independence, included in Table 1 are the present results for the separately used mesh sizes of $\Delta h = 1/30$ and $\Delta h = 1/40$. As can be seen in Table 1 for the cases of Re = 20 and 40, the converged solutions in L_w and C_D are obtained, and they are in good agreements with other published results. By refining the computational grids, it is shown to be able to improve the solutions for the flow conditions at a larger Reynolds number, such as Re = 100 or 200; this confirms the global accuracy of the present method. Comparisons between the present and other numerical results [43,46–50] of the drag and lift coefficients (C_D and C_L), and Strouhal number (St) for the cases of Re = 100 and 200 are summarized in Table 2. The present results were computed by using a refined grid size, i.e., $\Delta h = 1/50$. It can be seen the present results are shown to have very good agreements with others given in the literature. For the cases of Re = 100 and 200, the transient solution procedure was carried out to reach the prescribed time, i.e., t = 200. Time variations of the dimensionless drag and lift coefficients at Re = 200 are presented in Figure 3. Due to the induced vortex motions, the transition processes until the C_D and C_L approach to the stabilized periodic variations are clearly depicted in Figure 3. Figure 4 exhibits the instantaneous distribution patterns of pressure and vorticity at t = 200 for various Reynolds numbers of 40, 80, 100, and 200. The patterns of vortex shedding for the cases of Re = 80, 100, and 200 can be clearly observed. The predicted results of vortex shedding demonstrate the proposed method can predict accurately the transient flow patterns. As can be seen, the wake triggered at the cylinder surface is asymmetric in reference to the direction of flow. The formation of vortices on either side of the flow direction does not simultaneously occur. In fact, the shedding of vortices alternates from side to side, which had been observed in experiments. Thus, the pressure distribution around the obstacle is asymmetric about the flow direction.

Table 1. Drag coefficient and reattachment length for flow past a stationary cylinder at Re = 20 and Re = 40.

Authors	Re = 20		Re = 40	
	C_D	L_w	C_D	L_w
Tritton [42]	2.22	-	1.48	-
Calhoun [43]	2.19	0.91	1.62	2.18
Russel and Wang [44]	2.22	0.94	1.63	2.35
Silva et al. [45]	2.04	1.04	1.54	2.55
Xu and Wang [46]	2.23	0.92	1.66	2.21
Kolahdouz et al. [47]	2.10	0.93	1.58	2.31
Present $\Delta h = 1/30$	2.15	0.94	1.57	2.25
Present $\Delta h = 1/40$	2.15	0.94	1.57	2.25

Table 2. Drag and lift coefficients and Strouhal number for flow past a stationary cylinder at Re = 100 and Re = 200.

Authors	Re = 100			Re = 200		
	C_D	C_L	St	C_D	C_L	St
Calhoun [43]	1.330 ± 0.014	±0.298	-	1.172 ± 0.058	±0.668	-
Xu and Wang [46]	1.423 ± 0.013	±0.340	0.171	1.420 ± 0.040	±0.66	0.202
Kolahdouz et al. [47]	1.370 ± 0.015	±0.351	0.168	1.390 ± 0.060	±0.75	0.198
Liu et al. [48]	1.350 ± 0.012	±0.339	0.164	1.170 ± 0.058	±0.67	0.202
Choi et al. [49]	1.340 ± 0.011	±0.315	0164	1.360 ± 0.048	±0.64	0.191
Griffith and Luo [50]	-	-	-	1.360 ± 0.046	±0.70	0.195
Present $\Delta h = 1/50$	1.400 ± 0.014	±0.341	0.166	1.380 ± 0.050	±0.67	0.199

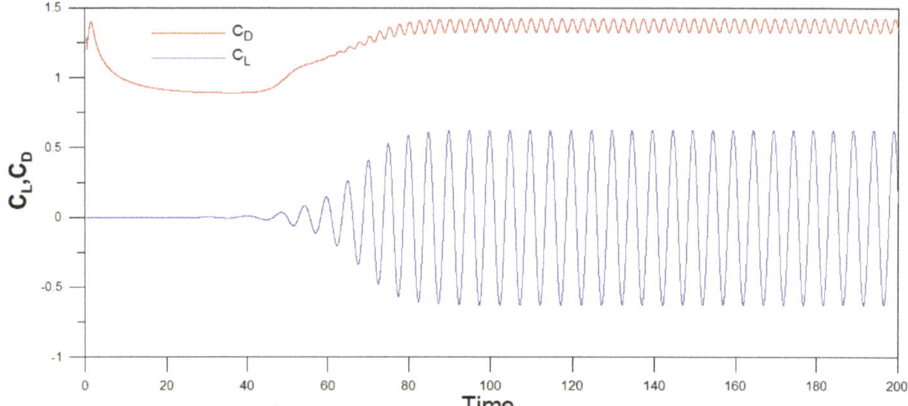

Figure 3. Drag and lift coefficients at Re = 200.

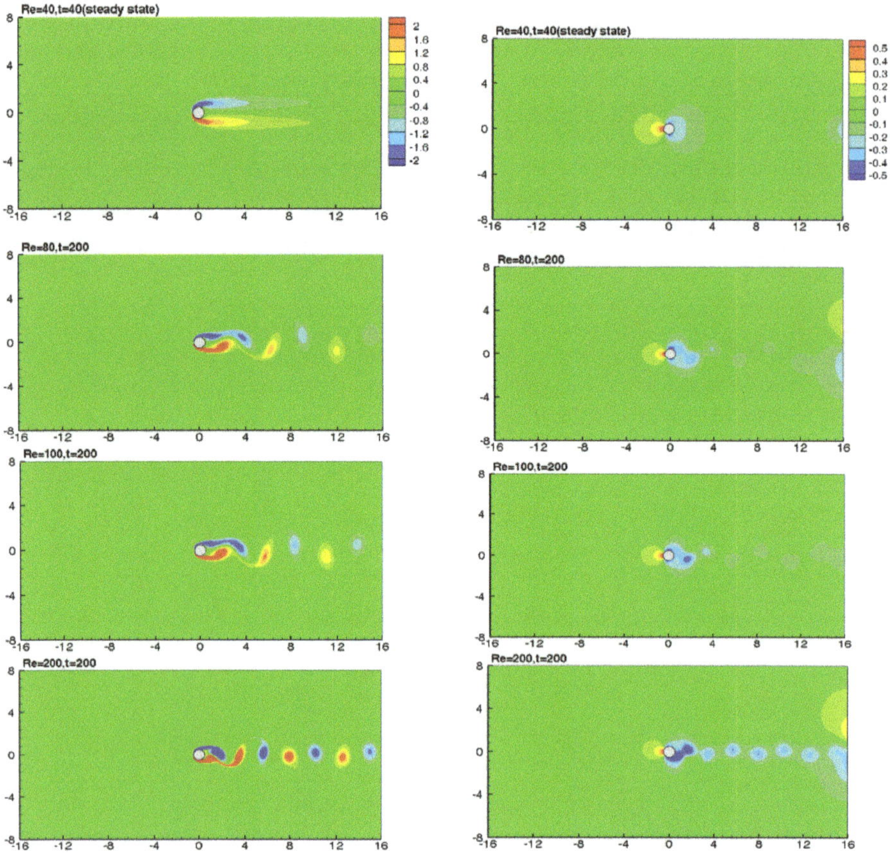

Figure 4. Instantaneous distribution patterns of (**a**) vorticity and (**b**) pressure at $t = 200$ for a uniform flow past a cylinder at Re = 40, 80, 100, and 200.

3.2. Two Cylinders Moving Against Each Other in Viscous Fluid

The present model is further tested to examine the hydrodynamic effect induced by two cylinders moving in opposite directions, in a domain of viscous fluid. Again, each of the two moving cylinders has a diameter of 1 unit, and the computational domain is 32 × 16. The constant velocities of the upper and lower moving cylinders are set respectively as −1 and 1 units. As shown in Figure 5, the centers of the two cylinders are initially placed at the locations of (0, − 0.75) and (16, 0.75), separately, within the domain ranges of −8 ≤ x ≤ 24 and −8 ≤ z ≤ 8.

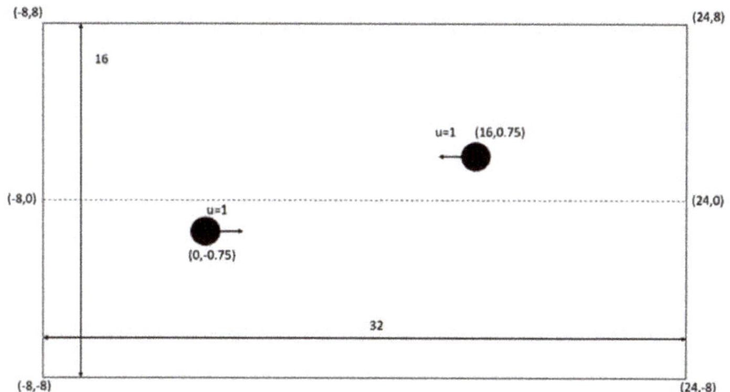

Figure 5. Schematic diagram of two moving cylinders in a viscous fluid domain.

To test the sensitivity of the grid size on the model solutions for this moving boundary problem, the set computational grid systems include 640 × 320, 960 × 480, and 1280 × 640. The time step was selected to satisfy the Courant–Friedrichs–Lewy (CFL) condition [51]. The CFL for n-dimensional domain can be generalized as $C = \Delta t \sum_{i}^{n} \frac{u_{x_i}}{\Delta x_i} \leq C_{max}$. Here, C is the Courant number, Δx_i, $i = 1, \ldots, n$, is the grid size in the ith dimension, which can be a varied value in a given dimension, u_{x_i} denotes the magnitude of velocity in the ith dimension (A two-dimensional case is used here), and Δt represents the fractional time-step. C_{max} is typically set to be equal to 1. Figure 6 presents the time variations of the dimensionless lift coefficient for the upper cylinder at Re = 40. The results from Russel and Wang [44] are also included in Figure 6, for comparisons. Russel and Wang [44] presented an efficient method based on an underlying Cartesian grid for solving 2D incompressible viscous flows around multiple moving objects. It can be seen the present results agree well with those from [44], especially under the grid system of 1280 × 640 (i.e., Δh = 1/40). The time varying drag coefficient (C_D) and lift coefficient (C_L) for the upper cylinder under the conditions of various Reynolds numbers are shown in Figure 7. From Figure 7, it is noted that, for all cases, the minimum values of the drag coefficient occur roughly at $\widetilde{t} = 16$, where $\widetilde{t}$ = Ut/(D/2); meanwhile, for the lift coefficient, the minimum values take place at a delayed time of around $\widetilde{t} = 18$. In general, the results show that the drag coefficient and the positive portion of the lift coefficient decrease with an increase in the Reynolds number. The transitions of the drag and lift coefficients become particularly pronounced at the timeframe from $\widetilde{t} = 14$ to $\widetilde{t} = 20$. In addition, to demonstrate that the present model is capable of successfully simulating the flow fields, Figure 8 depicts the instantaneous vorticity contours of two moving cylinders for Re = 100 at four selected instants. The color-coded vorticity contours range from −5 to 5. In this transient simulation study, the results of hydrodynamic effect and vortex shedding phenomena under the condition of two cylinders moving against each other in viscous fluid are found to be completely different when compared to those for a single moving body at the same Reynolds number. Due to the viscous effect within the boundary layers of the solid bodies, the moving rigid bodies are noticed to play an important role in generating the vortex flow in the immersed viscous fluid domain.

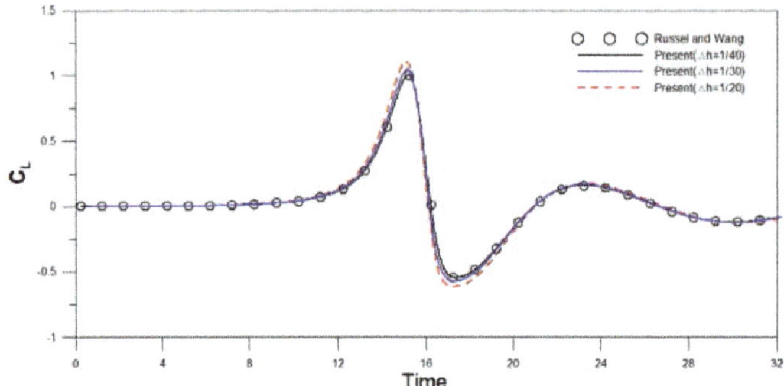

Figure 6. Comparisons of lift coefficient on the upper cylinder at Re = 40.

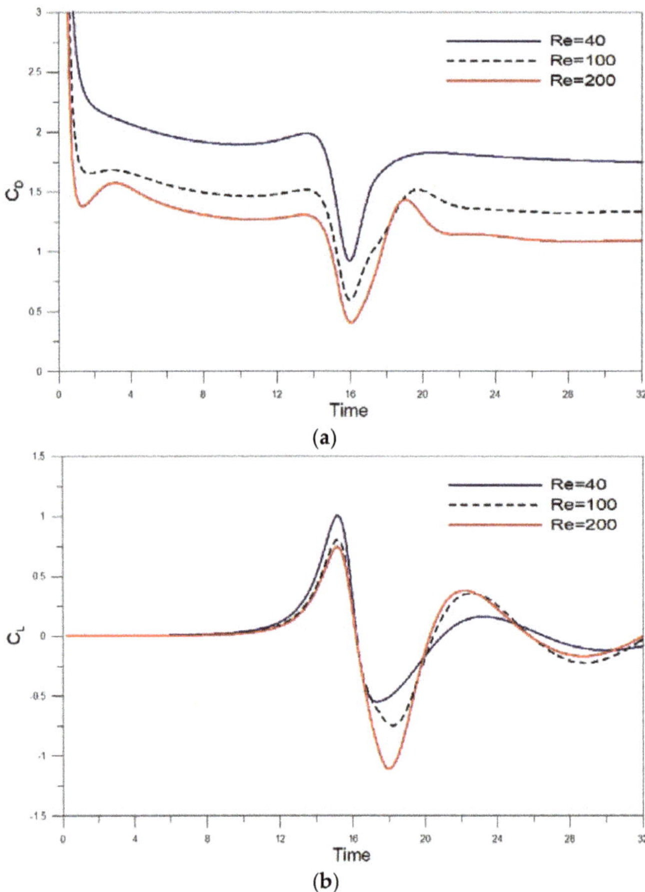

Figure 7. Comparisons of drag and lift coefficients on upper cylinder at various Reynolds numbers: (**a**) drag coefficient and (**b**) lift coefficient.

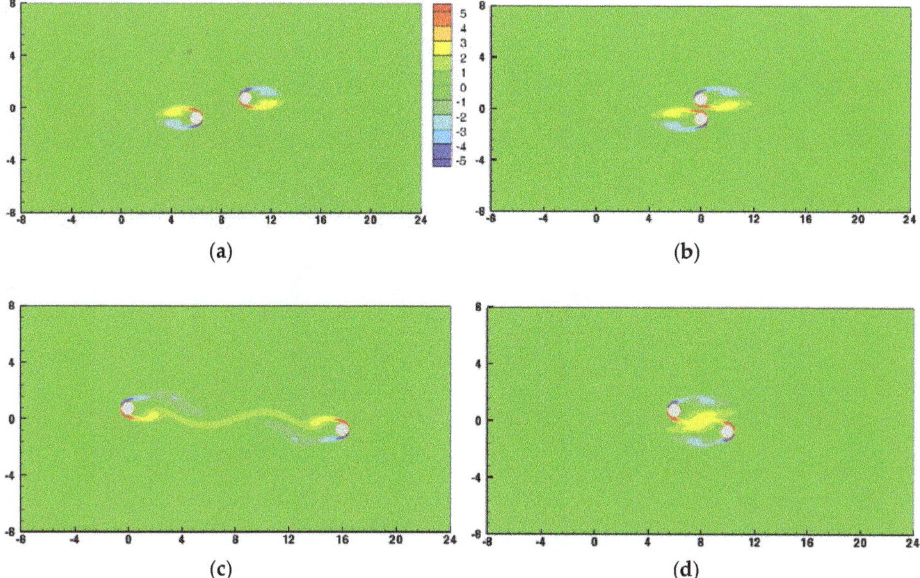

Figure 8. Instantaneous vorticity contours of two moving cylinders at Re = 100, at different times: (a) $\hat{t} = 12$ (b) $\hat{t} = 16$ (c) $\hat{t} = 20$ (d) $\hat{t} = 32$.

3.3. Dam-Break Problems

Simulations of selected dam-break problems are used to verify the free-surface capturing approach. The first example computes the position of the leading edge of a dam-break wave and compares the results with the experimental data of Martin and Moyce [52] and numerical results from Nomeritae et al. [53]. The other cases include a dam-break wave propagating over a submerged semicircular, or a rectangular, cylinder.

Following Martin and Moyce's [52] experimental study, the initial condition considered is a 0.057 m square water column (or H = 0.057 m) arranged behind a dam face in a 3 m long and 1 m high container. At $t > 0$, the dam is removed to allow the water to freely move toward downstream. Three uniform grid systems with 150×10 ($\Delta x = \Delta z = 0.02$ m), 300×20 ($\Delta x = \Delta z = 0.01$ m), and 600×40 grids ($\Delta x = \Delta z = 0.005$ m) and time step (Δt) of 5×10^{-4} sec were used to simulate the dam-break flows for the determination of the nondimensionalized leading-edge position. Figure 9 shows the comparisons between the present solutions on the time varying leading-edge position and those from measurements [52] and other numerical results [53]. For the Martin and Moyce [52] dam-break flow case, the corresponding Reynolds number was about 42,792. The computed surge front positions versus the nondimensionalized time reveal a good agreement with the data from Martin and Moyce [52] and the numerical results from Nomeritae et al. [53]. Moreover, the tests of the grid size effect indicate that the converged and nicely fit solutions are obtained when the refined grid size of $\Delta x = \Delta z = 0.005$ m is used.

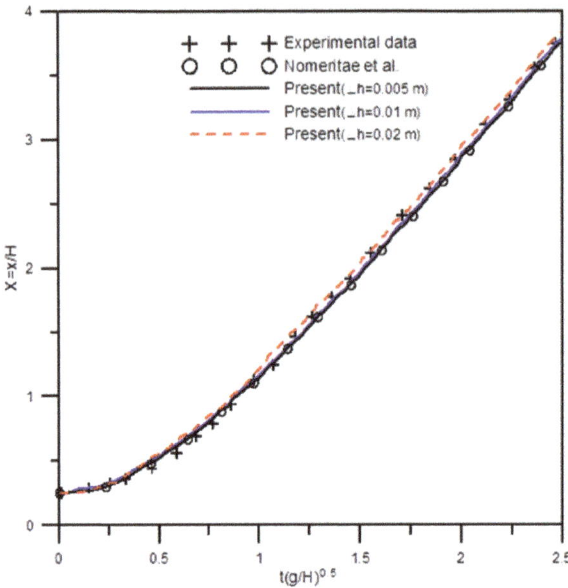

Figure 9. Comparisons between measured [52], EISPH [53] model, and present simulated leading-edge position at different times, with initial water volume of 0.057 m in length and height.

The present two-phase flow model is also simulated for the cases with dam-break waves over a submerged structure that is placed in an enclosed channel downstream of a dam face. Illustrated in Figure 10 are the schematic diagrams of the dam-break flow problem with either a completely submerged semicircular or a rectangular cylinder. The domain of the water column upstream of the dam face is 2 m by 1 m and the downstream standing water has a depth of 0.3 m. The dimensions of the channel in length and height are, respectively, 10 and 2 m. The diameter of the semicircular cylinder is 0.4 m, and the dimensions of the rectangular cylinder are 0.4 m in length and 0.2 m in height. Two different grid sizes, $\Delta x = \Delta z = \Delta h = 0.025$ m and $\Delta x = \Delta z = \Delta h = 0.033$ m, and $\Delta t = 0.005$ s are adopted for model simulations. Here, the interfacial parameter (ε) and the influential radius (r) of the Lagrangian marker points of the IB are, respectively, $2\Delta h$ and $1.5\Delta h$. The time varying velocities and pressures at three selected locations, namely P_1 at (2 m, 0.16 m), P_2 at (4 m, 0.16 m), and P_3 at (5 m, 0.26 m) (see Figure 10), are compared with the Flow-3D model results. The results obtained from the Flow-3D was based on the two-phase laminar flow model with the consideration of water and air phases. Figure 11 presents the comparisons between the present model results (velocity in x direction and pressure) and those from Flow-3D for the case of a submerged semicircular cylinder at point P_1. The time variations of both the velocity and pressure, except at the time near the end of the interaction process, are in good agreements with the Flow-3D results. The results obtained from the arrangements of two different grid sizes reveal the consistent and reliable model predications. The comparisons of time varying pressures at points P_2 and P_3 under the cases of either a semicircular or a rectangular cylinder are presented in Figure 12. It is found the variation trends of the pressure at P_2 and P_3 under the influence of a submerged semicircular cylinder are similar to those under the case of a submerged rectangular cylinder. However, due to a more severe geometry change that appeared in the rectangular cylinder, the pressures at P_2 and P_3 under the case of a submerged rectangular cylinder are greater than those with a semicircular cylinder after the surge waves impact on the cylinders.

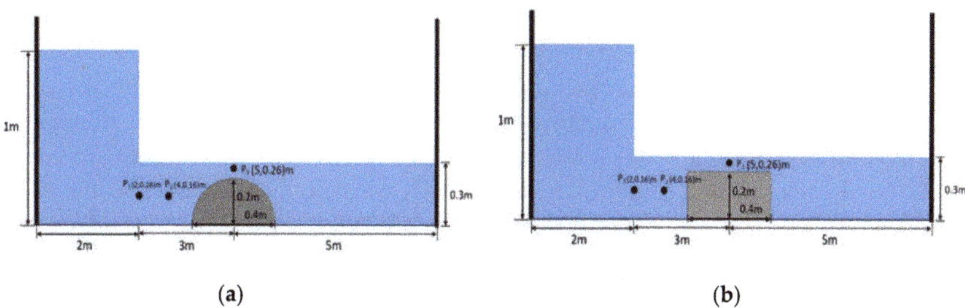

(a)   (b)

Figure 10. Schematic diagram of dam-break waves over a submerged structure: (**a**) semicircular cylinder and (**b**) rectangular cylinder.

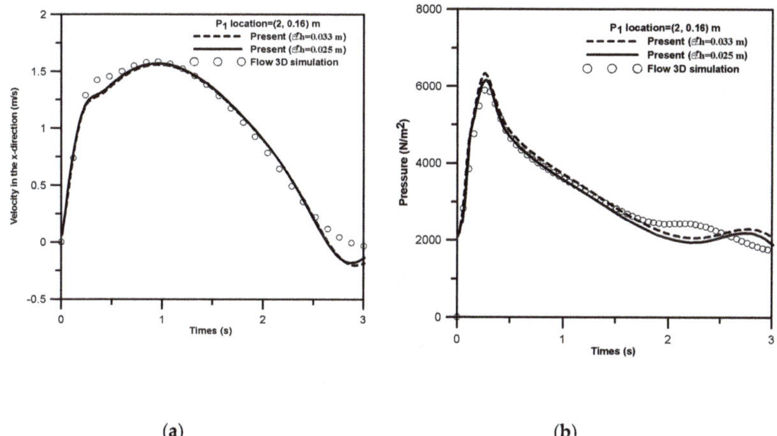

(a)   (b)

Figure 11. Comparisons between results (velocity and pressure) from the present model and Flow 3D simulations for the case of semicircular cylinder at point P1: (**a**) velocity and (**b**) pressure.

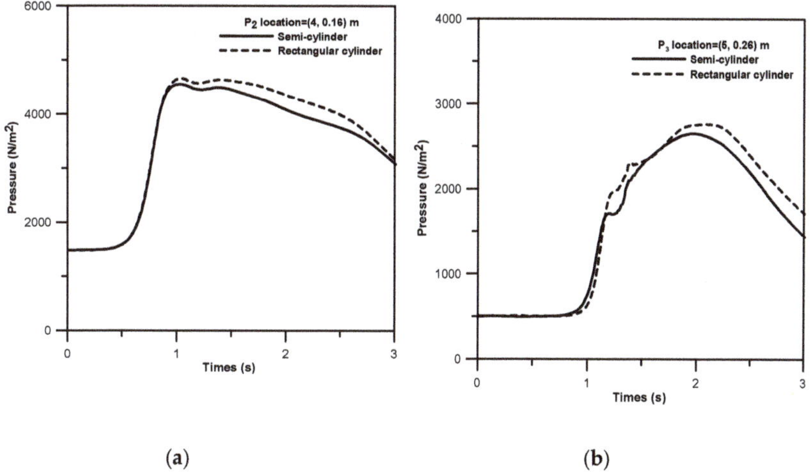

(a)   (b)

Figure 12. Comparisons of pressure variations for a dam-break flow over a semicircular and a rectangular cylinder at fixed points of (**a**) P2 at (4 m, 0.16 m) and (**b**) P3 at (5 m, 0.26 m).

In terms of force computations from IB approach, the time variations of drag forces acting respectively on a semicircular or a rectangular cylinder are presented in Figure 13. As a model verification, the present results for the case of a submerged rectangular cylinder are also compared with those obtained from the Flow-3D in Figure 13b. As indicated in Figure 13a, the maximum force induced by the leading surge wave on a semicircular cylinder (600 N) is less than that on a rectangular cylinder (850 N). For the case of a dam-break wave over a rectangular cylinder, as shown in Figure 13b, the predicted maximum in-line force from IB calculations is found to be slightly greater than that from the Flow-3D model. In addition, although the variation pattern is similar, the present model predicts higher values of the forces caused by the follow-up waves due to the slightly higher water-surface level predicted (See Figure 14). Regarding the case of a dam-break wave over a submerged rectangular cylinder, the sensitivity tests for force computation, using thee different input conditions, namely $\varepsilon = \Delta h$, $r = 1.5\Delta h$; $\varepsilon = 2\Delta h$, $r = 1.5\Delta h$; and $\varepsilon = 2\Delta h$, $r = 2\Delta h$, are presented in Figure 13b. Apparently, the difference of forces computed for cases of $\varepsilon = 2\Delta h$, $r = 1.5\Delta h$ and $\varepsilon = 2\Delta h$, $r = 2\Delta h$ is insignificant. In other words, the influential radius has reached its required converged value on the numerical simulation. However, the force predications have a noticeable difference when using two different interfacial thickness, i.e., $\varepsilon = \Delta h$ and $\varepsilon = 2\Delta h$, between two fluids. From results shown in Figure 13, it is evident that the values of hydrodynamic force are highly affected by the shape of the cylinder. Due to its smooth shape, the semicircular cylinder is subject to a relatively smaller hydrodynamic force than that to a rectangular cylinder. Figure 14 presents the spatial variations of the free-surface elevations at $t = 0.5$ s and $t = 2.0$ s, with the results obtained from both the present model and Flow-3D. Good agreement is noticed from the comparison plots. Due to the presence of a submerged structure, the predicted higher nonlinear growth of the free-surface flows has resulted in a more complex flow field with steep velocity gradient within the region of boundary layer.

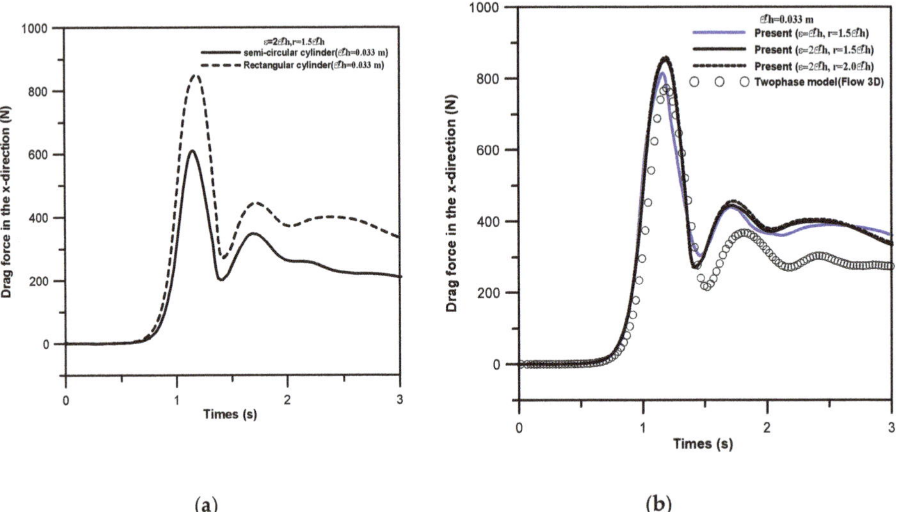

(a) (b)

Figure 13. Time variations of drag force (N): (**a**) semicircular cylinder vs. rectangular cylinder and (**b**) rectangular cylinder.

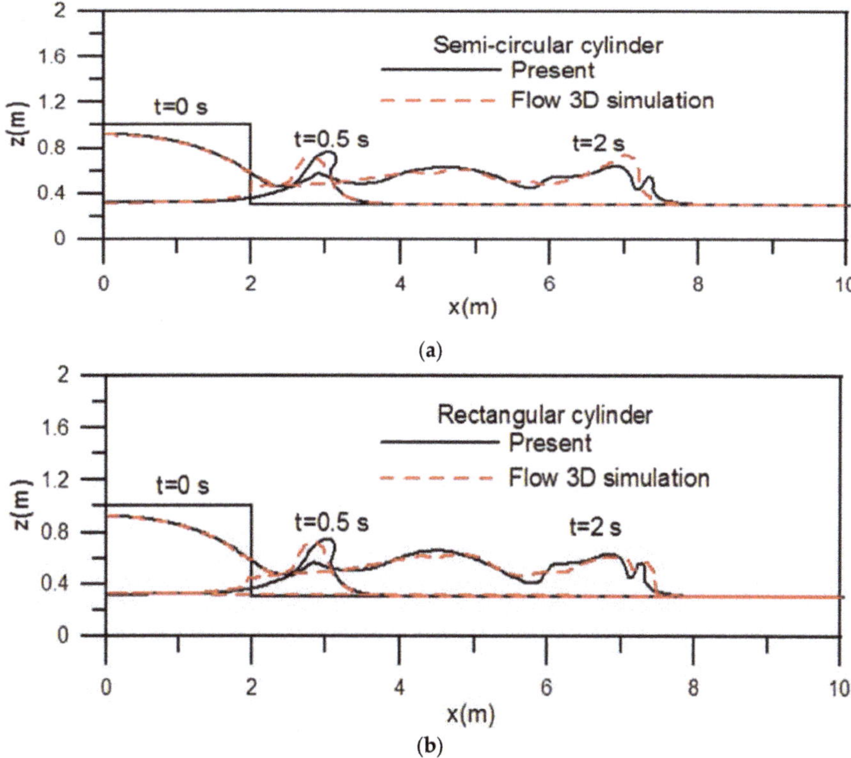

(a)

(b)

Figure 14. Comparisons of free-surface elevation at $t = 0.5$ s and $t = 2.0$ s: (**a**) semicircular cylinder and (**b**) rectangular cylinder.

In order to compare the flow patterns induced by either a submerged semicircular cylinder or a rectangular cylinder, the evolutions of density distributions of showing the water surface levels at various times are illustrated in Figure 15. Notably, a very similar variation trend of the free-surface elevation in cases with two tested submerged cylinders can be observed. Figure 16 presents the velocity vectors of a dam-break wave passing over the submerged structures of either a semicircular cylinder or a rectangular cylinder. The two-phase flow patterns between air and water phases are plotted particularly to show the motion of the vortices, which are generated during the process of wave-structure interaction. As seen in Figure 16, the plots also illustrate the local velocity distribution near the free-surface and around the submerged structures. The results obtained from the present two-phase flow model indicate that the shape of submerged rigid bodies has a noticeable effect on the vortex motion of the gas-phase fluid above the interface of the advancing waves. It should be noted that the results presented above for the dam-break wave problems were obtained from the simulations of the present two-phase flow model, involving the interfacial handing techniques between the air–water phase and liquid–solid phase that allow uniquely the use of a Cartesian-coordinate based mesh system. Though the single-phase flow model can be used to resolve the free-surface boundary condition of the Navier–Stokes equations, the numerical procedure adds further complexity for modeling free-surface flow problems.

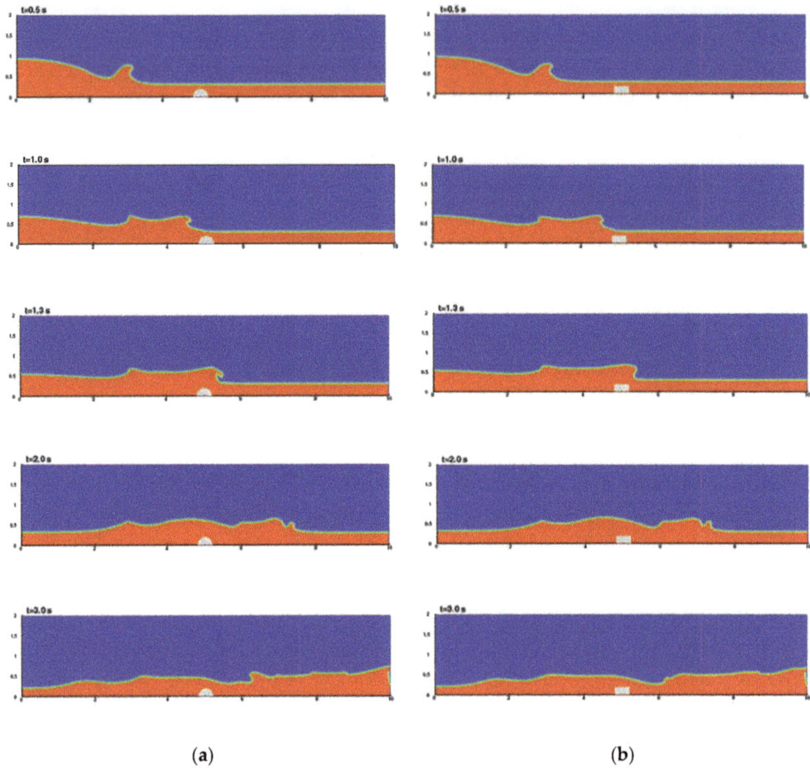

(a)　　　　　　　　　　(b)

Figure 15. Density distributions of a dam-break wave over submerged structures at different times: (**a**) semicircular cylinder and (**b**) rectangular cylinder.

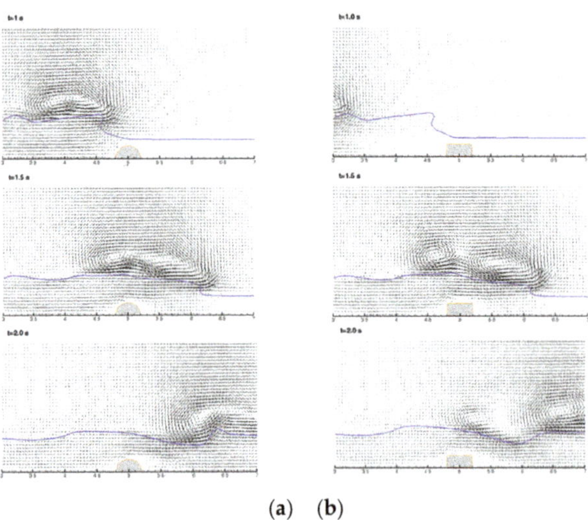

(a)　(b)

Figure 16. Velocity distributions of a dam-break wave over submerged structures at different times: (**a**) semicircular cylinder and (**b**) rectangular cylinder.

3.4. Wave Decomposition Process over a Submerged Trapezoid Breakwater

The model application is further extended to simulate wave generation, propagation, and interacting with a submerged trapezoidal bar. This case has been experimentally investigated by Beji and Battjes [54], where the measurements were performed in a 36 m long flume with a set water depth of 0.4 m. The numerical simulation was performed in a computational flume (or domain), which is 36 m long and 0.6 m high, and the water depth is 0.4 m. The spatial domain in this numerical wave tank (NWT) is divided into three zones: the wave generation zone, the wave propagation region, and the wave-absorbing layer. The length of wave generation zone is one wavelength long, and the center of the zone is located at x = 10.5 m. The numerical wave-absorbing layers are located in both the upstream and downstream domains with the ranges of 0 < x < 8 m and 28 < x < 36 m, respectively. The length of bar base is 11 m (between x = 14 m and x = 25 m), and its height is 0.3 m. The slopes of the inclined front and rear faces of the trapezoid breakwater are, respectively, 1/20 and 1/10.

The simulations start with a state of rest in water body subject to hydrostatic pressure. The free-surface elevation and velocity component for the incoming Stokes waves are specified at the inlet of the upstream boundary as the inflow conditions [55]. Separately, a total of 360,000 and 450,000 grid points together with $\Delta t = 0.002$ s were used in the numerical calculations. The mesh sizes Δx and Δz for 360,000 grid points are, respectively, 0.01 and 0.005 m, whereas, for cases using 450,000 grid points, they are $\Delta x = 0.01$ m and $\Delta z = 0.004$ m. In the experimental study by Beji and Battjes [54], an incident wave train with a wave height, H = 2 cm, a wave period of T = 2.02 s, and a wavelength of 3.73 m was generated to propagate over a submerged trapezoidal bar. The free-surface elevations computed with the same input conditions are compared with the measurements to demonstrate the performance of the developed model.

Presented in Figure 17 are comparison plots showing the time varying free-surface elevations at the locations of six gauges, identified as a, b, c, d, e, and f, with positions at x = 18.5 m, 20.5 m, 21.5 m, 22.5 m, 23.5 m, and 25.5 m, respectively. The results in Figure 17 exhibit the temporal free-surface evolution and the wave transformation processes due to the shoaling effect along the region of non-constant water depth. Because the decrease of the water depth along the upward slope of the bar, the wave elevation is shown to deform with a steepened wave amplitude and sawtoothed-like profile at x = 20.5 m and 21.5 m (See Figure 17b,c). Additionally, as shown in Figure 17d, high and sharp wave crests are formed over the bar crest at x = 22.5 m due to the wave shoaling and nonlinear interaction between waves and a submerged bar. Figure 17e reveals the decomposition of the wave with the development of higher harmonic components at regions from x = 23.5 m onward, as the wave propagates over and past the back face of the bar. It is noticed that the wave elevations obtained from the present model at the six-gauge locations match nicely with the experimental measurements [54] and the Flow-3D results. The simulated results of Flow-3D are based on a single phase with either laminar or turbulent $(k - \varepsilon)$ flow model. The comparisons of the present and Flow-3D wave elevations indicate that both results are well predicted when compared with the measurements at the gauge locations during the period from $t = 33$ s to $t = 39$ s. However, relatively, the results from the present model are shown to have slightly better agreements with the experimental measurements than those from the Flow-3D, considering either the laminar or the turbulent $(k - \varepsilon)$ flow module. It again demonstrates that the present two-phase flow model that combines LS and IB methods can be implemented with reasonable predictions for the study of hydrodynamic interactions between waves and structures.

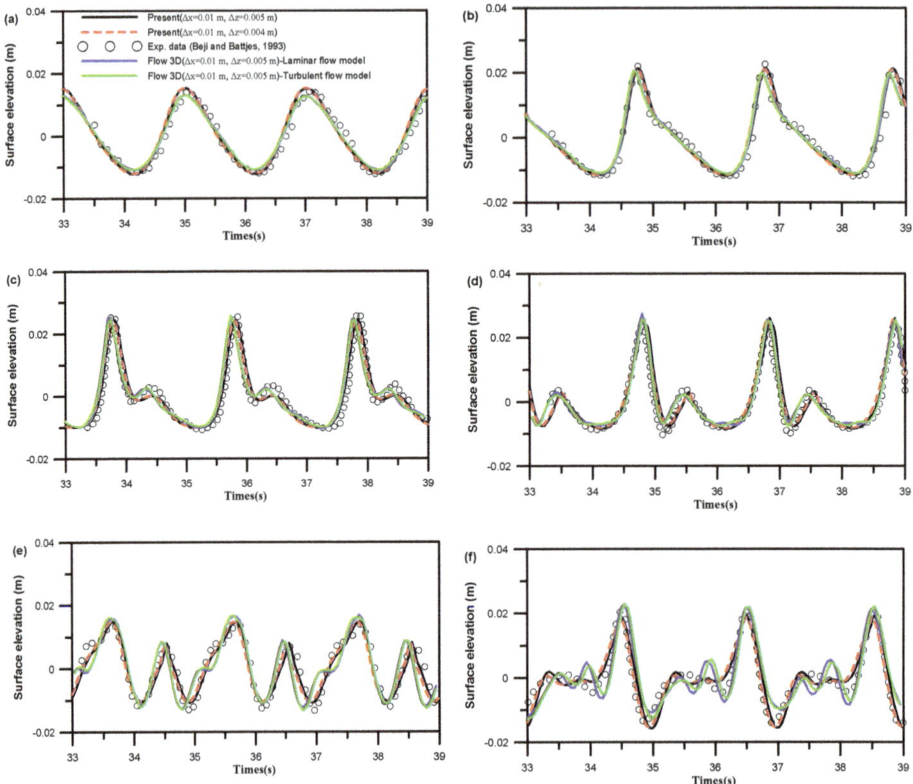

Figure 17. Free surface elevations on six wave gauges (**a**) x = 18.5 m; (**b**) x = 20.5 m; (**c**) x = 21.5 m; (**d**) x = 22.5 m; (**e**) x = 23.5 m; (**f**) x = 25.5 m.

3.5. Free Falling Wedge

In this section, with the added effect of the dynamic response of a moving solid body, a two-dimensional free-falling wedge that is subject to the motions of three degrees of freedom (DOF) after its entering a fluid domain is investigated with the developed numerical model. A solid symmetric wedge with a density of ρ_b = 466.6 kg/m^3 is selected for model simulations. It is positioned with its vertex of the obtuse angle pointing downward. The length of the wedge base is 1.2 m, and the side angle is 25°. The computational results in terms of the position of wedge's bottom vertex and its velocity are compared with the experimental data collected by Yettou et al. [56]. A 2D computational domain is considered, in which the length of the domain is 20 m and the height is 4 m, including a fluid domain with water depth of 1 m. Following the experimental setup, initially, the distance from the tip of the bottom vertex to the free-surface is 1.3 m. The wedge is suddenly released into the 20 m long tank. Two grid sizes as Δh = 0.04 m and Δh = 0.02 m are used for the numerical simulations. Figure 18 shows the induced velocity fields of the fluid domains at nine selected instants under the action of a free-falling wedge and its impact on the free surface. The variations of the interface also reflect the motions of the free-surface. The induced severe interfacial oscillations and coupled air motions after the wedge entering the water body can be observed. As the present numerical approaches treats more closely the two-phase flow conditions as a process of fully nonlinear fluid-structure interaction, it is believed more accurate and physically fitted results can be generated by the present model when compared to other models considering only the single-phase flow conditions without the inclusion of the effect of air phase. The use of orthogonal mesh system is highly essential to capture the vortex generation at the regions near or around a solid structure in the cases involving the interactions of air/water/solid. The comparisons between the simulated results and measured data from Yettou et al. [56] in terms of the time varying vertical position in z coordinate and vertical velocity of the bottom vertex for a free-falling wedge described above are presented in Figure 19a,b, respectively. The results obtained by using either the grid size of Δh = 0.04 m or Δh = 0.02 m are concluded to have nearly identical values, confirming the converged solutions. In terms of the model performance, the present results are shown to have good agreements with the published numerical solutions [36] and experimental data [56]. The maximum vertical velocity reaches 4.8 m/s at t = 0.56 s. This computational case demonstrates the model's capability in simulating the coupled motions between the dynamic responses of a free-falling wedge and the induced free-surface waves.

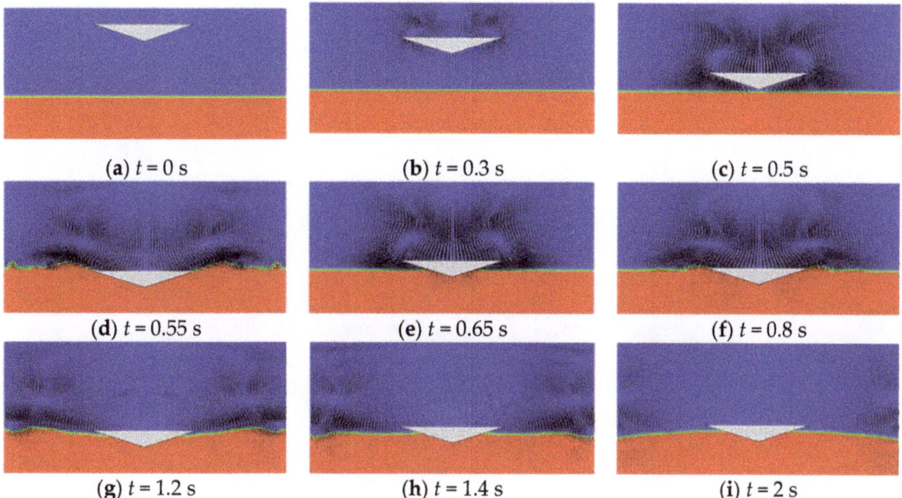

Figure 18. Velocity vector of free-falling wedge; the contour shows the density distribution of two-phase at different times: (**a**) t = 0 s; (**b**) t = 0.3 s; (**c**) t = 0.5 s; (**d**) t = 0.55 s; (**e**) t = 0.65 s; (**f**) t = 0.8 s; (**g**) t = 1.2 s; (**h**) t = 1.4 s; (**i**) t = 2 s.

3.6. Wave Packet Interacting with a Floating Body

As an extension of the modeling study given in Section 3.5, a case considering a wave-packet interacting with a floating body is numerically investigated. In this case, the numerical setup follows the experimental study performed by Hadzic et al. [57] in a small towing tank of the Berlin University of Technology. The tested floating object was a rectangular prism with the dimensions of 10 cm in length (along the x direction), 29 cm in width (along y direction), and 5 cm in high (along z direction). The density of the body was 680 kg/m^3, and the mass of the body was 0.986 kg. As shown in Figure 20, the center of the body was situated at a distance of 2.11 m away from the wavemaker. The computational domain is set as a 13 m long and 0.8 m high channel (water depth = 0.4 m). In Hadzic et al.'s [57] experiments, a flap-type wavemaker was controlled to produce a wave packet with a focusing point at the original location of the object. The grid sizes of Δx = 0.02 m and Δz = 0.005 m were used for the 2D numerical computations. Temporal variations of the body motions in three DOFs are computed during the interaction process between the wave packet and the floating body. Figure 21 depicts the time-based changes of the oscillating angle of the flap wavemaker that generate a wave packet. The simulated time varying wave elevations at the locations of x = 1.65 m and x = 2.66 m during the evolution of the wave packet are presented in Figure 22. The experimentally measured data from Hadzic et al. [57] are also included in Figure 22 for comparisons. The simulated wave packet is shown to fit nicely with the measured data. The maximum values of the wave elevation occur roughly at t =

5.8 s and $t = 7.7$ s at the locations of x = 1.65 m and x = 2.66 m, respectively. In general, the results show the increasing trend of the wave elevation until $t = 5.8$ s at the location of x = 1.65 but delayed to $t = 7.7$ s at the location of x = 2.66 m. Afterward, the wave elevation is shown to have a trend of periodic decay with time.

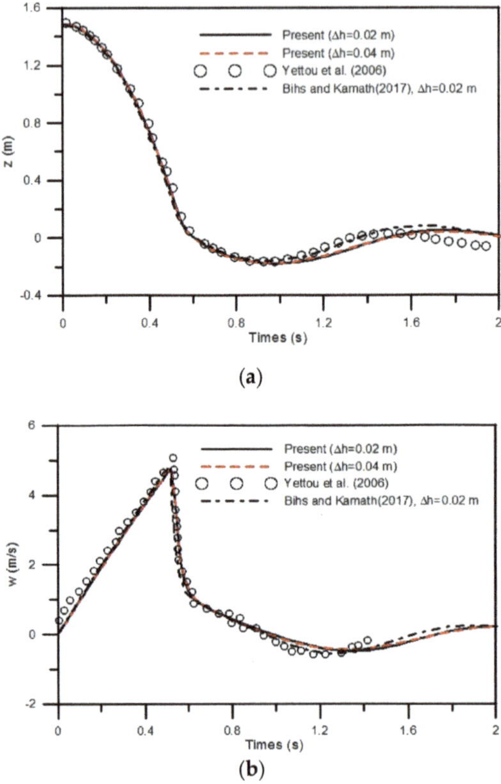

(a)

(b)

Figure 19. Motion of a free-falling wedge: (**a**) vertical position of the bottom vertex; (**b**) vertical velocity of the bottom vertex.

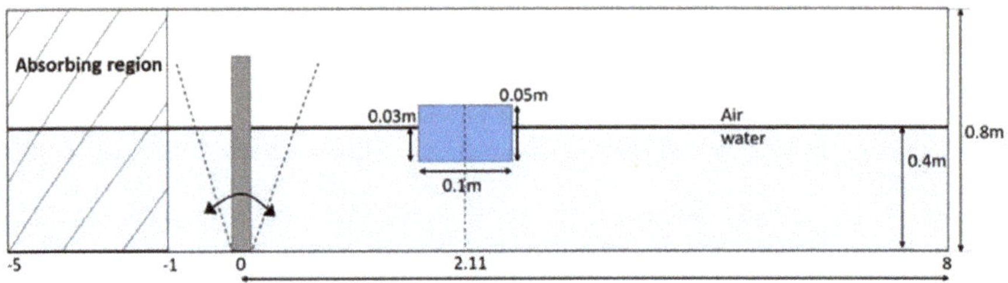

Figure 20. Layout of the freely floating box interacting with a wave packet.

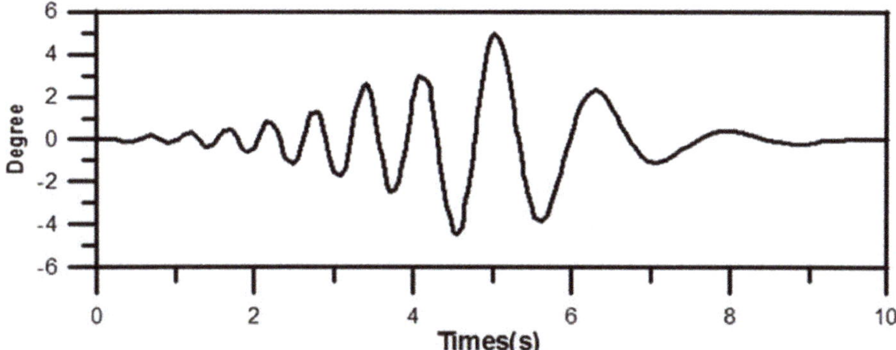

Figure 21. Time-based variations of the flap wavemaker angle used for the generation of a wave packet.

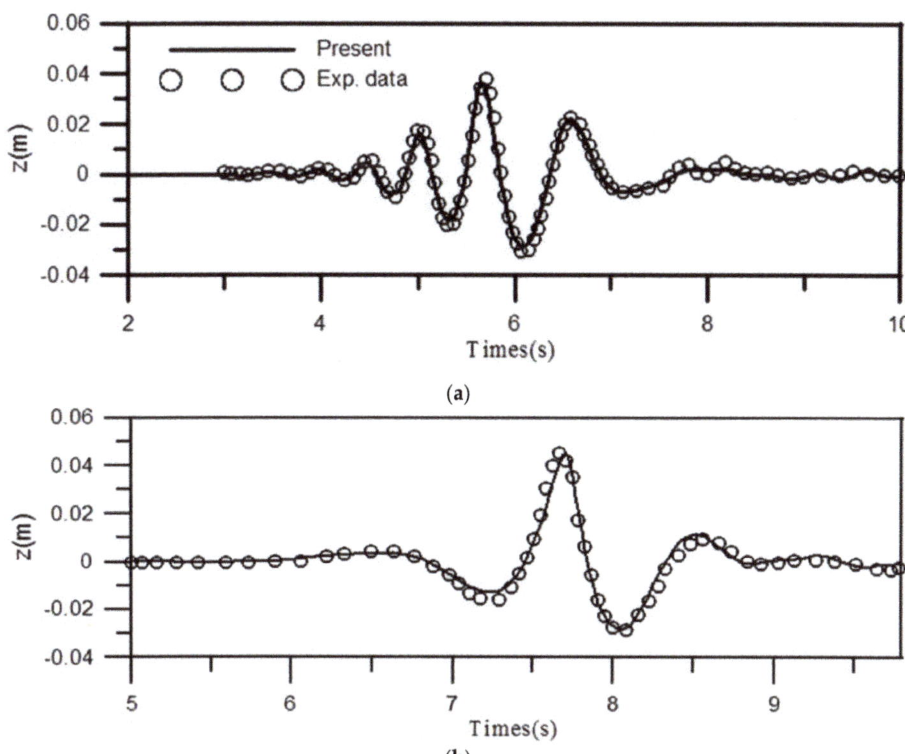

Figure 22. Time-based variations of wave elevation during the evolution of the wave packet at (**a**) x = 1.65 m and (**b**) x = 2.66 m.

In terms of the three DOF motions of a floating body, it is generally defined that the rotational motion that is referenced to its longest axis (in this case, the body width in y direction) as the rolling motion. Then, in this study, the translational motion along the x direction is defined as the sway motion. The vertical motion is as usual the heave motion. The time variations of computed sway, heave, and rolling motions with values referenced to its original body center location are plotted together with the measured data from Hadzic et al. [57] in Figure 23. The comparisons suggest that the predicted body motions in 3DOFs are in reasonably good agreements with measured ones. As shown in Figure 23a, the floating body's sway motion reaches the maximum value of 0.11 m at $t = 7.6$ s. The observed oscillatory heave motion is indicated in Figure 23b, where the motion varies within the range of −0.04 to 0.04 m, until it damps out at roughly $t = 8.5$ s. Figure 23c shows the time-based variations of the rolling angles of the floating body motion in which the angle of rolling motion ranges approximately from −20° to 20°. Finally, two snapshots showing the free-surface deformation and fluid velocity vectors surrounding the floating body at $t = 7.2$ s and $t = 7.6$ s are presented in Figure 24. Overall, the good matched comparisons between the simulated fluid and body motions with the experimental measurements demonstrate, again, that the developed two-phase viscous flow model can provide reasonable predictions on waves interacting with a floating body.

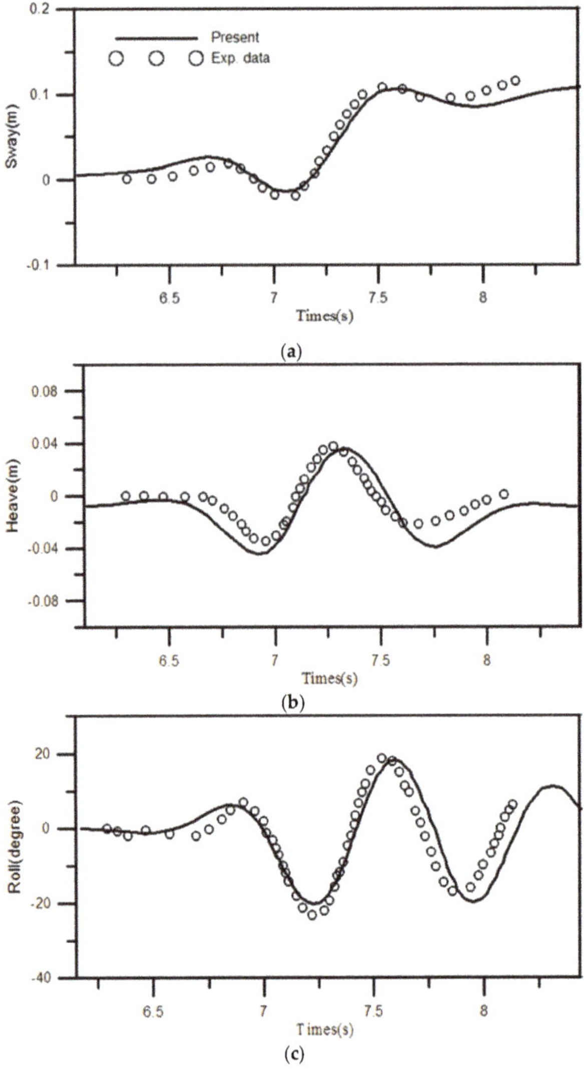

Figure 23. Time-based variations of the floating body motions during the interaction with the wave packet (**a**) sway, (**b**) heave, and (**c**) roll.

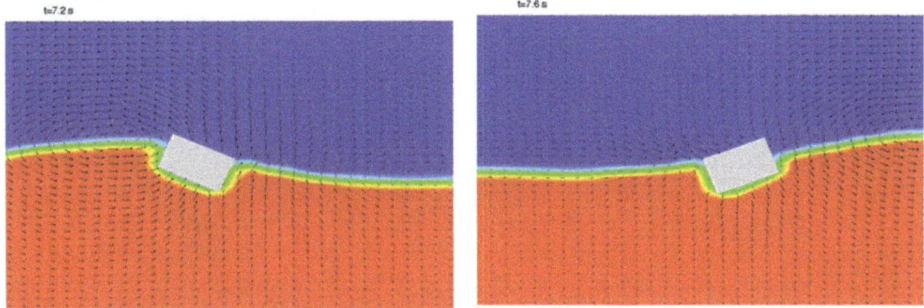

Figure 24. Density distributions and velocity vectors of floating body at $t = 7.2$ s (**left**) and $t = 7.6$ s (**right**).

The promising results obtained from the present 2D two-phase flow model suggest that the numerical approaches presented in this study can be extended to the development of a 3D model to study the three-dimensional two-phase flow problems. By using similar methodologies as those proposed in the present study, the Navier–Stokes-equations-based projection method, the LS method, and the IB approach can be extended to the domain of three dimension, including using a combined Eulerian Cartesian and Lagrangian grid system. Certainly, the complexity of the problems arises from modeling the induced 3D motions of a floating body, where six degrees of freedom should be considered. The movements of a rigid body subject to the external forces and moments will be governed by the equations of motion covering three linear translational and three rotational motions. Following the similar procedure of using a combined Cartesian and Lagrangian grid system, the movements of the 3D interfacial boundaries and the dynamic responses of a moving body can be simulated. As a future research direction, the present research group plans to extend the developed 2D two-phase free-surface flow model to an applicable 3D model for solving the 3D problems.

4. Conclusions

A compact interfacial method by solving the Navier–Stokes equations with added source terms of external forces from immersed bodies for modeling two-phase flow problems is presented in this study. The treatments of the moving free-surface and the immersed bodies with separately using the LS and IB methods are included in the developed model. It can be noted that the present study uses a combined Eulerian Cartesian and Lagrangian grid system to avoid the re-meshing procedure for two-phase flow modeling involving coupled fluid-structure interactions. A higher-order Runge–Kutta integration in time and WENO scheme in space are included to discretize the LS equation to track the interfacial positions with severe free-surface deformations. The proposed numerical technique does not require the conformation of the grid lines onto the boundary of an immersed object, and at the same time, the effect of an immersed object can be implemented through the right-hand side source term of the Navier–Stokes equations.

The study cases include both the moving objects in a viscous fluid and interfacial two-phase flows, such as dam-break flows and wave motions over a submerged structure and the motions of a floating body. Good accuracy of the computed results, as confirmed by the comparisons with published results in calculating the variations of velocity/pressure/vorticity field and the hydrodynamic forces, was achieved. It is demonstrated that the present numerical model can simulate reasonably well for the cases of a dam-break flow passing over either a submerged semicircular or a rectangular cylinder. The model was also tested successfully to simulate the periodic waves propagating over a coastal bar, where the predicted free-surface elevations reflect an improved resolution by comparing to the Flow-3D simulations, using the same grid size. The predicted results are shown to have slightly better agreements with the experimental measurements when comparing to those from Flow-3D, using either the laminar or turbulent ($k - \varepsilon$) flow module. Moreover, the use of an interfacial method in this study is capable of capturing directly the vortex formulations at the recirculation zone downstream of a submerged structure. With detailed calculations of the free-surface deformation, the present model can provide a better estimation of the free-surface elevation and flow patterns for strongly nonlinear interaction between waves and structures. It is concluded with certainty that the two-phase free-surface flow model developed in this study is able to successfully simulate nonlinear water wave problems, including the accurate predictions of severe wave deformation processes.

For cases with dynamic responses of moving bodies, the present model is also extended to compute with good comparison results the motions of a free-falling wedge as it is entering a water body and wave packet interacting with a 2D floating body. The results obtained from the present study suggest that a two-phase flow model with a coupled immersed body calculation can be recommended for the investigation of nonlinear water waves interacting with structures, and with the integration of equations of motion, to predict the responses of a floating body. Our findings also support the conclusions that the present model is an effective numerical tool for free-surface flow computations in a domain including gas, liquid, and solid phases, with the use of a combined Eulerian Cartesian and Lagrangian grid system.

Author Contributions: D.-C.L. and K.-H.W. devised the experimental strategy and carried out this numerical experiment; D.-C.L. wrote the manuscript, and K.-H.W. contributed to the revisions; T.-W.H. partially contributed to the experiment and analysis of the data. All authors have read and agreed to the published version of the manuscript.

Funding: This research was funded by the Ministry of Science and Technology, Taiwan, grant number No. MOST 107-2221-E-992-020.

Conflicts of Interest: The authors declare no conflicts of interest.

Nomenclature

English Symbols

k	curvature of an interface
f	hydrodynamic forces
A_b	wave absorbing coefficient
g	gravitational acceleration vector
t	Time
$\mathbf{u}$	velocity vectors
P	Pressure
x_{xt}	starting position of the absorbing region
x_{ab}	length of the absorbing region
$\mathbf{u}^*$	velocity vectors at each of the intermediate (between n and n+1) time level
$\hat{\mathbf{u}}$	velocity vectors without considering the effect of immersed boundary
$\mathbf{F}$	force density vector at the Lagrangian points
$\mathbf{X}_l$	Lagrangian points with coordinates
$\mathbf{U}^\Gamma$	Lagrangian point velocity of body
$\mathbf{U}_b$	translational velocity vectors of body
V_b	body volume
I_b	body moment of inertia
$\mathbf{r}$	position vector relative to the body centroid
D	cylinder diameter
U	uniform flow velocity
C_D	drag coefficient
C_L	lift coefficient
St	Strouhal numbers

Greek Symbols

δ	Dirac delta function
Σ	surface tension coefficient
ϕ	level set function
$H(\phi)$	Heaviside function
ρ	fluid density
μ	fluid viscosity
ε	interfacial thickness
ΔV_l	control volume defined about the m-th Lagrangian marker
ρ_b	body density
ω_b	body angular velocity vectors
Re	Reynolds number
υ	fluid kinematics viscosity

References

1. Koo, W.; Kim, M.-H. Freely floating-body simulation by a 2D fully nonlinear numerical wave tank. *Ocean. Eng.* **2004**, *31*, 2011–2046. [CrossRef]
2. Finnegan, W.; Goggins, J. Numerical simulation of linear water waves and wave–structure interaction. *Ocean Eng.* **2012**, *43*, 23–31. [CrossRef]
3. Halow, F.H.; Welch, J.E. Numerical calculation of time-dependent viscous incompressible flow of fluid with a free surface. *Phys. Fluids* **1965**, *12*, 2182–2189. [CrossRef]
4. Hirt, C.W.; Nicholas, B.D. Volume of fluid method for the dynamics of free boundaries. *J. Comput. Phys.* **1981**, *39*, 201–225. [CrossRef]
5. Anbarsooz, M.; Passandideh-Fard, M.; Moghiman, M. Fully nonlinear viscous wave generation in numerical wave tanks. *Ocean Eng.* **2013**, *59*, 73–85. [CrossRef]
6. Kang, L.; Jing, Z. Depth-averaged non-hydrostatic hydrodynamic model using a new multithreading parallel computing method. *Water* **2017**, *9*, 184. [CrossRef]
7. Gingold, R.A.; Monaghan, J.J. Smoothed particle hydrodynamics: Theory and application to non-spherical stars. *Mon. Not. R. Astron. Soc.* **1977**, *181*, 375–389. [CrossRef]
8. Ramaswamy, B.; Kawahara, M. Lagrangian finite element analysis applied to viscous free surface flow. *Int. J. Num. Meth. Fluids* **1987**, *7*, 953–984. [CrossRef]
9. Hansbo, P. Lagrangian incompressible flow computations in three dimensions by use of space-time finite elements. *Int. J. Num. Meth. Fluids* **1995**, *20*, 989–1001. [CrossRef]
10. Monaghan, J.J. Smoothed particle hydrodynamics and its diverse applications. *Annu. Rev. Fluid Mech.* **2012**, *44*, 323–346. [CrossRef]
11. Hirt, C.W.; Amsden, A.A.; Cook, J.L. An arbitrary Lagrangian Eulerian computing method for all flow speeds. *J. Comput. Phys.* **1974**, *14*, 227–253. [CrossRef]
12. Chan, R.K.C. A generalized arbitrary Lagrangian-Eulerian method for incompressible flows with sharp interfaces. *J. Comput. Phys.* **1975**, *17*, 311–331. [CrossRef]
13. Pracht, W.E. Calculating three-dimensional fluid flows at all speeds with an Eulerian-Lagrangian computing mesh. *J. Comput. Phys.* **1975**, *17*, 132–159. [CrossRef]
14. Hughes, T.J.R.; Liu, W.K.; Zimmerman, T.K. Lagrangian Eulerian finite element formulation for incompressible viscous flows. *Comput. Meth. Appl. Mech. Eng.* **1981**, *29*, 329–349. [CrossRef]
15. Hieu, P.D.; Tanimoto, K. Verification of a VOF-based two-phase flow model for wave breaking and wave-structure interaction. *Ocean Eng.* **2006**, *33*, 1565–1588. [CrossRef]
16. Osher, S.; Sethian, J.A. Fronts propagating with curvature-dependent speed: Algorithms based on Hamilton–Jacobi formulations (PDF). *J. Comput. Phys.* **1988**, *79*, 12–49. [CrossRef]
17. Osher, S.; Fedkiw, R.P. Level set methods: An overview and some recent results. *J. Comput. Phys.* **2001**, *169*, 463–502. [CrossRef]
18. Osher, S.; Fedkiw, R.P. *Level Set Methods and Dynamic Implicit Surfaces*; Springer: Berlin, Germany, 2002.
19. Unverdi, S.O.; Tryggvason, G. A front-tracking method for viscous, incompressible, multi-fluid flows. *J. Comput. Phys.* **1992**, *100*, 25–37. [CrossRef]

20. Agresar, G.; Linderman, J.J.; Tryggvason, G.; Powell, , K.G. An adaptive, Cartesian, front tracking method for the motion, deformation and adhesion of circulating cells. *J. Comput. Phys.* **1998**, *43*, 346–380. [CrossRef]

21. Bunner, B.; Tryggvason, G. Direct Numerical simulations of three-dimensional bubbly flows. *Phys. Fluids* **1999**, *11*, 1967–1969. [CrossRef]

22. Tryggvason, G.; Bunner, B.; Esmaeeli, A.; Juric, C.; Al-Rawahi, N.; Tauber, W.; Han, J.; Nas, S.; Jan, Y.-J. A front-tracking method for the computations of multiphase flow. *J. Comput. Phys.* **2001**, *169*, 708–759. [CrossRef]

23. Peskin, C.S. Flow patterns around heart valves: A numerical method. *J. Comput. Phys.* **1972**, *10*, 252–271. [CrossRef]

24. Peskin, C.S. Numerical-analysis of blood-flow in heart. *J. Comput. Phys.* **1977**, *25*, 220–252. [CrossRef]

25. Roma, A.M.; Peskin, C.S.; Berger, M.J. An adaptive version of the immersed boundary method. *J. Comput. Phys.* **1999**, *153*, 509–534. [CrossRef]

26. Iaccarino, G.; Verzicoo, R. Immersed boundary technique for turbulent flow simulations. *Appl. Mech. Rev.* **2003**, *56*, 331–347. [CrossRef]

27. Tseng, Y.-H.; Ferziger, J.H. A ghost-cell immersed boundary method for flow in complex geometry. *J. Comput. Phys.* **2003**, *192*, 593–623. [CrossRef]

28. Uhlmann, M. An immersed boundary method with direct forcing for the simulation of particulate flows. *J. Comput. Phys.* **2005**, *209*, 448–476. [CrossRef]

29. Yang, X.; Zhang, X.; Li, Z.; He, G.-W. A smoothing technique for discrete delta functions with application to immersed boundary method in moving boundary simulations. *J. Comput. Phys.* **2009**, *228*, 7821–7836. [CrossRef]

30. Simon, G.; Thomas, B.; Dominique, A. A coupled volume-of-fluid/immersed method for the study of propagating waves over complex-shaped bottom: Application to the solitary wave. *Comput. Fluids* **2016**, *131*, 56–65.

31. Lo, D.C.; Lee, C.-P.; Lin, I.-F. An efficient immersed boundary method for fluid flow simulations with moving boundaries. *Appl. Math. Comput.* **2018**, *328*, 312–337. [CrossRef]

32. Liu, X.-D.; Osher, S.; Chan, T. Weighted essentially non-oscillatory schemes. *J. Comput. Phys.* **1994**, *115*, 200–212. [CrossRef]

33. Carrica, P.M.; Wilson, R.V.; Noack, R.W.; Stern, F. Ship motions using single-phase level set with dynamic overset grid. *Comput. Fluids* **2007**, *36*, 1415–1433. [CrossRef]

34. Yang, J.; Wang, Z.; Stern, F. Sharp interface immersed-boundary/level-set method for wave-body interactions. *J. Comput. Phys.* **2009**, *228*, 6590–6616. [CrossRef]

35. Calderer, A.; Kang, S.; Sotiropoulos, F. Level set immersed boundary method for coupled simulation of air/water interaction with complex floating structures. *J. Comput. Phys.* **2014**, *277*, 201–227. [CrossRef]

36. Bihs, H.; Kamath, A. A combined level set/ghost cell immersed boundary representation for floating body simulations. *Int. J. Num. Meth. Fluids* **2017**, *83*, 905–916. [CrossRef]

37. Sethian, J.A.; Smereka, P. Level set methods for fluid interfaces. *Annu. Rev. Fluid Mech.* **2003**, *35*, 341–372. [CrossRef]

38. Sussman, M.; Smereka, P.; Osher, S. A level set approach for computing solutions to incompressible two-phase flow. *J. Comput. Phys.* **1994**, *114*, 146–159. [CrossRef]

39. Lin, P.; Liu, P.L.F. Discussion of vertical variation of the flow across the surf zone. *Coast. Eng.* **2004**, *50*, 161–164. [CrossRef]

40. Chorin, A.J. A numerical method for solving incompressible viscous flow problem. *J. Comput. Phys.* **1967**, *2*, 12–26. [CrossRef]

41. Suzuki, K.; Inamuro, T. Effect of internal mass in the simulation of a moving body by the immersed boundary method. *Comput. Fluids* **2011**, *49*, 173–187.

42. Tritton, D.J. Experiments on the flow past a circular cylinder at low Reynolds number. *J. Fluid Mech.* **1959**, *6*, 547–567. [CrossRef]

43. Calhoun, D.A. Cartesian grid method for solving the two-dimensional streamfunction-vorticity equations in irregular regions. *J. Comput. Phys.* **2002**, *176*, 231–275. [CrossRef]

44. Russel, D.; Wang, Z.J. A Cartesian grid method for modeling multiple moving objects in 2D incompressible viscous flow. *J. Comput. Phys.* **2003**, *191*, 177–205. [CrossRef]

45. Silva, A.A.; Neto, S.; Damasceno, J.J.R. Numerical simulation of two-dimensional flows over a circular cylinder using the immersed boundary method. *J. Comput. Phys.* **2003**, *189*, 351–370. [CrossRef]

46. Xu, S.; Wang, Z.J. An immersed interface method for simulating the interaction of a fluid with moving immersed boundaries. *J. Comput. Phys.* **2006**, *216*, 454–493. [CrossRef]

47. Kolahdouz, E.M.; Bhalla, A.P.S.; Craven, B.A.; Griffith, B.E. An immersed interface method for discrete surfaces. *J. Comput. Phys.* **2020**, *400*, 1–37. [CrossRef] [PubMed]

48. Liu, C.; Sheng, X.; Sung, C.H. Preconditioned multigrid methods for unsteady incompressible flows. *J. Comput. Phys.* **1998**, *139*, 35–57. [CrossRef]

49. Choi, J.I.; Oberoi, R.C.; Edwards, J.R.; Rosati, J.A. An immersed boundary method for complex incompressible flow. *J. Comput. Phys.* **2007**, *224*, 757–784. [CrossRef]

50. Griffith, B.E.; Luo, X.Y. Hybrid finite difference/finite element version of the immersed boundary method. *Int. J. Numer. Methods Biomed. Eng.* **2017**, *33*, e2888. [CrossRef]

51. Courant, R.; Fredrichs, K.O.; Lewy, H. Über die partiellen differenzengleichungen der mathematischen physik. *Math. Ann.* **1928**, *100*, 32. [CrossRef]

52. Martin, J.; Moyce, W. Part IV. An experimental study of the collapse of liquid columns on a rigid horizontal plane. *Philos. R. Soc. London. A Math. Phys. Eng. Sci.* **1952**, *244*, 312–324.

53. Nomeritae; Edoardo, D.; Stefania, G.; Ha, H.B. Explicit incompressible SPH algorithm for free-surface flow modelling: A comparison with weakly compressible schemes. *Adv. Water Resour.* **2016**, *97*, 156–167. [CrossRef]

54. Beji, S.; Battjes, J.A. Experimental investigation of wave propagation over a bar. *Coast. Eng.* **1993**, *19*, 151–162. [CrossRef]

55. Dean, R.G.; Dalrymple, R.A. *Water Wave Mechanics for Engineers and Scientists*; World Scientific Publishing Company: Singapore, 1984.

56. Yettou, E.M.; Desrochers, Y.; Champoux, Y. Experimental study on the water impact of a symmetrical wedge. *Fluid Dyn. Res.* **2006**, *38*, 47–66. [CrossRef]

57. Hadzic, I.; Hennig, J.; Peric, M.; Xing-Kaeding, Y. Computation of flow induced motion of floating bodies. *Appl. Math. Model.* **2005**, *29*, 1196–1210. [CrossRef]

 *water*

Article

10.3390/w12020427 Flooding of Piazza San Marco (Venice): Physical Model Tests to Evaluate the Overtopping Discharge

Piero Ruol *, Chiara Favaretto, Matteo Volpato and Luca Martinelli

ICEA Department, Padova University, V. Ognissanti 39, 35129 Padova, Italy;
chiara.favaretto@dicea.unipd.it (C.F.); matteo.volpato@unipd.it (M.V.); luca.martinelli@unipd.it (L.M.)
* Correspondence: piero.ruol@unipd.it

Received: 11 December 2019; Accepted: 1 February 2020; Published: 5 February 2020

Abstract: This paper aims at evaluating the wave overtopping discharge over the pavement of "Piazza S. Marco" (Venice) in order to select the best option to mitigate the risk of flooding of the Piazza and to protect the monuments and historic buildings, e.g., the "Basilica S. Marco". In fact, the MO.S.E. (MOdulo Sperimentale Elettromeccanico) system is designed to keep the water level below a certain value, for the safety of the lagoon, but it does not guarantee the defence of the Piazza, where flooding is still possible, being its pavement locally much lower than the maximum expected water level. To completely defend the Piazza, specific additional works are planned to prevent the back-flow through the natural drainage system (now the primary pathway) or by filtration or by overtopping. This paper investigates on the overtopping mechanism, under conditions compatible with a fully operational MO.S.E. system, through 2-D experiments. The pavement of the Piazza is gently sloping towards the masonry quay which, in some parts is formed by 5 descending steps, and in some other parts, is just a vertical wall. Close to the "Marciana" Library, a critical part is present, with a slightly lower crest freeboard. In total, three cross-sections were examined in the 36 m long wave flume of the Padova University. The test programme includes 10 irregular wave attacks and three different water levels. The test results differ considerably from the results of the available formulas, since the investigated cross-sections by far exceed their range of applicability. The presence of the steps affects only the reflection coefficient rather than the overtopping discharges. In general, if the waves incident to the Piazza are higher than 40 cm, which is a possible scenario, some other adaptation works must be considered, such as the pavement rise, temporary barriers or the reduction of the waves impacting the quay through, for instance, floating breakwaters.

Keywords: Venetian lagoon; flooding; wave overtopping; astronomical tide; storm surge; experimental investigation

1. Introduction

In order to cope with the expected sea-level rise Cazenave at al. [1] and subsidence Tosi et al. [2], and the consequent increasing frequency of flooding of Venice Lionello [3], the well-known MO.S.E. (MOdulo Sperimentale Elettromeccanico, www.mosevenezia.eu) system is designed to keep the water level below a certain value, thus preserving the people mobility and the economic activities. In particular, the maximum expected water level in Venice when the barriers are closed is +1.10 m relative to the local tidal reference (named ZMPS, Zero Mareografico Punta Salute). Obviously, in order to guarantee this maximum level, the MO.S.E. gates will close when the mean water level at the inlets is much below this value, in order to anticipate the predicted wind set-up, rainfall contribution, etc. Rinaldo et al. [4]. However, the gates are not intended to specifically protect "Piazza S. Marco", that needs additional works to avoid its flooding, e.g., regulation of the drainage system to avoid the water entry from the sewer drains and from seepage. In fact, the mean elevation of the square is ~0.95 m ZMPS, with some areas in front of the S. Marco church lower than 0.7 m ZMPS.

Another adaptation measure could be the reduction of the possible wave overtopping from the S. Marco Basin. In fact, waves generated by boats or by wind blowing from South-East can overtop the quay (named "Riva S. Marco", Figure 1). The evaluation of this discharge is not straightforward due to the complex topography of the site and to the uncertainties of the boundary conditions and forcing (water levels, wind velocities and directions). In particular, the Riva S. Marco is characterized by a natural stone pavement mildly sloping toward the vertical wall or, in some parts, toward a vertical wall with descending steps in front. In case of very high-water level inside the lagoon, the pavement is completely inundated.

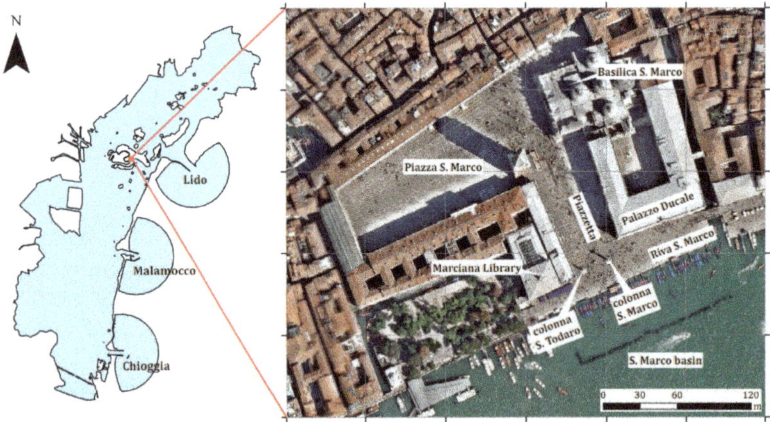

Figure 1. Location and orthophoto of the study site: Piazza S. Marco, Venice, Italy.

Wave overtopping is widely investigated in the literature since its evaluation is crucial for structures that protect the inland against flooding. There are many simple predictive tools such as the Goda's design charts [5], the "Overtopping of Seawalls: Design and Assessment Manual 2" [6], the "Coastal Engineering Manual" [7], and the EurOtop 2018 manual [8]. Almost all the empirical formulae proposed by these guidelines are derived from large datasets of laboratory and field measurements and take into account three types of structures: (i) sloping sea dikes; (ii) armoured rubble slopes and mounds; (iii) vertical walls. The experimental dataset address by the CLASH European project is described by Van der Meer et al. [9]. Recently, van Gent et al. [10] and Zanuttigh et al. [11] developed neural network systems that predict the overtopping for complicated structure geometries and variable wave conditions, on the basis of a homogeneous and wide database of measurements that trains the network Verhaeghe et al. [12].

The mildly sloping pavement over the quay in Piazza S. Marco could be assimilated to a promenade placed behind a vertical wall. However, the existing studies usually consider the presence of a promenade over smooth dikes. For instance, Van Doorslaer et al. [13] examined the typical defence structures of the Belgian coasts, characterized by smooth dikes, with a long and mildly sloping promenade above the still water level. From this study, the first set of formulae were established to parametrize the combined effect of crown wall, bullnose and promenades into a reduction coefficient γ^* to be included in the EurOtop formulae [8] for the prediction of the overtopping discharge. Pullen et al. [14] investigated, through field and laboratory measurements, the overtopping discharge at Samphire Hoe. The structure is a composite vertical wall comprising a rock toe berm and, on the top, a parapet and a 23 m large promenade, that during severe overtopping events, is completely inundated. Conversely to the S. Marco case, most of the discharge in Samphire Hoe fall in the area directly behind the parapet wall.

The particular geometry that characterizes the "Riva S. Marco" has never been investigated and it is out of the range of validity of the γ^* coefficient relative to the promenade effect. The present study aims at evaluating the wave overtopping, considering three representative cross-sections and several forcing conditions. The results are compared to the EurOtop formula for vertical wall and an influence factor accounting for the presence of the sloping pavement of the S. Marco quay is proposed. The investigation could be useful to select an appropriate adaptation measure that reduces overtopping, minimizing the impact and considering the historic and architectural constraints of the city of Venice. Among the measures and solutions that reduce the amount of overtopping, those that may be considered are, for instance, the addition of structural elements to the quay wall such as berms Burcharth et al. [15], crown walls Van Doorslaer et al. [13], Formentin et al. [16], steps Kerpen et al. [17], McCabe et al. [18], bullnoses Pearson et al. [19], Martinelli et al. [20]; the promenade upgrades Van Doorslaer et al. [21], De Finis et al. [22] combined to the addition of stilling wave basins Geeraerts et al. [23], Kisacik et al. [24] or overspill basin Grossi et al. [25], Cappietti et al. [26]; the reduction of the waves incident the quay by floating breakwaters Ruol et al. [27].

In order to meet the aims, physical model tests were performed in the wave flume (2D) of the Padova University.

In addition to this introduction, this paper includes two main sections and a concluding paragraph. The first section describes the study site, the laboratory facility, the tested configurations, the test programme and the types of analysis carried out. The second section presents the results in terms of waves and overtopping discharges, discussing the effect of the geometry of the quay and of the steps in front of the vertical quay. Lastly, conclusions useful for design purposes are drawn.

2. Materials and Methods

2.1. Study Site

The Venice lagoon is a shallow brackish water body (area ~550 km^2), located in the Northern part of the Adriatic Sea. The morphology consists of small islands, tidal flats, marshes and a complex network of channels and it is connected to the sea through three large inlets: Lido, Malamocco and Chioggia. The lagoon is characterized by a semidiurnal tidal regime with a spring tidal range of about 1 m D'Alpaos et al. [28] and the storm surge is an additional relevant component. The highest surge ever measured was on 4 November 1966 when the sea level rose approximately 194 cm above ZMPS and persisted above the 120 cm level for more than 15 h Canestrelli et al. [29]. More recently, on 29 October 2018 at 13 UTC, the fifth highest historical level was recorded in Venice (since 1872, starting year of the measurements) with a maximum sea level of 1.56 m ZMPS. A second peak (1.48 m ZMPS) followed this maximum, with an estimated storm surge of 1.30 m, that "luckily" was not in phase with the astronomical tide so that it was not reached a much more severe flooding Cavaleri et al. [30]. Very recently, on 12 November 2019 at 22.50 UTC, the water level reached 1.87 m ZMPS, the second highest value ever measured, due to an "unlucky" combination of an astronomical tidal peak with a severe storm surge generated by a strong wind (up to 30 m/s) and a sudden pressure drop down to 987 h Pa.

The North Adriatic Sea, and consequently the Venice lagoon, is characterized by two prevailing wind regimes, the Bora and the Scirocco, which blow from North-East and South-East respectively Ruol et al. [31]. The Scirocco wind regime (SE) is responsible for the highest storm surge in front of the Piazza S. Marco Mel et al. [32]. For instance, the direction measured during the 29 October 2018 storm was approximately from 100° N to 150° N and the direction during 12 November 2019 quickly turned from 100° N to 230° N.

The MO.S.E. was designed to mitigate the effect of the increasing number of high tide conditions Trincardi et al. [33] and to maintain the water level in front of Venice below a certain threshold (1.10 m ZMPS in extreme cases, more frequently in the range 0.9 m–1.0 m ZMPS).

The "Piazza S. Marco", constructed in the ninth century as a small square, was significantly enlarged in 1174 and it is dominated at its eastern end by the "Basilica S. Marco". The "Piazzetta" S. Marco is an extension of the Piazza towards S. Marco basin in its South-East border (Figure 1). The Piazzetta lies between the "Palazzo Ducale" (on the east side) and the "Marciana" library (West). The Piazza was paved in the late 12th century with bricks and in 1723 the bricks were replaced with masonry blocks, named "masegni".

The "Riva S. Marco" (historic quay of Venice—Figure 2) is characterized by a pavement mildly sloping toward the lagoon. The quay can be subdivided into three portions, hereafter named A, B and C (Figure 2), characterized by different cross-sections. To describe these cross-sections, three distinctive points are defined: $z1$ is the edge of the quay, $z2$ is the point where the pavement height is equal to 1.1 m ZMPS (maximum expected level with MO.S.E. gates in operation) and $z3$ is the crest height (highest point of the cross-section). Type A and B have a vertical wall (Figure 3a). Type A cross-section, placed in proximity of the "Marciana" library is the most critical since the crest height ($z3$) is 1.10 m ZMPS, distant 5 m from the edge. Type B, placed between the "Colonna San Todaro" and "Colonna San Marco", has a crest at height $z3 = 1.17$ m ZMPS, distant 20 m from the edge. Type C (Figure 3b) is characterized by 5 descending steps, 15 cm high. The crest and its distance from the edge are the same of type B cross-section. Table 1 summarizes the characteristics of the three cross-sections. In front of the quay, a small floating breakwater is placed in a water depth of 3.5 m, that attenuates the waves generated by small boats and protects the "gondole", docked at the Riva. This structure is not intended to be efficient for wind waves (longer period).

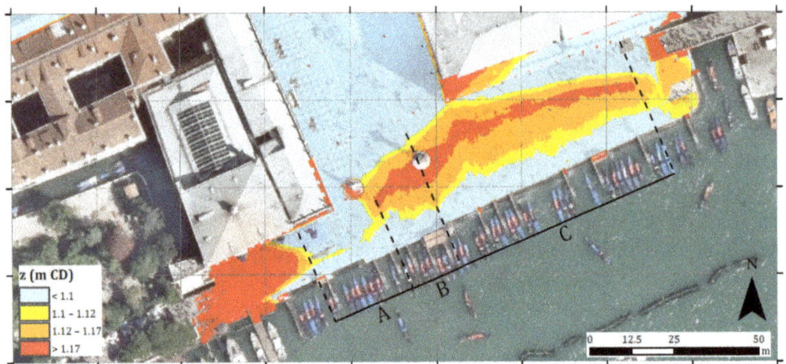

Figure 2. Location and orthophoto of the study site: Piazza S. Marco, Venice, Italy.

Table 1. Summary of the cross-section types that characterize the Riva S. Marco.

No.	Type	Length	z1	z2	z3	L1	L2
A	vertical wall (Figure 3a)	25 m	1.00 m	1.10 m	1.10 m	5 m	0 m
B	vertical wall (Figure 3a)	15 m	1.00 m	1.10 m	1.17 m	5 m	15 m
C	with steps (Figure 3b)	85 m	1.00 m	1.10 m	1.17 m	5 m	15 m

A dataset of waves relative to 2003–2013 measured at the "Punta della Salute" station (situated approximately in front of the square) by the "Centro Previsioni e Segnalazioni Maree—CPSM" institution, allows to analyse the typical wave characteristics in the S. Marco basin. In the same station, a very long dataset of sea level is also available. The wave dataset includes: significant wave height Hs (m), maximum wave height H_{MAX} (m), mean period Tm (s), peak period Ts (s) and water level z (m ZMPS). Moreover, the CPSM institution collected, since 1983, wind and sea level data both inside and outside the lagoon Mel et al. [32].

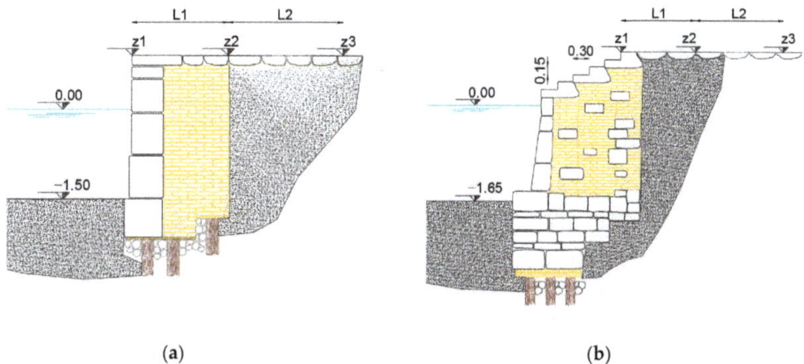

(a)   (b)

Figure 3. Cross-section of the quay at Riva S. Marco, in correspondence of the vertical wall (**a**) and of the descending steps (**b**). The water depths are referred to ZMPS.

It was found that the waves approaching the quay (generated both from ships and from winds) are characterized by significant wave heights in the range 0.1–0.5 m and peak periods in the range 1.5–3.5 s. Figure 4 shows an example of measurements during an extreme event occurred in February 2012, when the wind speed at the Lido inlet remained over 15 m/s for 24 h, reaching a maximum of 25.5 m/s. This event was characterized by a SE direction (Scirocco), generating waves up to 38 cm in front of the Piazza. This value is considered to have a return period of approximately one year. The maximum water level in the S. Marco basin was 0.76 m ZMPS. The plot of *Hs* shows also the daily average. This value is obtained by averaging the wave heights at the same hour of every day, considering all the available measurements. The trend highlights that, during the night, the waves are characterized by lower *Hs* than during the daytime hours since the ship traffic is less intense.

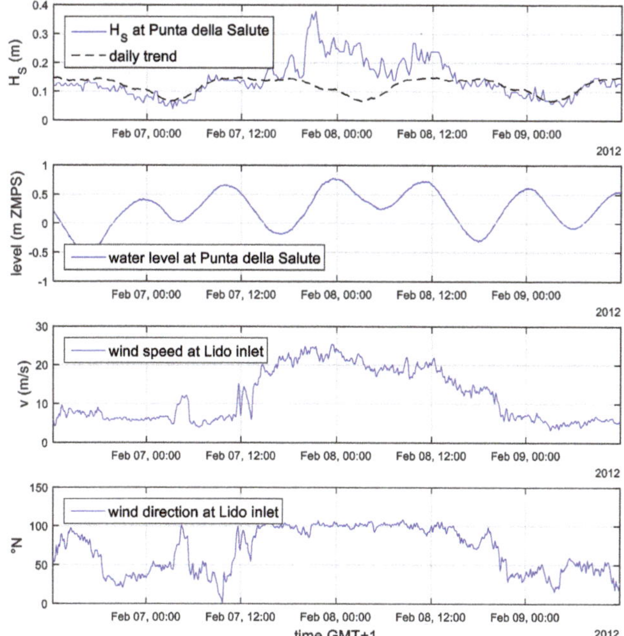

Figure 4. Example of measurements (February 2012), from top to bottom: significant wave height (m), sea level (m ZMPS), wind speed (m/s) and wind direction (°N).

In general, if the wind blows with a speed of 15 m/s from South-East (100° N–140° N), i.e., the directions of the maximum fetch in front of the square, waves higher than 30 cm are generated in the lagoon and therefore they can easily overtop the quay. In order to statistically characterise the occurrence probability of the waves, a statistical analysis of the winds measured at the Lido inlet was carried out. Winds blowing from SE with intensity 15 m/s were observed three times per year.

2.2. Experimental Set-Up and Test Programme

Physical model tests were performed in the wave flume of the Padova University, that is 36.0 m long, 1.0 m wide and 1.30 m deep (Figure 5), and is equipped with a dual piston-flap type wavemaker, capable of generating regular and irregular waves, with active wave absorption. A series of thin metallic plates, parallel to the side walls, are placed in front of the wavemaker to avoid transverse oscillations. The bed is made of lean concrete and the sides are glass made. A recirculation system is present.

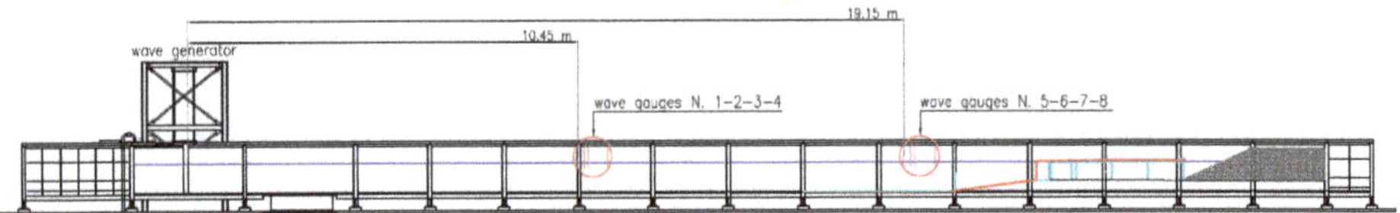

Figure 5. Wave flume and test setup with wave gauges and structure positions.

In this study, the bottom of the model is fixed and non-erodible and reproduces a simplified bathymetry. The bed is horizontal, except for the zone in front of the structure that has a slope of 1:10. The piston-type mode was used.

Tests were carried out in geometrical scale 1:5, according to Froude similarity. The three cross-sections previously described were tested and their geometries were accurately reproduced (Figure 6). A vertical structure simulates the vertical wall of the "Riva S. Marco" and 4 plastic plates were placed behind it, to mimic the presence of the pavement over the quay (Figure 7). During the tests for the cross-section C, 5 metallic descending steps were added in front of the vertical wall. The distance of the model structure from the wave generator was 22.7 m.

Two arrays of four wave gauges were used to measure the incident and reflected waves in the offshore and nearshore zones, and with a sampling frequency of 100 Hz. The offshore array is placed at 10.45 m from the wave generator and the nearshore one is placed at 19.15 m (Figure 5). In both arrays, the intervals between wave gauges are, from offshore to onshore: 19 cm, 11 cm and 53 cm.

The test programme (Table 2) includes irregular wave attack characterized by 5 significant wave heights Hs ranging from 0.2 to 0.75 m, 2 wave steepness ($Hs/L = 0.048$ and 0.063) and three different tidal levels (+0.9 m, +1.0 m and +1.1 m ZMPS). A more detailed investigation of the effect of the wave steepness was not carried out, after observing that the effect of this variable did not affect significantly the results to have practical design value. All the 28 Wave States (WS given in Table 1) are generated for each of the three cross-sections, reaching a total of 84 tests. The white noise filter method was used to generate the irregular wave attacks, lasting ~15 min (i.e., number of waves for each test ~1000), aiming at reproducing JONSWAP spectra with standard peak enhancement factor = 3.3. The random seed used to test the different sections is common for the same wave attack.

Due to the high wave reflection induced by the vertical wall, the efficiency of the wavemaker absorption system is critical. The hardware system used in the laboratory is a consolidated active wave absorption from HR Wallingford. The system efficiency was proven during the phase of transfer function calibration. For the tests under analysis, it was simply verified that the incident wave field is stationary even for long runs.

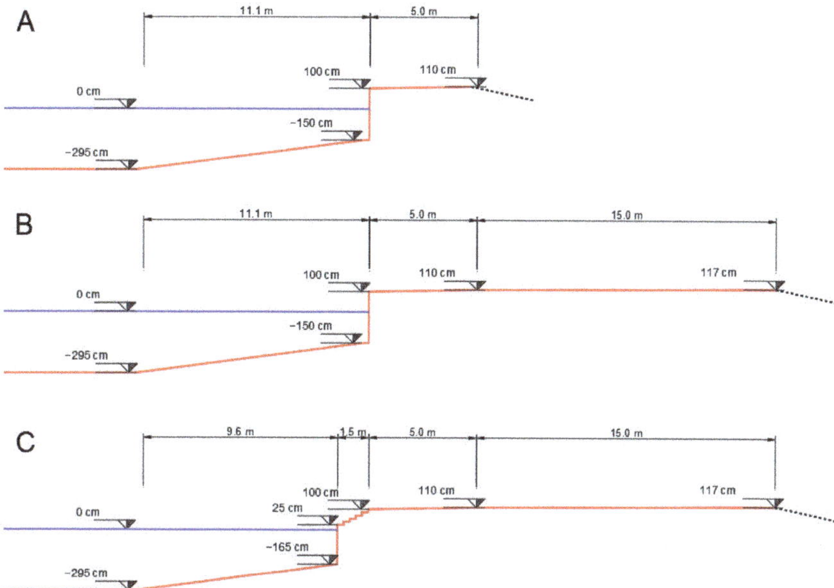

Figure 6. Geometries of the tested sections: type **A**, **B** and **C**.

(a)   (b)

Figure 7. Front view of two tested types of cross-section: vertical wall type **B** (a), and wall with steps type **C** (b).

Table 2. Test programme repeated for each tested cross-section.

WS	Tide (m ZMPS)	Hs (m)	Tp (s)	WS	Tide (m ZMPS)	Hs (m)	Tp (s)
1	0.9	0.20	1.57	15	0.9	0.20	1.79
2	0.9	0.30	1.92	16	0.9	0.30	2.19
3	0.9	0.40	2.21	17	0.9	0.40	2.53
4	0.9	0.50	2.47	18	0.9	0.50	2.83
5	0.9	0.75	3.03	19	0.9	0.75	3.46
6	1.0	0.20	1.57	20	1.0	0.20	1.79
7	1.0	0.30	1.92	21	1.0	0.30	2.19
8	1.0	0.40	2.21	22	1.0	0.40	2.53
9	1.0	0.50	2.47	23	1.0	0.50	2.83
10	1.0	0.75	3.03	24	1.0	0.75	3.46
11	1.1	0.20	1.57	25	1.1	0.20	1.79
12	1.1	0.30	1.92	26	1.1	0.30	2.19
13	1.1	0.40	2.21	27	1.1	0.40	2.53
14	1.1	0.50	2.47	28	1.1	0.50	2.83

The overtopping discharge was measured for all the configurations. Three different setups were used, each appropriately sized to collect the maximum expected volume of overtopped water preserving an accuracy of at least 1%. The first setup was a tray that redirected the overtopping flow to a large bucket, and the volume was measured at the end of the test. The second was similar, but with a narrower tray width (10 cm), so that only the fraction at mid-section (considered representative of the full width) was collected. The last, suited to very large discharges, also comprised a pumping system, placed in the bucket, a flowmeter, and a non-return valve along the outlet pipe.

2.3. Analysis Methods

The same analysis methods described in Martinelli et al. [20] was used. In short, the water elevation signals were band-pass filtered in the range 0.1–20 Hz to remove the average and possible noise. Incident and reflected waves were identified using the method described in Zelt et al. [34]. Typical time-domain analysis were carried out to find the significant (*Hs*) and maximum (*Hmax*) values of the wave height and the significant (*Ts*) and mean (*Tm*) wave period. Spectral analysis allowed to find the peak wave period *Tp* and the estimate of the significant wave height *Hm0*.

The measured value of the mean overtopping discharge was obtained by dividing the accumulated overtopping volume by the test duration and by the collecting tray width, so to obtain unit values. The measurements were compared to the predictions of the formula (Equation (1)) given by the EurOtop Manual [8]. This equation is relative to a vertical wall considering non-impulsive conditions and influencing foreshore, since the bed in the laboratory in front of the breakwater has a slope of 1:10.

$$\frac{q}{\sqrt{gH_{m0}^3}} = 0.05exp\left(-\frac{2.78}{\gamma}\frac{R_C}{H_{m0}}\right) \tag{1}$$

where γ takes into account the presence of a berm, oblique waves, a crown wall, etc. The effect of the presence of a sloping pavement at the top of the vertical wall can be assimilated to a promenade and its effect could be included using an influence factor γ_{prom} given by the EurOtop Manual (Equation (2)).

$$\gamma_{prom} = 1 - 0.47\frac{G_C}{L_{m-1,0}} \tag{2}$$

where the $L_{m-1,0}$ is the deepwater wavelength and Gc is the promenade width, valid in the range $G_C/L_{m-1,0} = 0.05$–0.5.

At S. Marco quay, Gc is equal to (L1 + L2) for the tests with 0.9 m and 1.0 m ZMPS water levels and Gc is equal to L2 for tests with 1.1 m ZMPS water level (L1 and L2 are defined in Figure 3 and in Table 1). However, the coefficient γ_{prom} is not applicable for this investigation, since the ratio between G_C and $L_{m-1,0}$ is largely outside the range of application, $G_C/L_{m-1,0} = 0.6$–4, and the formula gives inaccurate or even negative results. Therefore, in the following, a new coefficient is proposed that fits the measures, in order to have a simple tool for the prediction of the overtopping within the experimentally investigated range.

3. Results

The results are given in terms of reflection coefficients and mean overtopping discharges. Comparisons with available formulas are also proposed that enhanced the need for physical model tests for the analyzed geometries.

3.1. Waves

This sub-section presents the main results in terms of incident and reflected wave characteristics, summarized in Tables 3–5. Figure 8 shows the wave reflection coefficient k_R (ratio between reflected and incident *Hs*) for the three cross-sections (A, B and C) against the wave period. In the figure, crosses (x) are relative to the wave steepness $S_{op} = 0.063$ and circles (o) are relative to $S_{op} = 0.048$.

Table 3. Test results for cross-section type A.

WS	H_{sI} (m)	H_{sR} (m)	H_{m0} (m)	$T_{m-1,0}$ (s)	$q/\sqrt{gH_{m0}^3}$	WS	H_{sI} (m)	H_{sR} (m)	H_{m0} (m)	$T_{m-1,0}$ (s)	$q/\sqrt{gH_{m0}^3}$
1	0.22	0.17	0.22	1.70	0.0011	15	0.22	0.17	0.22	1.67	0.0011
2	0.31	0.22	0.31	1.75	0.0031	16	0.32	0.25	0.32	2.06	0.0040
3	0.43	0.31	0.44	2.07	0.0087	17	0.43	0.33	0.44	2.48	0.0078
4	0.54	0.39	0.54	2.50	0.0129	18	0.54	0.39	0.55	2.49	0.0131
5	0.78	0.55	0.81	2.86	0.0243	19	0.78	0.56	0.77	3.27	0.0272
6	0.22	0.13	0.22	1.70	0.0061	20	0.22	0.13	0.22	1.70	0.0064
7	0.31	0.18	0.32	1.75	0.0113	21	0.32	0.21	0.32	2.07	0.0129
8	0.43	0.26	0.42	2.06	0.0224	22	0.43	0.28	0.43	2.47	0.0234
9	0.53	0.34	0.54	2.48	0.0277	23	0.52	0.34	0.53	2.48	0.0288
10	0.79	0.51	0.79	2.85	0.0400	24	0.77	0.52	0.78	3.24	0.0435
11	0.22	0.09	0.22	1.69	0.0302	25	0.21	0.09	0.22	1.67	0.0255
12	0.31	0.13	0.31	1.76	0.0399	26	0.32	0.16	0.31	2.07	0.0416
13	0.43	0.20	0.43	2.09	0.0499	27	0.43	0.23	0.42	2.44	0.0495
14	0.53	0.29	0.53	2.45	0.0543	28	0.52	0.29	0.52	2.48	0.0531

Table 4. Test results for cross-section type B.

WS	H_{sI} (m)	H_{sR} (m)	H_{m0} (m)	$T_{m-1,0}$ (s)	$q/\sqrt{gH_{m0}^3}$	WS	H_{sI} (m)	H_{sR} (m)	H_{m0} (m)	$T_{m-1,0}$ (s)	$q/\sqrt{gH_{m0}^3}$
1	0.22	0.18	0.22	1.70	0.0000	15	0.21	0.18	0.22	1.70	0.0000
2	0.31	0.24	0.31	1.79	0.0000	16	0.32	0.27	0.32	2.06	0.0000
3	0.42	0.34	0.44	2.11	0.0005	17	0.42	0.35	0.43	2.47	0.0007
4	0.52	0.41	0.53	2.50	0.0020	18	0.53	0.41	0.54	2.52	0.0025
5	0.78	0.59	0.81	2.87	0.0115	19	0.80	0.61	0.80	3.25	0.0129
6	0.21	0.14	0.22	1.67	0.0000	20	0.21	0.14	0.21	1.68	0.0000
7	0.31	0.19	0.30	1.80	0.0002	21	0.31	0.22	0.33	2.08	0.0004
8	0.42	0.29	0.44	2.13	0.0027	22	0.42	0.31	0.43	2.47	0.0032
9	0.51	0.36	0.52	2.48	0.0082	23	0.53	0.38	0.54	2.49	0.0077
10	0.78	0.54	0.79	2.92	0.0186	24	0.80	0.59	0.79	3.24	0.0192
11	0.21	0.10	0.21	1.65	0.0000	25	0.20	0.10	0.21	1.68	0.0000
12	0.30	0.15	0.30	1.79	0.0015	26	0.31	0.17	0.33	2.07	0.0025
13	0.41	0.24	0.43	2.14	0.0065	27	0.41	0.27	0.42	2.45	0.0090
14	0.51	0.32	0.52	2.44	0.0169	28	0.53	0.33	0.54	2.46	0.0183

Table 5. Test results for cross-section type C.

WS	H_{sI} (m)	H_{sR} (m)	H_{m0} (m)	$T_{m-1,0}$ (s)	$q/\sqrt{gH_{m0}^3}$	WS	H_{sI} (m)	H_{sR} (m)	H_{m0} (m)	$T_{m-1,0}$ (s)	$q/\sqrt{gH_{m0}^3}$
1	0.21	0.05	0.21	1.66	0.0000	15	0.21	0.05	0.21	1.68	0.0000
2	0.30	0.07	0.30	1.74	0.0000	16	0.30	0.08	0.31	2.05	0.0000
3	0.41	0.11	0.42	2.10	0.0004	17	0.39	0.10	0.40	2.42	0.0009
4	0.50	0.11	0.50	2.40	0.0034	18	0.51	0.11	0.53	2.45	0.0035
5	0.73	0.17	0.75	2.85	0.0167	19	0.74	0.25	0.72	3.29	0.0185
6	0.21	0.05	0.20	1.66	0.0000	20	0.21	0.05	0.21	1.68	0.0000
7	0.29	0.07	0.30	1.75	0.0000	21	0.30	0.08	0.31	2.03	0.0000
8	0.41	0.10	0.42	2.10	0.0019	22	0.39	0.09	0.41	2.42	0.0025
9	0.50	0.10	0.51	2.41	0.0068	23	0.51	0.10	0.53	2.45	0.0076
10	0.73	0.14	0.76	2.86	0.0225	24	0.73	0.20	0.72	3.25	0.0268
11	0.21	0.04	0.21	1.65	0.0000	25	0.20	0.04	0.21	1.68	0.0000
12	0.30	0.06	0.30	1.76	0.0009	26	0.29	0.06	0.31	2.05	0.0028
13	0.41	0.08	0.42	2.10	0.0079	27	0.39	0.07	0.41	2.42	0.0095
14	0.50	0.09	0.51	2.41	0.0145	28	0.51	0.09	0.52	2.45	0.0151

For the vertical wall sections (A and B), k_R decreases with the water level, due to the increasing overflow. For the lowest water level (i.e., 0.9 m ZMPS, red symbols), the promenade does not play a relevant role and k_R is very close to 1.0 and independent from Hs and Tp, since all the tested waves are almost completely reflected by the wall. For the highest water level 1.1 m ZMPS (black symbols), at rest, the promenade is partially submerged. The smaller/shorter waves run up the promenade and the resulting k_R is only ~0.5; the higher/longer waves are more effectively reflected (k_R ~ 0.8) by the vertical wall below the mean sea level. Therefore, in this case, k_R increases with wave height/period. The case with water level = 1.0 m ZMPS (blue symbols) is intermediate: at rest, the still water level reaches the edge of the promenade, and the reflection is not affected by Hs or Tp since the vertical wall and the promenade affect in the same measure all the waves.

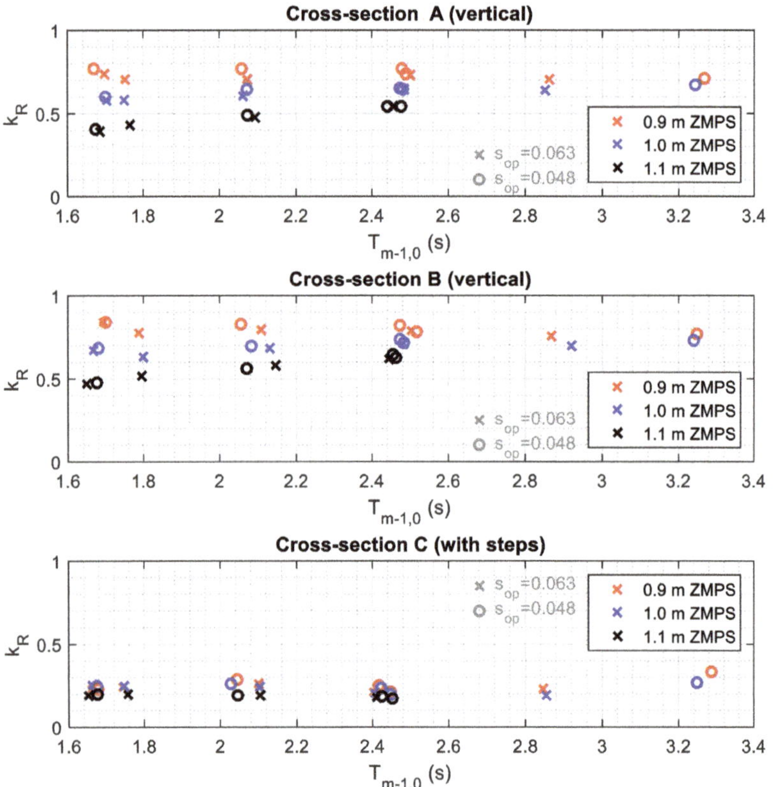

Figure 8. Wave reflection coefficient for the three cross-sections, from top to bottom: type **A**, **B** and **C**.

For the third analysed cross-section (type C, with steps) k_R is significantly lower ($k_R \sim 0.3$). In fact, the presence of the 5 steps can be compared to a "rough slope" that diminishes the reflection and facilitates the run-up over the quay. Clearly, the number of steps, as well as their dimensions, influence the energy dissipation. This agrees with the results of Kerpen et al. [17]. The pictures in Figure 9 give a qualitative description of the wave-structure interaction at the quay border. The same wave is selected from the videos of test no. 7 for two configurations, with and without steps.

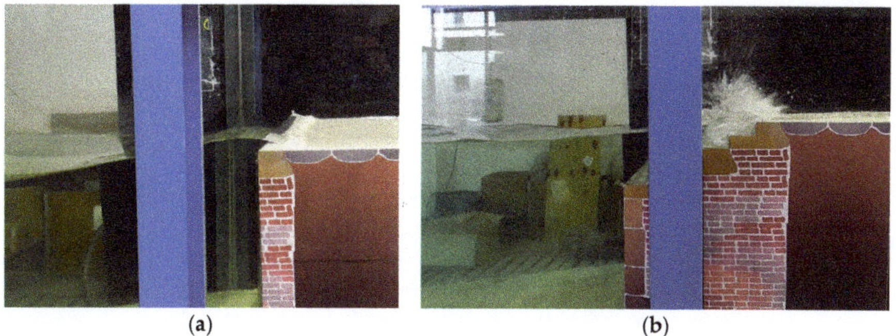

Figure 9. Snapshot of the same wave approaching two different cross-sections: type **B** (**a**), type **C** (**b**).

3.2. Mean Overtopping Discharge

The mean overtopping discharge was measured for all the tests and all the cross-sections. Figure 10 shows the wave overtopping results for the three tested cross-sections against the measured significant wave height.

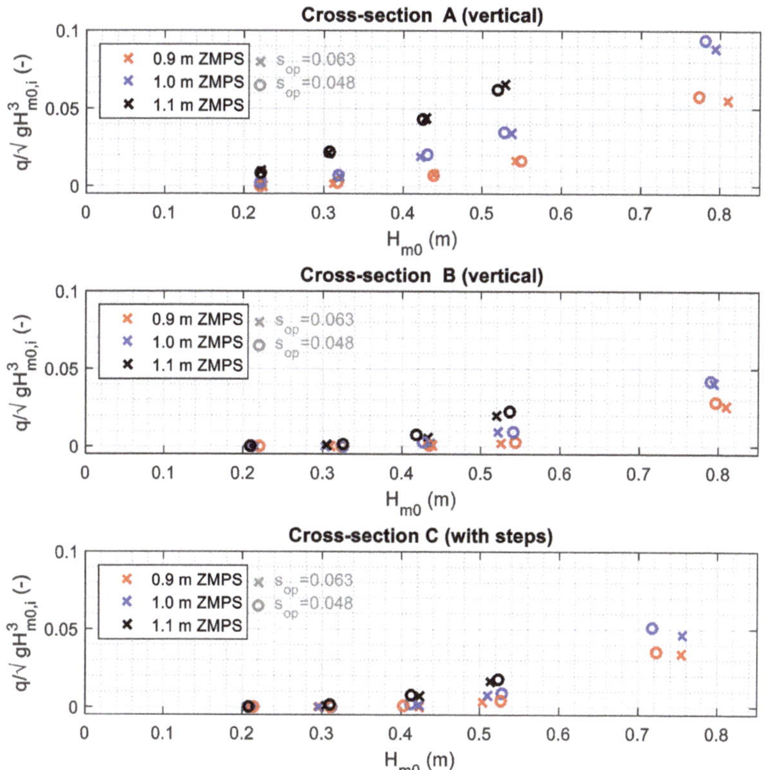

Figure 10. Wave overtopping results for the three tested cross-sections, from top to bottom: type **A**, type **B** and type **C**.

As anticipated, the cross-section A represents a critical issue since the maximum level of the pavement is only 5 m apart the quay and its elevation is only 10 cm higher than the border (Figure 6A).

In this section, with the highest water level, almost all the waves cause overtopping. Therefore, this portion of the quay requires a specific design to reduce flooding considering multiple mitigation strategies, e.g., provisional barriers to be temporarily placed on the pavement to reduce overtopping volumes, and/or the installation of an efficient floating breakwater to reduce the incident wave heights. Moreover, in this area, a significant rise of the pavement is not feasible since the slope between the crest of the quay and the square cannot be too steep.

Focusing on the cross-sections B and C, for which the maximum level of the pavement is about 20 m apart the quay and its elevation is 17 cm higher than the border, it was found that only the waves higher than 0.3 m are responsible for considerable discharges over the quay, and therefore can cause, or exacerbate, the flooding of Piazza S. Marco. The differences in terms of overtopping between sections B and C (without and with the steps) are negligible. In fact, the vertical wall (B) causes a high reflection that almost doubles the wave height increasing the overtopping, whereas the series of small steps (C) behaves more or less like a ramp that facilitates run-up and consequently the overtopping. Both these behaviours were observed to provide quantitatively similar results.

Figures 11a, 12a and 13a compare the measured dimensionless overtopping discharge to the EurOtop [8] prediction using Equation (1), without any influence factor. As expected, the predictions largely overestimate the results, since the presence of the promenade is ignored. Moreover, for the type A under the highest water level condition, the formula gives a single result since the freeboard is zero.

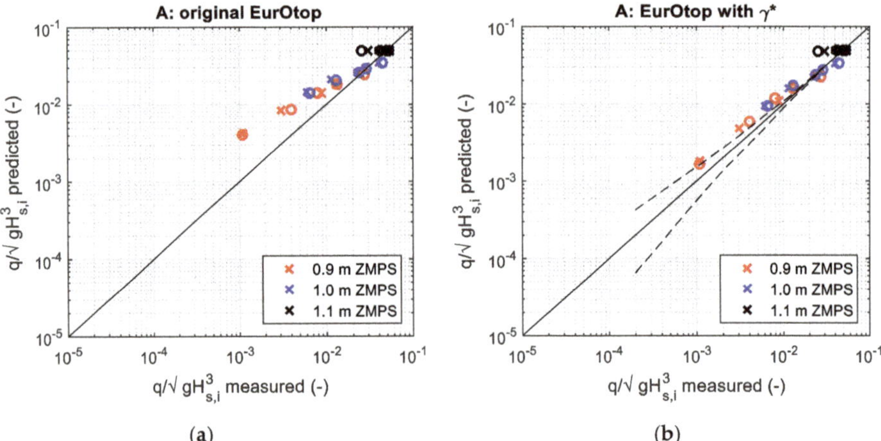

(**a**) (**b**)

Figure 11. Measured overtopping and predictions for the cross-sections A: according to EurOtop (**a**), applying γ^* (**b**). Crosses (x) are relative to $S_{op} = 0.063$, circles (o) are relative to $S_{op} = 0.048$. Dashed lines are the upper and lower 95% prediction bounds.

To overcome this issue, a new influence factor accounting for the presence of a mildly sloping pavement is proposed. The presence of the steps is ignored since it has a minor and uncertain influence on the measured overtopping discharge. The reduction effect found by Kerpen et al. [17] is relative to a stepped revetment that continues down to the bottom with no vertical wall, i.e., a geometry significantly different from section C.

The formulation (Equation (3)) follows the one proposed by the EurOtop manual, i.e., taking into account the dimensionless promenade width ($G_C/L_{m-1,0}$).

$$\gamma^* = a\left[1 - b\frac{G_C}{L_{m-1,0}}\right] \tag{3}$$

The best fit of experimental data was achieved through the least-squares method, obtaining b equal to 0.2251 (with 95% confidence bounds = [0.2128–0.2374]). The a coefficient takes into account the wall slope at the free surface (α) and it was set equal to $a = (a_1 + \alpha/180°)$, where a_1 is equal to 0.5 (with 95% confidence bounds = [0.4513–0.5487]). For the S. Marco quay, α is 90° for water levels equal to 0.9 and 1.0 m (i.e., $a \sim 1$) and α is $\sim 0°$ for the water level 1.1 m, i.e., $a \sim a_1$ when the promenade is partially submerged. The investigated range for Equation (3) is $G_C/L_{m-1,0} = 0.6\text{–}4$.

Figures 11b, 12b and 13b compare the measured dimensionless overtopping discharges with the EurOtop prediction using (Equation (1)), introducing the new proposed influence factor (Equation (3)). These figures show also an empirical indication of the upper and lower 95% prediction bounds, based on the coefficient variability. The prediction bounds are numerically obtained through a MonteCarlo simulation that randomly samples the coefficients, assumed to be Gaussian distributed. The 95% bounds are evaluated for a limited set of conditions. Since different conditions give potentially the same expected overtopping but different confidence intervals, an average value is assumed. Finally, the curve is smoothed.

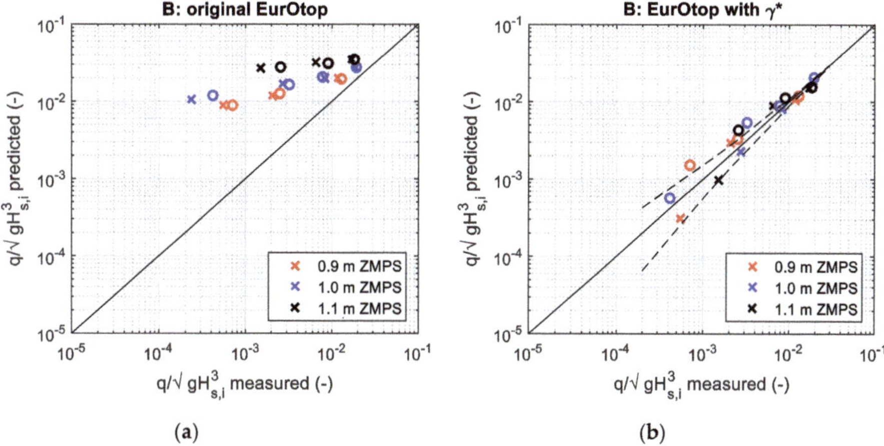

(a)   (b)

Figure 12. Measured overtopping and predictions for the cross-sections B: according to EurOtop (**a**), applying γ^* (**b**). Crosses (x) are relative to $S_{op} = 0.063$, circles (o) are relative to $S_{op} = 0.048$. Dashed lines are the upper and lower 95% prediction bounds.

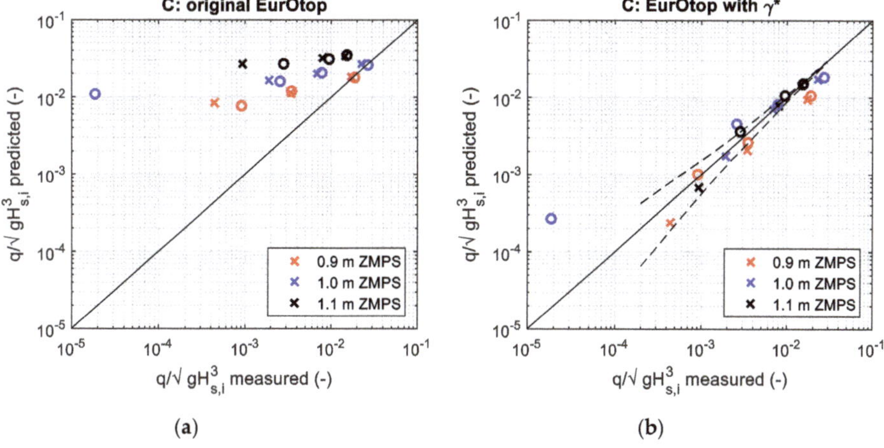

(a)   (b)

Figure 13. Measured overtopping and predictions for the cross-sections C: according to EurOtop (**a**), applying γ^* (**b**). Crosses (x) are relative to $S_{op} = 0.063$, circles (o) are relative to $S_{op} = 0.048$. Dashed lines are the upper and lower 95% prediction bounds.

It may be observed that the structure of the γ-formula proposed by the Eurotop Manual (Equation (2)) is effective also for the prediction of the overtopping measurements of this investigation, although a different calibration of the parameter (Equation (3)) is necessary to achieve a good agreement with the results. Clearly, γ cannot effectively represent the effect of the promenade when Rc tends to zero, since the overtopping in Equation (1) becomes independent from. Therefore, as expected, for cross-section A the tests with water level equal to 1.1 m ZMPS ($Rc = 0$) remain in disagreements with the predictions.

To assess the quality of the estimates, three performance metrics were calculated (Table 6): the coefficient of efficiency *NSE* Nash et al. [35], the index of agreement *D* Willmott et al. [36] and the square of the correlation coefficient r^2. Complete disagreement is described by $D = 0$, $r^2 = 0$ and negative *NSE*. All indexed are $= 1$ for perfect agreement. Applying the original EurOtop formula (Equation (1)), the *NSE* is equal to 0.7488, *D* is equal to 0.9336 and r^2 is equal to 0.8632. Considering the new influence factor γ^*, the *NSE* reaches the value of 0.9471, *D* is equal to 0.9847 and r^2 is equal to 0.9655.

Table 6. Test results for cross-section type C.

	EurOtop Original Formula			EurOtop with γ^*		
	NSE	*D*	r^2	*NSE**	*D**	r^{2*}
section A	0.9540	0.9862	0.9822	0.9542	0.9865	0.9826
section B	−0.0908	0.8404	0.9187	0.9889	0.9972	0.9891
section C	0.4280	0.8756	0.7994	0.8614	0.9521	0.953
all section	0.7522	0.9341	0.8632	0.9504	0.9856	0.9655

4. Conclusions

This paper discusses the wave overtopping that can flood the "Piazza S. Marco", Venice, under particular combinations of water level and incident wave height. Currently, the inundation may be caused by several factors: the water inlet from the sewer drains, the seepage, the overflow from the bounds and the wave overtopping over the "Riva S. Marco" border. Today, the first mechanism is dominant.

The MO.S.E. system is expected to guarantee that the water level remains below 1.1 m ZMPS in the whole lagoon, but the "Piazza S. Marco" would be slightly flooded in these conditions. Works are planned to completely defend the "Piazza", avoiding the back-flow through the drainage system, and reducing the overflow and overtopping volumes.

The first problem will be solved using special devices, operated by motorized sluice gates, able to close the drainage network towards the lagoon when necessary. The overflow from the lower boundaries along the smaller channels (not affected by waves) will be avoided with a local increase of the pavements. In light of this, the major problem remains the estimate of the overtopping volumes entering "Piazza S.Marco" from its quay (Riva San Marco) caused by the wave action.

This phenomenon was faced up by means of physical model tests, performed in the wave flume of Padova University: 84 tests have been carried out, considering 3 water levels, 2 wave steepness and up to five wave heights. Three representative cross-sections have been analyzed to represent the dis-homogeneity of the quay. EurOtop empirical formula can describe the effect of the mildly sloping pavements through the promenade influence factor. However, the range of validity of the formula does not include the investigated case. The structure of the γ-formula was found to be still effective also for our case, although a different calibration of the coefficient is necessary (Equation (3)). The coefficients and their confidence interval are given. The obtained formula is a useful design tool able to predict the overtopping for the S. Marco quay, under a wide range of wave heights and water levels.

In general, the experiments show that, under waves higher than 30 cm (an event that is expected to occur a few times per year), the overtopping is significant for all the tested water levels. For rare case of waves higher than 40 cm, the overtopping is critical and some mitigation measure is required since the discharge alone exceeds the limits of the drainage system envisaged for the rainfall.

One of the possible mitigation measures consists of an efficient floating breakwater, to be placed in front of the "Piazza", able to significantly reduce the transmitted waves. The quay in front of the "Marciana" library needs a specific solution, e.g., a temporary barrier, since its crest freeboard may fall below the level in the lagoon, even when the MO.S.E. is operating.

Author Contributions: Conceptualization, P.R., C.F., M.V. and L.M.; methodology, P.R., C.F., M.V. and L.M.; formal analysis, C.F., M.V. and L.M.; investigation, M.V.; writing—original draft preparation, C.F.; writing—review and editing, C.F. and L.M.; supervision, P.R. and L.M. All authors have read and agreed to the published version of the manuscript.

Funding: This research received no external funding.

Acknowledgments: The fruitful discussion with the colleagues of ICEA Department, Consorzio Venezia Nuova, Thetis, Konstructiva, and Mate Eng., regarding the flood protection of Piazza San Marco, is greatly acknowledged.

Conflicts of Interest: The authors declare no conflict of interest.

References

1. Cazenave, A.; Cozannet, G.L. Sea level rise and its coastal impacts. *Earth's Future* **2014**, *2*, 15–34. [CrossRef]
2. Tosi, L.; Strozzi, T.; Da Lio, C.; Teatini, P. Regional and local land subsidence at the Venice coastland by TerraSAR-X PSI. *Proc. Int. Assoc. Hydrol. Sci.* **2015**, *372*, 199–205. [CrossRef]
3. Lionello, P. The climate of the Venetian and North Adriatic region: Variability, trends and future change. *Phys. Chem. Earth. Parts A/B/C* **2012**, *40*, 1–8. [CrossRef]
4. Rinaldo, A.; Nicótina, L.; Celegon, E.A.; Beraldin, F.; Botter, G.; Carniello, L.; Cecconi, G.; Defina, A.; Settin, T.; Uccelli, A.; et al. Sea level rise, hydrologic runoff, and the flooding of Venice. *Water Resour. Res.* **2008**, *44*. [CrossRef]
5. Goda, Y. *Random Seas and Design of Maritime Structures*; World Scientific Publishing Co Pte Ltd.: Singapore, 2010.
6. Environment Agency. *Wave Overtopping of Seawalls, Design and Assessment Manual*; R&D Technical Report W178; HR Wallingford Ltd.: Wallingford, UK, 1999.
7. US Army Corps of Engineers. *Coastal Engineering Manual-Part VI*; USACE: Washington, DC, USA, 2002.
8. van der Meer, J.W.; Allsop, N.W.H.; Bruce, T.; De Rouck, J.; Kortenhaus, A.; Pullen, T.; Schuuttrumpf, H.; Troch, P.; Zanuttigh, B. EurOtop: Manual on Wave Overtopping of Sea Defences and Related Structures. Available online: www.overtopping-manual.com. (accessed on 28 December 2018).
9. Van Der Meer, J.W.; Verhaeghe, H.; Steendam, G.J. The new wave overtopping database for coastal structures. *Coastal Eng.* **2009**, *56*, 108–120. [CrossRef]
10. Van Gent, M.R.; Boogaard, H.F.V.D.; Pozueta, B.; Medina, J.R. Neural network modelling of wave overtopping at coastal structures. *Coast. Eng.* **2007**, *54*, 586–593. [CrossRef]
11. Zanuttigh, B.; Formentin, S.M.; Van Der Meer, J.W. Prediction of extreme and tolerable wave overtopping discharges through an advanced neural network. *Ocean Eng.* **2016**, *127*, 7–22. [CrossRef]
12. Verhaeghe, H.; Van Der Meer, J.W.; Steendam, G.-J.; Besley, P.; Franco, L.; Van Gent, M. Wave Overtopping Database As the Starting Point for a Neural Network Prediction Method. In Proceedings of the Coastal Structures, Portland, OR, USA, 26–30 August 2003; pp. 418–430.
13. Van Doorslaer, K.; De Rouck, J.; Audenaert, S.; Duquet, V. Crest modifications to reduce wave overtopping of non-breaking waves over a smooth dike slope. *Coast. Eng.* **2015**, *101*, 69–88. [CrossRef]
14. Pullen, T.; Allsop, W.; Bruce, T.; Pearson, J. Field and laboratory measurements of mean overtopping discharges and spatial distributions at vertical seawalls. *Coast. Eng.* **2009**, *56*, 121–140. [CrossRef]
15. Burcharth, H.F.; Andersen, T.L.; Lara, J.L. Upgrade of coastal defence structures against increased loadings caused by climate change: A first methodological approach. *Coast. Eng.* **2014**, *87*, 112–121. [CrossRef]
16. Formentin, S.M.; Zanuttigh, B. A Genetic Programming based formula for wave overtopping by crown walls and bullnoses. *Coast. Eng.* **2019**, *152*, 103529. [CrossRef]
17. Kerpen, N.B.; Schoonees, T.; Schlurmann, T. Wave Overtopping of Stepped Revetments. *Water* **2019**, *11*, 1035. [CrossRef]
18. McCabe, M.; Stansby, P.; Apsley, D. Random wave runup and overtopping a steep sea wall: Shallow-water and Boussinesq modelling with generalised breaking and wall impact algorithms validated against laboratory and field measurements. *Coast. Eng.* **2013**, *74*, 33–49. [CrossRef]

19. Pearson, J.; Bruce, T.; Allsop, W.; Kortenhaus, A.; Van Der Meer, J.; Smith, J.M. Effectiveness of recurve walls in reducing wave overtopping on seawalls and breakwaters. *Coast. Eng.* **2004**, *4*, 4404–4416.

20. Martinelli, L.; Ruol, P.; Volpato, M.; Favaretto, C.; Castellino, M.; De Girolamo, P.; Franco, L.; Romano, A.; Sammarco, P. Experimental investigation on non-breaking wave forces and overtopping at the recurved parapets of vertical breakwaters. *Coast. Eng.* **2018**, *141*, 52–67. [CrossRef]

21. Van Doorslaer, K.; Romano, A.; De Rouck, J.; Kortenhaus, A. Impacts on a storm wall caused by non-breaking waves overtopping a smooth dike slope. *Coast. Eng.* **2017**, *120*, 93–111. [CrossRef]

22. De Finis, S.; Romano, A.; Bellotti, G. Numerical and laboratory analysis of post-overtopping wave impacts on a storm wall for a dike-promenade structure. *Coast. Eng.* **2020**, *155*, 103598. [CrossRef]

23. Geeraerts, J.; De Rouck, J.; Beels, C.; Gysens, S.; De Wolf, P.; Smith, J.M. Reduction of wave overtopping at seadikes: stilling wave basin (SWB). *Coast. Eng.* **2006**, *5*, 4680–4691.

24. Kisacik, D.; Tarakcioglu, G.O.; Baykal, C. Stilling wave basins for overtopping reduction at an urban vertical seawall—The Kordon seawall at Izmir. *Ocean Eng.* **2019**, *185*, 82–99. [CrossRef]

25. Grossi, A.; Kortenhaus, A.; Romano, A.; Franco, L. Wave Overtopping Prediction for Sloping Coastal Structures with Overspill Basins at the Crest. In Proceedings of the Coastal Structures and Solutions to Coastal Disasters, Boston, MA, USA, 9–11 September 2015; pp. 743–753.

26. Cappietti, L.; Aminti, P.L. Laboratory investigation on the effectiveness of an overspill basin in reducing wave overtopping on marina breakwaters. *Coast. Eng. Proc.* **2012**, *1*, 20. [CrossRef]

27. Ruol, P.; Martinelli, L.; Pezzutto, P. Formula to Predict Transmission for π-Type Floating Breakwaters. *J. Waterw. Port Coast. Ocean Eng.* **2013**, *139*, 1–8. [CrossRef]

28. D'Alpaos, A.; Carniello, L.; Rinaldo, A. Statistical mechanics of wind wave-induced erosion in shallow tidal basins: Inferences from the Venice Lagoon. *Geophys. Res. Lett.* **2013**, *40*, 3402–3407. [CrossRef]

29. Canestrelli, P.; Mandich, M.; Pirazzoli, P.A.; Tomasin, A. Wind, Depression and Seiches: Tidal Perturbations in Venice (1951–2000). Centro Previsioni e Segnalazioni Maree. Comune Venezia 2001. pp. 1–104. Available online: http://93.62.201.235/maree/DOCUMENTI/Venti_depressioni_e_sesse.pdf (accessed on 19 November 2018).

30. Cavaleri, L.; Bajo, M.; Barbariol, F.; Bastianini, M.; Benetazzo, A.; Bertotti, L.; Chiggiato, J.; Davolio, S.; Ferrarin, C.; Magnusson, L.; et al. The October 29, 2018 storm in Northern Italy—An exceptional event and its modeling. *Prog. Oceanogr.* **2019**, *178*, 102178. [CrossRef]

31. Ruol, P.; Martinelli, L.; Favaretto, C. Vulnerability Analysis of the Venetian Littoral and Adopted Mitigation Strategy. *Water* **2018**, *10*, 984. [CrossRef]

32. Mel, R.; Carniello, L.; D'Alpaos, L. Addressing the effect of the MO. SE barriers closure on wind setup within the Venice lagoon. *Estuar. Coast. Shelf Sci.* **2019**, *225*, 106249. [CrossRef]

33. Trincardi, F.; Barbanti, A.; Bastianini, M.; Benetazzo, A.; Cavaleri, L.; Chiggiato, J.; Papa, A.; Pomaro, A.; Sclavo, M.; Tosi, L.; et al. The 1966 Flooding of Venice: What Time Taught Us for the Future. *Oceanography* **2016**, *29*, 178–186. [CrossRef]

34. Zelt, J.A.; Skjelbreia, J.E. Estimating incident and reflected wave fields using an arbitrary number of wave gauges. In Proceedings of the 23rd International Conference on Coastal Engeering, Venice, Italy, 4–9 October 1992; pp. 777–789.

35. Nash, J.E.; Sutcliffe, J.V. River Flow forecasting through conceptual models-Part I: A discussion of principles. *J. Hydrol.* **1970**, *10*, 282–290. [CrossRef]

36. Willmott, C.J.; Ackleson, S.G.; Davis, R.E.; Feddema, J.J.; Klink, K.M.; LeGates, D.R.; O'Donnell, J.; Rowe, C.M. Statistics for the evaluation and comparison of models. *J. Geophys. Res. Space Phys.* **1985**, *90*, 8995. [CrossRef]

Article

Green Water on A Fixed Structure Due to Incident Bores: Guidelines and Database for Model Validations Regarding Flow Evolution

Jassiel V. Hernández-Fontes [1,2,*], Paulo de Tarso T. Esperança [2], Juan F. Bárcenas Graniel [3], Sergio H. Sphaier [2] and Rodolfo Silva [1]

1   II-UNAM, Instituto de Ingeniería, Universidad Nacional Autónoma de México, Edificio 17, Ciudad Universitaria, Mexico City 04510, Mexico; rsilvac@iingen.unam.mx

2   LabOceano (Laboratório de Tecnologia Oceânica), Programa de Engenharia Oceânica, COPPE/UFRJ, Universidade Federal do Rio de Janeiro, Parque Tecnológico do Rio, Rio de Janeiro 21941-907, Brazil; ptarso@laboceano.coppe.ufrj.br (P.T.T.E.); shsphaier@oceanica.ufrj.br (S.H.S.)

3   Departamento de Ciencias Básicas e Ingeniería, Universidad del Caribe, SM. 78, Manzana 1, Lote 1, Esq. Fraccionamiento Tabachines, Cancun 77528, Quintana Roo, Mexico; jbarcenas@ucaribe.edu.mx

*   Correspondence: jassiel.hernandez@gmail.com

Received: 19 November 2019; Accepted: 3 December 2019; Published: 7 December 2019

Abstract: This paper presents a two-dimensional experimental study of the interaction of wet dam-break bores with a fixed structure, regarding the evolution of the incident flows and the resultant green water events on the deck. The study employs image-based techniques to analyse flow propagation from videos taken by high-speed cameras, considering five different shipping water cases. The features of small air-cavities formed in some green water events of the plunging-dam-break type were analysed. Then, the spatial and temporal distribution of water elevations of the incident bores and green water were investigated, providing a database to be used for model validations. Some guidelines for the selection of the freeboard exceedance, which is of relevance for green water simulations, were provided. Finally, the relationship between the incident bore and water-on-deck kinematics was discussed. The proposed study can be used as a reference for performing simplified and systematic analyses of green water in a different two-dimensional setup, giving high-resolution data that visually capture the flow patterns and allow model validations to be performed.

Keywords: image analysis; green water; wet dam-break bore; 2D experimental study; water elevation database

---

1. Introduction

Green water events, defined as compact masses of water shipping on the deck of marine structures [1], are phenomena of importance in applications involving the interaction of waves with fixed, moored, free-floating, or advancing devices. Detailed analyses of such events can help to improve the design of these structures.

The study of green water events has employed methodologies that are analytical [2–4], numerical [1,5,6], and experimental [7–9]. The latter methodology type is important since it can be used to validate analytical and numerical models. However, at the model scale, green water events occur in a short time and must be captured at adequate sampling rates to give sufficiently detailed information to validate the models.

Experiments on green water have been carried out using wave trains in wave flumes [1,7,10] and ocean basins [2,11]. Sometimes, these experiments take long periods of time, requiring a reduction in the sampling rates of sensors and cameras to capture enough details of the green water events generated. Recently, the need for more detailed investigations to better understand the types of green water events that impinge on structures has been noted in the literature [12]. It was seen as necessary to employ high-resolution tecniques in the study of green water patterns and their evolution over the deck. Some research that has treated these topics includes the works of [5,7]. They presented two-dimensional experimental studies of green water using regular wave trains, reporting several types of green water events. However, the water elevations of the incoming wave and green water were monitored by obstructive wave probes at very few positions [1], thus limiting a more detailed understanding of the water evolution along the deck.

Later on, Ryu et al. [10,13] also used a two-dimensional experimental setup to study green water, considering a broken incident wave. However, the focus of their works was the study of the wave kinematics by applying an image-based method, disregarding the identification of flow patterns and the evolution of water in terms of elevations.

In order to understand details of different types of green water events, Hernández-Fontes et al. [14,15] proposed an alternative approach to generate systematic experiments of green water in a controlled environment, investigating the resulting patterns. They used the wet dam-break method in a small rectangular tank (~1 m length) to generate incident bores, which produced isolated green water events on a fixed rectangular structure. The wet dam-break method considers the flow generated from the sudden interaction of a water volume located upstream of a vertical gate (initial water depth, h_1) with a water volume located downstream of the gate (initial water depth, h_0), taking into account that h_1 is always higher than h_0 ([16]). From a high-speed video, Hernández-Fontes et al. [14,15] presented snapshots and descriptions of events that had some resemblance with the dam-break (DB), plunging-dam-break (PDB), and hammer-fist (HF) types reported by [7]. In the DB, the flow flooding the deck resembles a dam-break type flow. In the PDB, a cavity forms at the beginning of the deck followed by a DB flow. In the HF-type, a block of water rises at the bow, forming a suspended arm that directly hits the deck.

The use of the wet dam-break approach to study the green water problem in a simplified way is a recent proposal [14,15]. However, the study of the physics and applications of such a method is not new. Early experimental research was performed by [17], who investigated the resulting flows based on their kinematic characteristics. Some years later, Stoker [16] established well-known analytical procedures to represent the physics of dry ($h_0 = 0$, see also [18]) and wet ($h_0 > 0$, $h_0 < h_1$) dam-break phenomena, considering infinite conditions in the flow propagation domains. Later, Nakagawa et al. [19] performed several experimental tests to characterize the flow resulting from the dry and wet dam-break cases in a 0.5 m wide and 30 m long wave flume, employing the Stoker approach to analyse the resulting flows. From their observations, they proposed three types of wet dam-break flow based on the h_0/h_1 ratio: A wave of uniform and progressive features with a breaking front, similar to a moving hydraulic jump ($0 < h_0/h_1 \leq 0.4$); a wave resembling an undular bore of unstable dynamics with its front partially broken ($0.4 < h_0/h_1 \leq 0.56$); and a wave similar to an undular bore with its front unbroken ($0.56 \leq h_0/h_1 < 1.0$). Later on, in a flume 0.4 m wide and 15.24 m long, Stansby et al. [20] performed a numerical and experimental study, considering the ratios $h_0/h_1 = 0.1$ and 0.45 to investigate the initial stages of the wet dam-break problem. They reported that a mushroom-like jet is formed just after gate release, which interacts with the downstream, quiet water in a complex way, entraining air before a wave resembling a bore or a spilling breaker is developed. These findings were further confirmed numerically by [21]. Other studies have employed dam-break flows to study their interactions with structures located downstream. For example, Oertel and Bung [22], through experimental and numerical tests, evaluated the effects of a rectangular obstacle on the flow derived from the dry dam-beak case ($h_0 = 0$, $h_1 > 0$). Also, using numerical and experimental methods, Kocaman and Ozmen-Cagatay [23] investigated the impacts of wet dam-break waves on a vertical wall. More recently, Hernández-Fontes et al. [24,25] used the experimental wet dam-break approach to investigate green water problems, validating an analytical model for green water elevations and loading over the deck of a fixed structure, respectively.

The use of the wet dam-break method to investigate the green water problem on fixed structures can still be extended to explain in more detail the interaction of the incident flow with the structure and to increase the understanding of the physics of different green water patterns. This paper thus extends the experimental wet dam-break application of [14,15], described above, to investigate details of the interaction of bores with a fixed rectangular structure, using image-based analyses to study the flow evolution in time and space. The evolution of the incoming bores and the generated green water events, in terms of water elevations, are included in this research. As an alternative to previous two-dimensional green water studies that have been performed with other methods, such as regular waves, this paper offers an approach that allows some details in green water research to be investigated in a systematic and practical way. The main objectives are described as follows:

(a) To analyse features of the air cavities formed at the beginning of the deck during the initial stages of some green water events that occur in the form of a plunging-dam-break.

(b) To investigate the spatial and temporal evolution of the incident flow and the resultant water on deck of the structure, providing a database of time series of water elevations for all the study cases performed in this work, which can be employed by other authors to validate analytical or numerical models. The procedure followed to obtain these data can be very helpful to analyse green water elevations in other two-dimensional applications, which until now, have generally been measured at only a few positions, by obstructive wave probes.

(c) To analyse the difference between selecting the freeboard exceedance at the bow or an upstream position, including the relationship existing between the freeboard exceedance and the incident bores. This parameter is of significant relevance in performing model implementations, then the approach followed here may be useful to evaluate adequate safety factors to predict the real amount of water on deck.

(d) To verify the relationship between the kinematics of green water with that of the incident bore, including the influence of the presence of the structure.

These topics were investigated in a dam-break installation of approximately 2 m in length, considering the interaction of five different wet dam-break bores with a fixed rectangular structure located downstream. The bores were generated with a wet dam-break ratio of $h_0/h_1 = 0.6$ and five freeboards (0.006 m $\leq FB \leq 0.042$ m). Experiments with and without the fixed structure were performed. High-speed cameras and conductive wave probes were employed to monitor flow propagation. The videos were analysed with open-source image-based methods to extract water elevation measurements at several positions along the installation, allowing a detailed visualization of flow evolution in time and space.

The paper is organized as follows: Section 2 briefly presents the theoretical wet dam-break approach. The experimental methods are shown in Section 3 and the comparisons of a conventional technique with VWPs (virtual wave probes, as named by [26]) are shown in Section 4. Then, the incoming wave and green water results are presented in Sections 5 and 6, respectively. Finally, Section 7 summarizes the main conclusions and future works.

2. The Theoretical Wet Dam-Break Approach

In the present experimental investigation, the isolated green water events were reproduced using the wet dam-break approach as the mechanism to generate the incident waves (i.e., bores). In this approach, a dam (gate) separates two sides of a horizontal tank of constant cross section at $x = 0$. Upstream and downstream of the dam, there is a volume of water with an initial water depth, h_1 and h_0, respectively. It is assumed that the tank extends to infinite in upstream and downstream directions, and that h_1 is always higher than h_0.

At the initial condition, the water at both sides of the dam is assumed as undisturbed (i.e., at rest position). Then, the gate is suddenly removed ($t = 0$). In this stage, the rapid formation of a wave in the form of a bore or a hydraulic jump [16] is expected in the downstream region, propagating with theoretical front velocity, U_0, and elevation, h_2, over the lower layer of water. These parameters were estimated from the graphical solution proposed by [16], verifying its applicability to the present experimental application. To apply this theoretical model, the effects of the finite length of the water reservoirs were neglected, assuming the infinite domains of propagation described by Stoker's theory.

3. Experimental Methods

3.1. Experimental Set-Up

The wet dam-break experiments were carried out at the Ocean Technology Laboratory (LabOceano/COPPE) facilities of the Federal University of Rio de Janeiro, Brazil.

The general arrangement of the experiment can be seen in Figure 1a. A prismatic tank, made of polymethyl methacrylate (PMMA) plates, 25 mm thick, to reduce hydroelastic effects, was mounted on a structural arrangement. This tank had a fixed rectangular box installed inside, representing the deck of a coastal, naval, or offshore structure. At the left side of the tank, a vertical gate separating an upstream volume of water from a downstream volume can be observed. This was made from a PMMA plate, 15 mm thick. The gate was pulled upwards (dam-break action) by a release mechanism mounted in a vertical structure separated from the tank support. The release mechanism was formed by a 16 kgf weight, held by an electromagnet at the beginning of the experiments. When the trigger was activated ($t = 0$ s), the weight fell down and opened the gate by means of a pulley arrangement in the external structure.

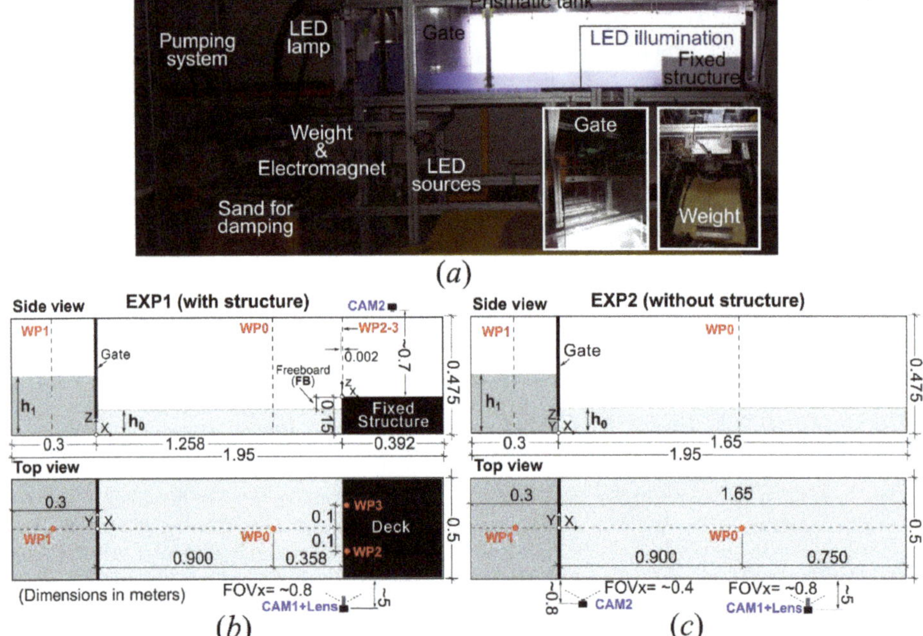

Figure 1. Experimental setup. (**a**) General arrangement. (**b**) Side and top views for the experiment with structure (EXP1). (**c**) Side and top views of the experiment without structure (EXP2).

In the present investigation, two types of experiments were carried out, considering the dam-break installation with (EXP1) and without (EXP2) the internal fixed structure. The main dimensions of the tank as well as the sensor positions for EXP1 and EXP2 are shown in Figure 1b,c, respectively.

3.2. Study Cases

In this work, five different cases of green water on a fixed structure were considered (C1–C5), as shown in Table 1. These were generated with the wet dam-break aspect ratio, $h_0/h_1 = 0.6$, and five different freeboards (*FB*, Figure 1b). The study case, C1, was employed as representative of the other cases for the main descriptions of the topics of the present work. However, all the cases were employed for comparison and discussion purposes.

Table 1. Cases of study and design initial conditions for the experiments.

Case	h_0/h_1	h_0 (in m)	h_1 (in m)	FB (in m)
C1	0.6	0.108	0.180	0.042
C2	0.6	0.120	0.200	0.030
C3	0.6	0.126	0.210	0.024
C4	0.6	0.132	0.220	0.018
C5	0.6	0.144	0.240	0.006

The ratio $h_0/h_1 = 0.6$ was chosen to generate unbroken incident flows, as suggested by [19]. Hernández-Fontes et al. [14], obtained unbroken incident flows considering the same wet dam-break ratio, with a smaller tank than that of this study. In that work, such a ratio caused isolated green water events that generated small cavities at the beginning of the deck, which were also expected in the present study.

The aim was to generate unbroken incident flows in order to apply a two-dimensional image-based methodology for the investigation of flow evolution (Section 3.4). It is important to mention that the features of the resultant wet dam-break wave might change according to the experimental setup. Yeh et al. [27] reported that for a tank 1.2 m wide and 9 m long, with a sloped beach 0.4 m downstream of the gate, the flow obtained with ratios $h_0/h_1 > 0.5$ had the form of an undular bore, but with its leading wave breaking at its crest. On the other hand, the results obtained by [19] in a 0.5 m wide and 30 m long wave flume, without downstream obstacles near the gate, suggest that it is possible to obtain unbroken bores for ratios $h_0/h_1 > 0.56$. The study cases were applied for the experiments with (EXP1) and without (EXP2) the structure. Each experiment was repeated five times to calculate the mean and standard deviation of the measured data.

3.3. Conventional Wave Probe Measurements

Two conductive wave probes (WP0 and WP1) were employed in the experiments EXP1 and EXP2 to monitor the initial conditions (i.e., the initial water levels) in the upstream (WP0) and downstream (WP1) water reservoirs (Figure 1b,c). Moreover, two shorter wave probes (WP3 and WP4) were installed above the deck to monitor the freeboard exceedance (Figure 1b). These sensors were used to compare the vitual wave probe measurements obtained with the image-based methodology described below. Details of the wave probe sensors as well as the measurement procedure are fully described in [28].

3.4. Image-Based Measurements

To capture the evolution of water in the stages of wave propagation downstream of the tank and green water on the structure, a high-speed digital camera (CAM1), model QUALISYS Oqus 310, was employed. This camera was used with an additional lens, AF DC-NIKKOR 105 mm f/2 D. The camera was set to 500 fps, with a resolution of 1284 × 1024 pixels. The camera was located to capture the side view of the tank, with the center view aligned to the bow edge of the structure, as shown in Figure 1b, for the experiment with the internal structure (EXP1). For the experiment without the structure (EXP2, Figure 1c), CAM1 was at the same place. A second camera (CAM2), model QUALISYS Oqus 110 was used in EXP1 and EXP2 to record the green water wavefront displacement and the wave generation stages during the gate aperture, respectively. For both cases, it was set at 200 fps, with a resolution of 640×480 pixels (Figure 1b,c). The software QUALISYS QTM was used to produce the videos.

3.4.1. Water Elevation Measurements

Measurements of water elevation were obtained from several virtual wave probes (VWPs), from the video made by the cameras that were parallel to the side view of the tank, that is, CAM1 for EXP1 and EXP2, and CAM2 for EXP2. To obtain the water elevation measurements, the open-source image-based methodology developed by [29] was employed. The methodology includes the three main stages of an artificial vision system: Image acquisition, processing, and analysis. These stages were implemented using open-source ImageJ software. Overall, gray-scale images were obtained from the video, then processed by intensity modulation and pseudo-color segmentation to obtain binarized images. Finally, through basic morphological operations, image analysis was carried out from the binary images to obtain water elevations at different regions of interest, which were recorded by the virtual wave probes (VWPs). See the work of [29] for more details regarding the procedures for image calibration, processing, and analysis to obtain the water elevation measurements.

For EXP1, virtual wave probes (VWPs) were located along the incident wave and green water propagation domains defined from CAM1, as shown in Figure 2a. This figure also defines the field of view (FOV) of the camera; ~0.83 and ~0.3 m in the x- and z-directions, respectively. A reference system of coordinates (xyz) was located at the beginning of the structure to locate the VWPs. Two VWPs were taken to measure the freeboard exceedance; VWPw0 and VWPd0, located at $x = -0.005$ m and $x = 0.005$ m from the origin. To monitor the water elevations on the deck, 33 VWPs, separated by a distance of 0.01 m, from $x = 0.01$ m (VWPd01) to $x = 0.33$ m (VWPd33), were used. To obtain the water elevations in the domain for the incident wave propagation, 42 VWPs, separated by a distance of 0.01 m were used, ennumerated as VWPw01 ($x = -0.01$ m) to VWPw42 ($x = -0.42$ m).

For EXP2, measurements were obtained for the same positions: VWPd01-VWPd33 and VWPw01-VWPw42, using CAM1. Furthermore, for CAM2 in this experiment, several VWPs were located along the domain for the wave generation and propagation defined in Figure 2b. In this case, the allowable FOV was ~0.63 m and ~0.3 m in the horizontal and vertical directions, respectively. A reference system of coordinates (XYZ), with origin at the gate, was considered to set 60 VWPs from $X = 0.01$ m (VWPg01) to $X = 0.6$ m (VWPg60), separated by 0.01 m.

3.4.2. Wavefront Velocity

As stated before, CAM2 was installed parallel to the deck in EXP1 to capture the green water wavefront displacement (Figure 3). From the videos, the velocity of the wavefront (U_{front}) was measured manually, using ImageJ visualization tools. The position of the wavefront edge was tracked every 0.01 s as water propagated over the deck. The movement of the green water wavefront may show cross-sectional variations along the width of the deck. Thus, for a practical measuring of its displacement, the spatial-average displacement over the cross-section line was monitored, starting from the instant at which the wavefront was first visible, close to the structure edge (Figure 3).

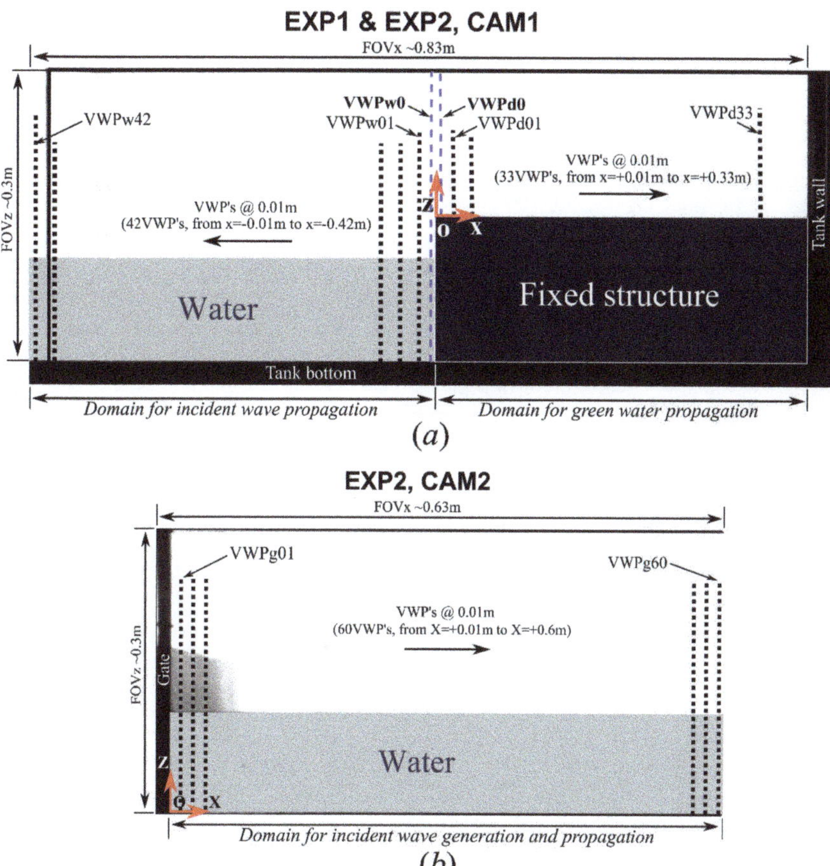

Figure 2. Scene field of view (FOV) of cameras' and virtual wave probes' (VWPs) positions. (**a**) CAM1, EXP1 and EXP2. (**b**) CAM2, EXP2.

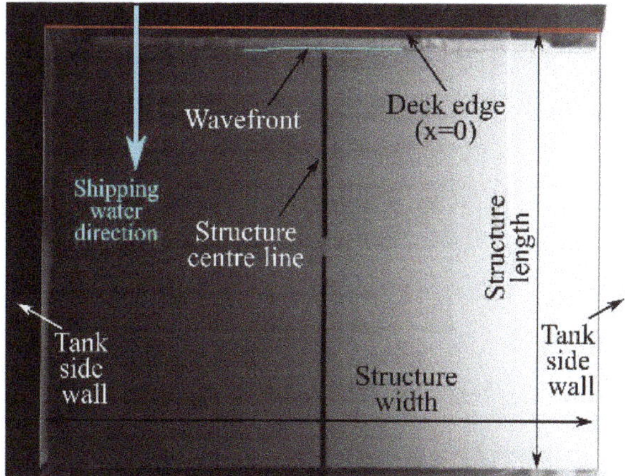

Figure 3. Typical frame for wavefront displacement measurements.

4. Comparison of Conventional and Virtual Wave Probes

The performance of the image-based approach to measure water elevations was verified by comparing the virtual results with the measurements given by conventional wave probes in the experiment with the internal structure (EXP1), considering C1 and C5 as representative of the other cases. Figure 4a,b show the comparison for the incident bore elevations between WP0 and a virtual wave probe, located at the same WP0 position, for C1 and C5, respectively. The signals shown in the figures are given in terms of mean and standard deviation values, obtained from the five repetitions. The VWP and WP0 signals have excellent agreement in measuring the static water level. A small shift in time is observed in the time series of the VWP with respect to WP0 for both cases, which can be partially related to the position of the camera, which was centered with respect to the bow edge of the structure. It is important to consider that WP0 measured the water elevations at the center of the tank, whereas the VWP monitored the elevations in the tank wall. Then, small 3D effects on flow observed in the wall might overestimate or underestimate the measurements at the center of the tank, which can explain the differences observed in Figure 4a,b. The agreement of the VWP with respect to WP0 is better for C5, which presented higher water elevations, following the trends of the incident and reflected waves.

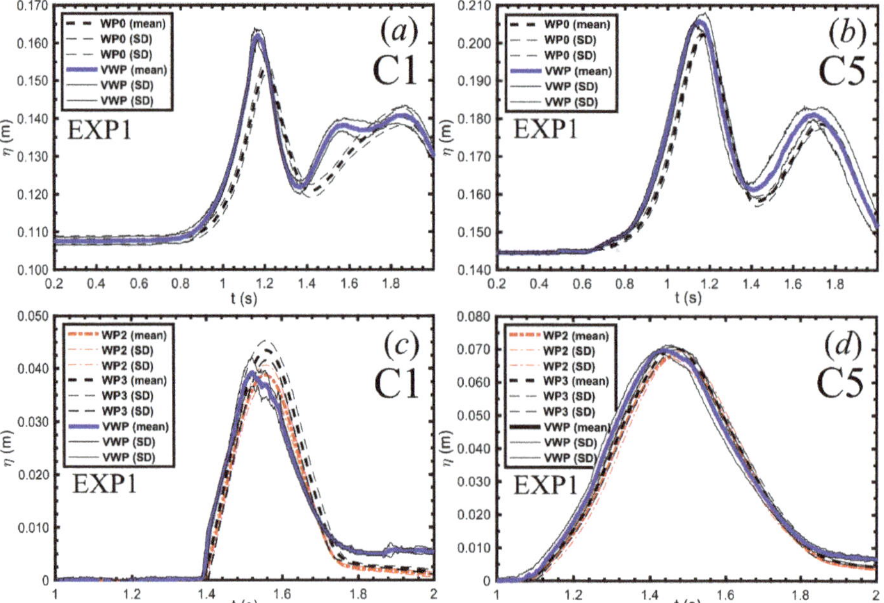

Figure 4. Comparison of water elevations obtained with virtual and conventional wave probes (mean and standard deviation values for the five repetitions) for C1 and C5. (**a**) Incoming bore elevations measured by WP0 and a VWP located at the same position for C1. (**b**) Incoming bore elevations measured by WP0 and a VWP located at the same position for C5. (**c**) Freeboard exceedance elevations measured by WP2, WP3, and a VWP located 1 mm upstream their position for C1. (**d**) Freeboard exceedance elevations measured by WP2, WP3, and a VWP located 1 mm upstream their position for C5.

On the other hand, Figure 4c,d show the comparison of the VWP measurements with those obtained with the conventional wave probes WP2 and WP3 installed above the deck, for C1 and C5, respectively. The VWPs were located about 1 mm upstream from the WP2 and WP3 position to allow image analysis, avoiding the obstruction of the sensors in the images. Note that the conventional wave probes indicate small 3D effects of the shipping flow, particularly for C1, as can be inferred from the differences in their maximum elevations (~0.004 m). These may be due mainly to installation issues of the conventional probes or a non uniform invasion of water over the deck. Overall, the VWP in C1 and C5 have reasonable agreement with water elevations over the deck measured by conventional wave probes.

Considering that the conventional wave probes can be subjected to different sources of error due to installation and performance (see, for instance, [1,29]), the applicability of the image-based approach was considered acceptable for the purposes of the present work.

5. Propagation and Characterization of the Bores

In this section, the results regarding the incoming bores are presented. These include the gate aperture, features presented during the generation and propagation of the bores, and a theoretical characterization of their steepness.

5.1. Gate Aperture

In the experimental dam-break approach, the gate aperture time (t_r) should be as short as possible, to replicate a sudden gate release. To attain a suitable gate aperture time, the condition proposed by [30,31] for a dry dam-break experiment (i.e., $h_0 = 0$, $h_1 > 0$) was considered: $t_r < \sqrt{2}/\sqrt{g/h_1}$, where t_r is the time for gate release, g is the acceleration due to gravity, and h_1 is the initial water depth in the volume of water upstream of the gate.

From the video obtained with CAM2 in the experiment without structure (EXP2), the experimental gate release times, t_r, were approximately 0.12, 0.13, 0.14, 0.15, and 0.16 s for cases C1, C2, C3, C4, and C5, respectively. These times were lower than the expected (i.e., design condition) t_r values, indicating that the gate opening times for all the experiments were within the range defined above. It is important to mention that for the analyses described hereafter, the time $t \approx 0$ s was considered as the time at which the trigger was activated. This was done to try to begin all sensor measurements almost at the same time. However, after triggering, the release of the gate took some milliseconds to occur, beginning its vertical displacement between 0.16 and 0.18 s in all the cases.

5.2. Bore Propagation

In the present methodology, when the gate opens, unbroken bores that propagate downstream are formed. For h_0/h_1 ratios lower than that of the present study (i.e., $h_0/h_1 < 0.6$), Stansby et al. [20] observed a mushroom-like jet after gate release. In the present case, some turbulence caused by the wave breaking over the downstream water volume and some drops of water falling from the gate were observed. Shigematsu et al. [32] suggest that the turbulence seen after gate release increases as the h_0/h_1 ratios drop. Furthermore, Liu and Liu [33] described that the resultant flow just after the gate release is commonly higher than the one observed when it developes farther from the gate, where water levels vary gradually.

To examine the transition of the bore development, from the gate area to some distance downstream, a spatial distribution of the water elevations was obtained from virtual wave probes from CAM2 (VWPg01-VWPg60, Figure 2b), as shown in Figure 5. This figure presents a comparison of bore profiles (i.e., water elevations, η, against the longitudinal positions from the gate, X) taken at various points in time ($t_1 = 0.35$ s, $t_2 = 0.40$ s and $t_3 = 0.75$ s). The theoretical value, h_2, estimated with the approach of Stoker (see Section 2) has also been included to make comparisons with the obtained results. The bore profile at t_1 resembles a steep hydraulic jump, with amplitudes close to the theoretical ones of the Stoker model. Conversely, for later stages (t_2), the shape of the profile changes their appearance to become an undular bore of greater amplitude. The crest of the bore was clearer as it propagated downstream. When its maximum elevation appeared, near $X \approx 0.42$ m, the bore developed a trough and a second crest of smaller amplitude formed behind the first ($t_3 = 0.75$ s). In this stage, the maximum elevation of the fully developed bore overestimated ~18% of the theoretical value, h_2, obtained with the Stoker approach. These results showed that Stoker's theory is more suitable for the hydraulic jump (shock wave) that appears after the gate release than for the fully-developed bore. This was also confirmed for the study cases C2–C5.

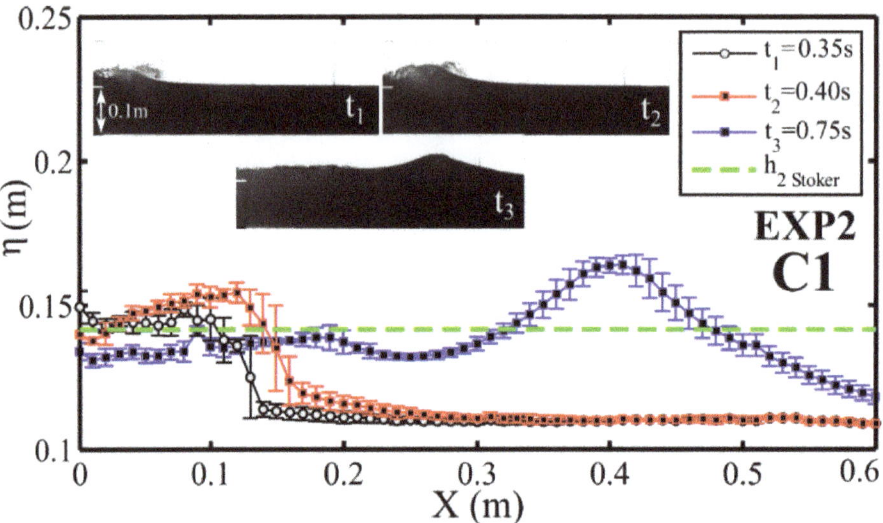

Figure 5. Comparison between spatial distributions of bore elevations (η vs. X) at different points in time ($t_1 = 0.35$ s, $t_2 = 0.40$ s, and $t_3 = 0.75$ s) for C1 (EXP2, CAM2). The dashed line corresponds to the elevation, h_2, given by the theoretical Stoker model.

Considering C1, water elevation time series for the incoming bore were obtained from CAM1 at the positions defined in Figure 2a (VWPw01-VWPw42), allowing a two-dimensional flow visualization in time (t) and space (x), as shown in Figure 6. Figure 6a,b show the time–space distributions of water elevation at several positions (VWPw01-VWPw42) obtained from EXP1 (with structure) and EXP2 (without structure), respectively. For the case with structure, it can be noted that the bore propagated with an almost constant elevation (see the constant color region of the incoming bore until $x \approx -0.160$ m), until the structure influenced it, increasing the elevations. These maximum elevations occurred just before the shipping of water onto the structure, in a stage known as bow run-up. Maximum values of ~0.196 m are seen close to the structure ($x \approx -0.01$ m), defining the maximum freeboard exceedance of the event, which is one of the main parameters in green water analyses (see Section 6.2). Next, a reflected wave is seen to be generated by the structure just after the incoming bore reached its maximum value, during the bow run-up stage (for $t > 1.6$ s at $x \approx -0.2$ m). The reflected wave has a maximum elevation of ~0.167 m, which is ~15% lower than that of the incident wave. In contrast, for the case without a structure (Figure 6b), specific regions of maximum elevations due to bow run-up and backflow are not seen. Instead, continuous color regions for the incident bore and a wave reflected by the tank wall are observed, suggesting that they propagate with almost constant elevation and velocity (see the slope of the colored regions). It is important to comment that in studies perfomed with wave trains, the backflow generated after bow run-up may significantly affect the features of after-coming waves. In these cases, the resultant green water events may be dependent on the backflow generated from previous wave interactions with the structure [7].

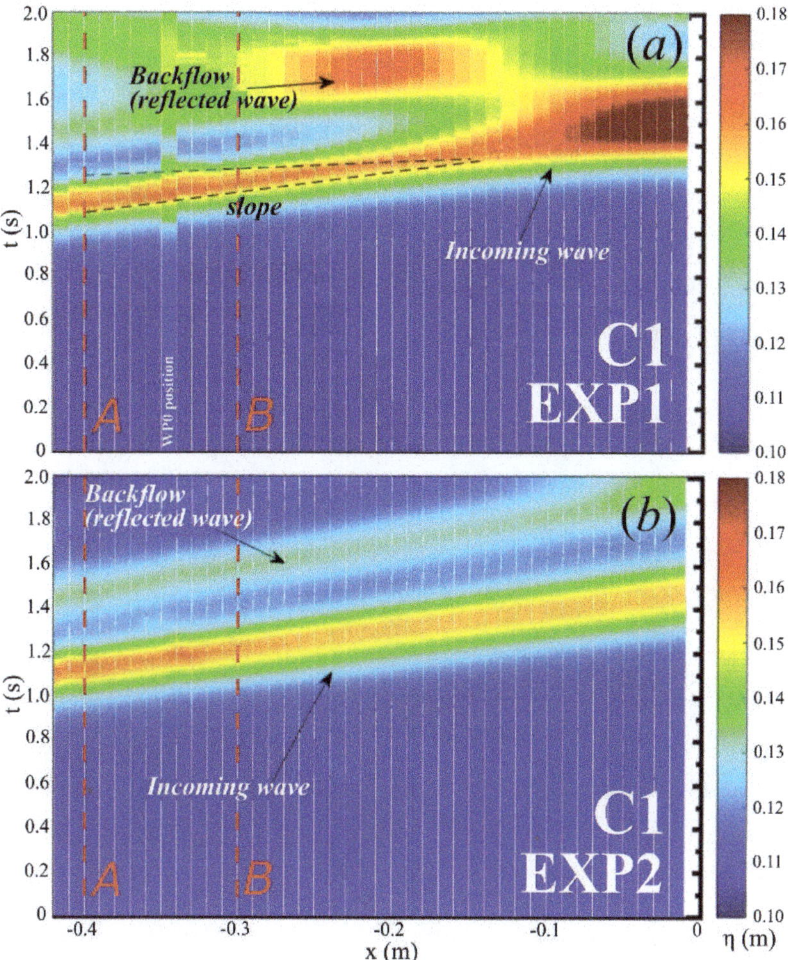

Figure 6. Time–space flow visualization of the water elevations of the incoming bore for C1. The bow edge of the structure was taken as the origin ($x = 0$). (**a**) t vs. x comparisons for the experiments with structure (EXP1, CAM1). (**b**) t vs. x comparisons for the experiments without the structure (EXP2, CAM1). A and B define a domain where the bore elevations are less influenced by the presence of the structure.

Results indicate that the structure presence has an influence on the features of the incoming bore, which is more noticeable close to the structure. Similar features were observed for C2–C5, differing mainly in the magnitudes of the incoming and reflected waves. Relationships between these waves and the resultant water on deck are further described in Section 6.

5.3. Bore Steepness

In order to characterize the incoming bores, their steepness has been estimated theoretically. To do this, they were assumed as solitary waves, which can be described by two parameters: Wave height and water depth [34]. For all the study cases, these parameters were considered from the mean values obtained in the time series of water elevations measured at point A for EXP1 (VWPw40, Figure 6a), considering the five repetitions. From these time series, the water depth parameter was considered as the mean value of the measured h_0 values (i.e., $\overline{h}_o$), whereas the wave height parameter was considered as the difference between the mean maximum amplitude in the VWPw40 time series and $\overline{h}_o$ (i.e., $\overline{H}$). The steepness of each bore was calculated as $\epsilon = \overline{H}/L_T$, where L_T is the theoretical length of the solitary wave, which was estimated as [35,36]: $L_T = 1.5\overline{h}_0\left(\overline{H}/\overline{h}_0\right)^{-0.5}$. The $\overline{H}$ values measured for C1, C2, C3, C4, and C5 were ~0.055, ~0.056, ~0.058, ~0.059, and ~0.060 m, respectively, whereas the corresponding steepnesses were ~0.241, ~0.216, ~0.209, ~0.203, and ~0.183. These values suggest that the bores have almost the same height, $\overline{H}$ (varying ~5 mm from C1 to C5), differing in length, which increases from C1 to C5. The incoming bore in C1 is the steepest of all cases; however, it has to be noted that all the bores used in this work present a close variation in steepness ($0.183 < \epsilon < 0.241$), that is, the interaction of very steep waves with the structure was not considered.

6. Interaction of Bores with the Structure: Green Water

This section presents the experimental results of the green water events originated from the interaction of the incident bores with the structure. First, main stages of green water for the representative case C1 are illustrated, emphasizing details of an air cavity occurring in the early stage of the event. Then, the evolution of green water elevations is treated similarly to that in Section 5, where temporal and spatial distributions were given for the incoming bore. Finally, some relevant parameters in green water research, such as freeboard exceedance and wavefront velocity over the deck, including their relationship with the incident bores, were analysed.

6.1. The Green Water Events

The green water events obtained for C1, C2, C3, and C4 presented the formation of a small air cavity at the beginning of the deck, whereas C5 did not presented such a cavity (see the initial stages shown in Figure 7a). The cavity size reduced from C1 to C4. From a qualitative point of view, the events obtained in C5 and C1–C4 can be classified as dam-break (DB) and plunging-dam-break (PDB) types of green water, resembling the ones obtained with regular waves on a fixed structure, as reported by [7]. In this section, the event of C1 presented the largest air cavity of all the PDB events found, then it was considered for reference to describe some details of water behaviour on deck. Figure 7b shows some stages of the green water event found in C1. Snapshots capture the flow at different points in time, including the initial (bow run-up) and final (backflow) stages of green water.

The features observed in the event found for C1 (Figure 7b) were also found by [15], using the same initial conditions for green water generation ($h_0/h_1 = 0.6$, $FB = 0.042$ m), in a smaller experimental installation (~1 m long, ~0.335 m wide). The installation of [15] had the same upstream reservoir length, and the same height for the fixed structure. The main differences are related to the distance available for the bore development and the tank width (~2.49 and ~1.49 times greater in the present work, respectively). Figure 8 presents the comparison between the green water patterns obtained in these two works. The patterns were captured 0.12 s after the water level reached the height of the deck. Note that both cases resemble the PBD-type of green water described by [1,7]; where an air cavity is produced while two small jets are formed as the wave front touches the deck. However, in [15], the cavity was larger (~1.7 and ~1.5 times in the horizontal and vertical directions, respectively) and the maximum elevation of the wave was higher (~30%) and occurred closer to the bow (~38% closer) than in the present work. Reduction in the cavity sizes with respect to the cases presented in [15] were also verified for cases C2, C3, and C4. The shorter distances for bore development due to the smaller tank used in that work may have influenced the increase in wave amplitudes because of the reflection effects, and in turn, the generation of larger cavities.

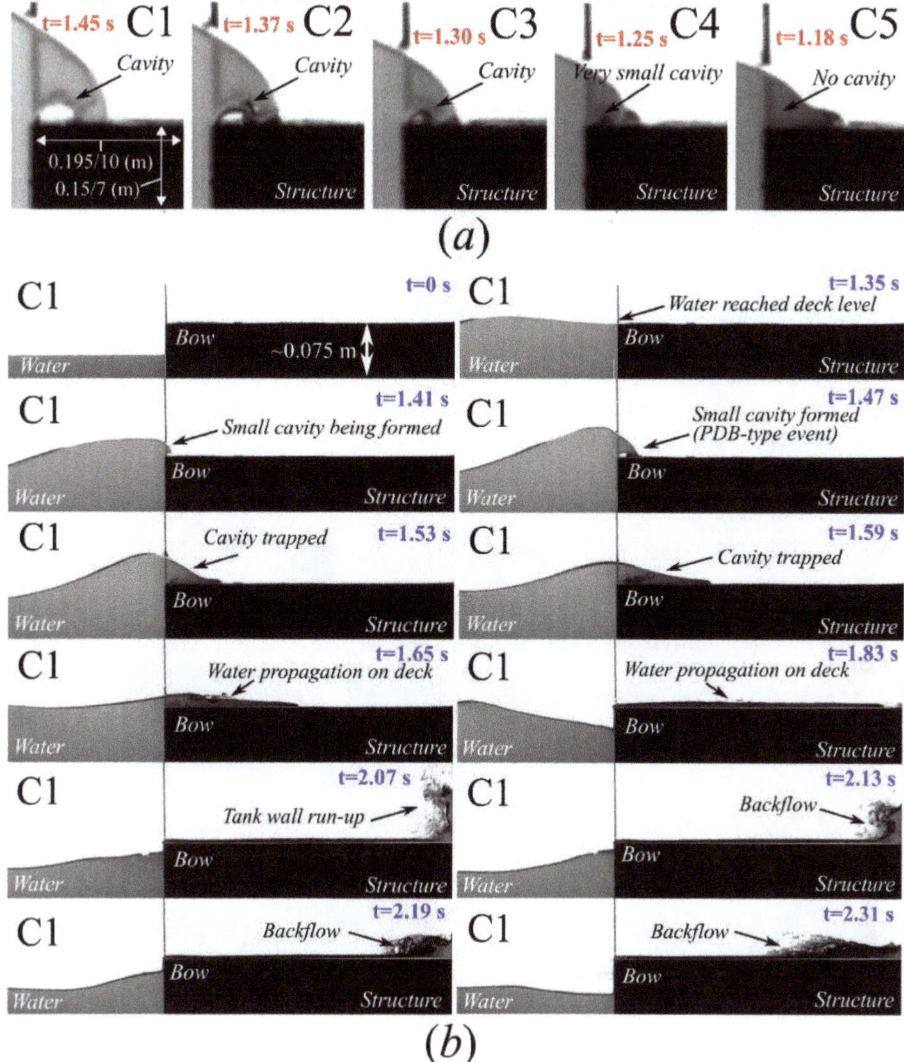

Figure 7. Representative snapshots of the green water events (EXP1, CAM1). (**a**) Features of the initial stages of green water for all the study cases (C1–C5). (**b**) Evolution of the green water event of C1.

Air Cavity Analysis

The study of the evolution of air cavities formed during wave interactions with marine structures is important since it may yield information about the induced loading on the structure. Some authors have stated that the presence of trapped air increases the loading on the structure, whereas others claim that the trapped air delays the impact [37]. For the case of cavities formed during green water events, Colicchio et al. [38] performed a detailed analytical and numerical investigation, evaluating scale effects in the cavity formed during PDB events obtained with regular waves. Alternatively, in this work the air cavities formed during the present PDB-type green water events (C1–C4) are analysed in a simplified way through analogies made with the air cavity formed in a vertical wall during some types of flip-through events (i.e., flip-through "Mode B", [37,39,40]). In these events, there is the formation of an air cavity due to the overturning of a wave crest approaching the vertical wall (the focusing stage). The cavity is subsequently closed, due to the generation of an upward jet, which meets the wave crest. Then, the cavity collapses and the jet rises above the wave crest [37]. Despite the kinematics of the entrainment mechanism of both the flip-through and green water problems is quite different, it is possible to relate some concepts from that phenomenon [39] to partially describe the evolution of the cavities formed in the present green water events as follows:

(1) The air cavity is entrapped against a horizontal surface rather than a vertical wall (Figure 9a).
(2) The deformation of the air cavity (i.e., compression and expansion) is very similar in the horizontal and vertical direction, that is, it suffers an *isotropic* compression and expansion (Figure 9b).
(3) The air-cavity deformation occurs mainly in the horizontal direction, that is, it presents an anisotropic compression/expansion due to the increase in the water column above it. However, in the present case, the cavity is also reduced by the effect of the backward jet that is formed as water propagates down the deck (Figure 9c).
(4) The air cavity collapses and fragments in small bubbles that mix downstream with the advancing flow (Figure 9d).

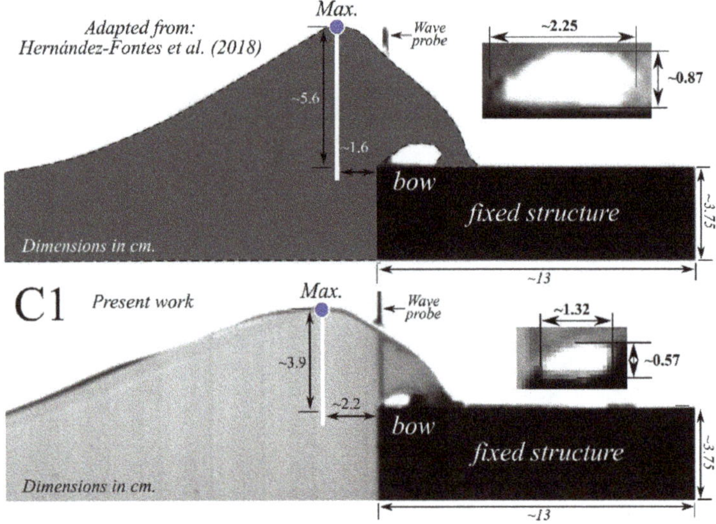

Figure 8. Comparison of the green water pattern obtained by [15] (**above**) and that obtained in the present work for C1 (**below**). Both snapshots were taken at 0.12 s from the instant at which the water reached the deck.

Lugni et al. [37] suggest the importance to evaluate the dynamic behavior of the air cavity, which shows high frequency oscillations before it collapses. This may be relevant to study structural loading effects, since structural vibrations, ventilation, and cavitation may be induced [39]. Thus, it is important to estimate its resonance frequency. The effect of this frequency in loading is out of the scope of this work. However, we practically evaluated it for each study case from an analogy made from the flip-though events. This can be done approximately by considering the air cavity as a two-dimensional bubble of semicircular cross section with a radius, R_c, during the initial stage of its formation, as soon as the tip of the plunging wave reaches the deck, as illustrated in Figure 9a. It is assumed that this cavity behaves like a harmonic oscillator that vibrates at a resonance frequency, f_r, when it is subjected to an impulsive force. Moreover, effects due to surface tension, dissipation, and damping must be ignored [37]. With these assumptions, f_r can be estimated as [37,41]: $f_n^2 = -\gamma p/2\pi^2 \rho R_c^2 \, log(R_c/2h)$, where p is the atmospheric pressure (~101.3 kPa), ρ is the water density (1000 kg/m^3), $\gamma \approx 1.4$ is the specific heat ratio (disregarding the effect of heat conduction from the bubble), and h is the distance from the center of the cavity at the vertical wall to the free surface (in the flip-through problem). For the present application to green water, h is assumed as the distance from the center of the cavity at the deck to the free surface ($h \approx 0.02$ m, Figure 9a).

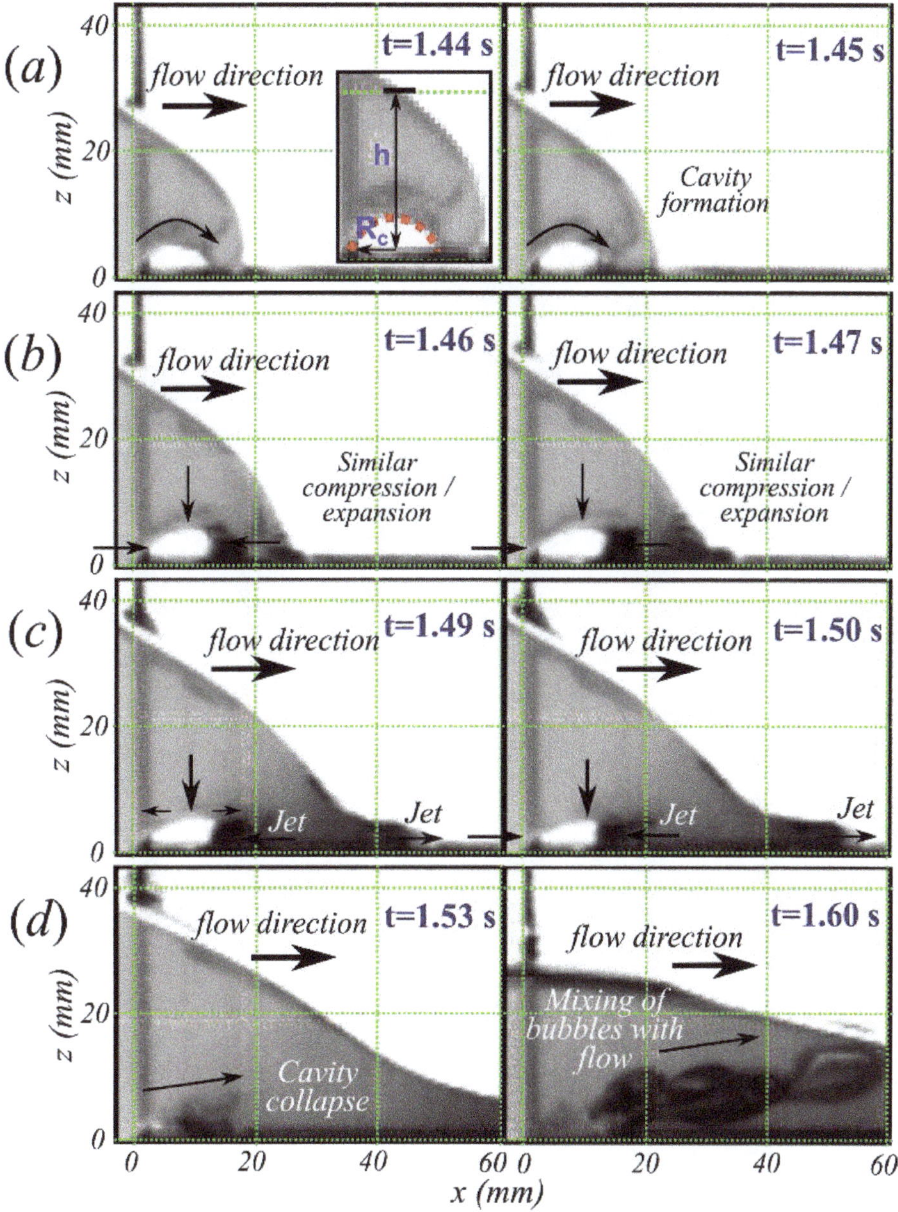

Figure 9. Main stages of air-cavity evolution for C1. (**a**) Cavity formation. (**b**) Isotropic compression/expansion. (**c**) Cavity deformation mainly in the horizontal direction (Anisotropic compression/expansion). (**d**) Cavity collapse and mixing of bubbles with flow.

Figure 10 presents the ratios, R_c/h, and the theoretical f_r values obtained for different steepnesses (C1–C4). The results are shown in terms of mean and standard deviation values. Note that for lower steepnesses, the cavity radius, R_c, is smaller in relation to the height, h. Conversely, for these steepnesses (cases with smaller cavities), the theoretical resonance frequency, f_n, is larger than those obtained for the cases with larger cavities, which were generated with steeper bores and higher freeboards (e.g., C1 and C2).

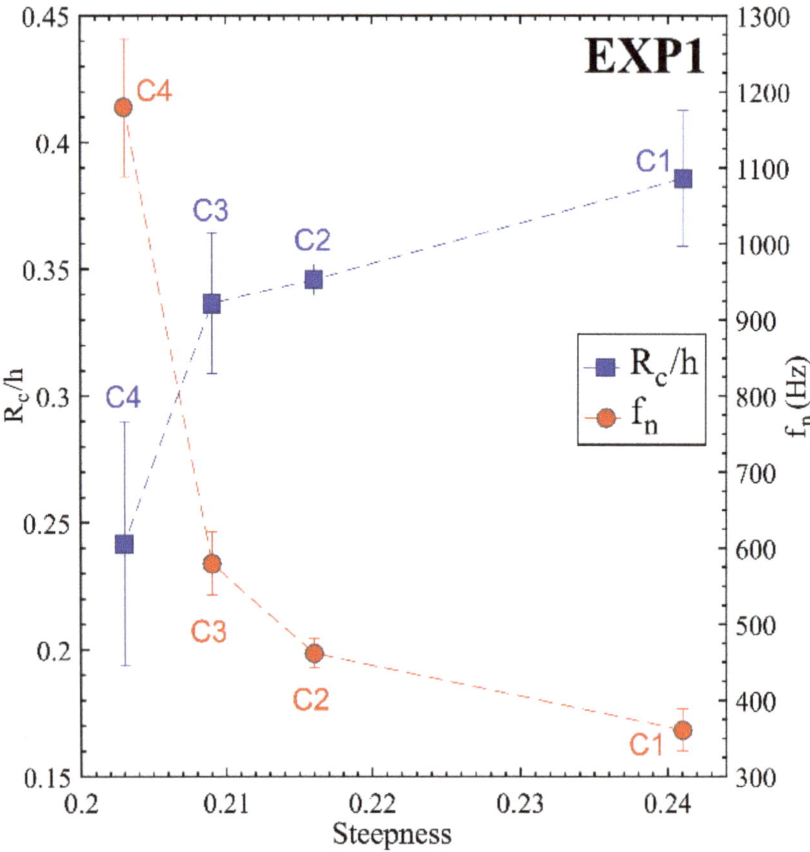

Figure 10. Ratio R_c/h (cavity parameters) and theoretical resonance frequency of the cavity (f_n) for different wave steepnesses (C1–C4). Errorbars indicate the standard deviation for each case, considering the five repetitions.

Although the physics of the cavity formed during a green water event is different from that of a flip-through phenomenon in a vertical wall, the present analysis could be considered as a practical alternative to analyse the dynamic behavior of the air cavity formed in green water events of the PDB-type.

6.2. Green Water Elevations

There are few experimental measurements available in the literature of the time series of green water elevations on the deck of a structure, considering several positions. To obtain these measurements it is common practice to use some conventional wave probe arrangements to extract water elevation data. Even in a two-dimensional setup, green water elevations have been measured by obstructive wave probes (e.g., [1,7]). Besides interfering with the flow propagation, these sensors yield information for very few positions on the deck, limiting the understanding of the water´s spatial evolution. This is valuable information for comparison or validation of analytical and numerical models.

The time series of green water elevations on the deck for C1 are shown in Figure 11. These were obtained for one of the five repetitions from virtual wave probes located along the deck (VWPd01-VWPd33, Figure 2a). Water evolution is shown for the complete duration of the experiment (3 s). However, it is important to consider that the range of applicability of the 2D image-based procedure is until ~2 s, before the tank wall run-up, where 3D effects are presented. Data processed for $t > 2$ s are shown for illustrative purposes to have an idea of flow behaviour during tank wall run-up and backflow. The evolution of the cavity that is formed in the present green water event was disregarded in the analyses.

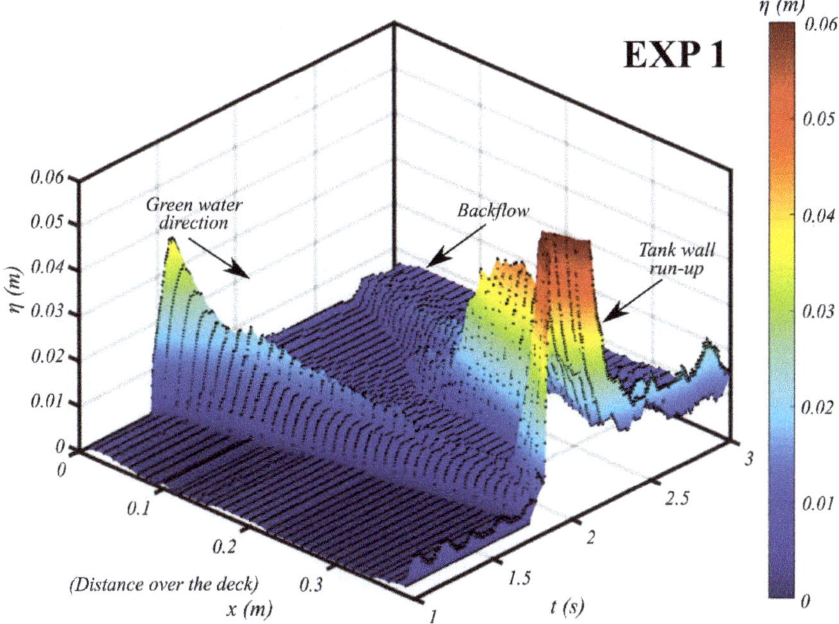

Figure 11. Green water evolution on deck for C1 (EXP1, CAM1).

Three main stages can be identified in Figure 11: First, the deck was completely dry ($x = 0$, $0 < t <$ 1.2 s), until a finite amount of water was shipped onto it and propagated to its end ($x = 0.392$ m, $t \approx 1.8$ s). The water then interacted with the tank wall, causing run-up (t > 1.8 s). Finally, the run-up water fell down due to gravity and backflow occurred on the deck (see the stages shown in Figure 7b). The stages observed in this case were also identified for the other cases, which presented larger amounts of water over the deck.

As an alternative to previous two-dimensional techniques to study water evolution on deck (e.g., [1]), the present approach can be extended to to acquire time series of green water elevations at several positions over the deck for different setups.

Freeboard Exceedance

One of the most important parameters in the study of green water events is the freeboard exceedance time series, or the effective water elevations from the incident wave that represent the volume of water that ships onto the deck. The maximum value of these series is known as the maximum freeboard exceedance (η_0, Figure 12a), and it is commonly considered as input of analytical (e.g., [24,42]) or numerical (e.g., [43]) models to study green water propagation over the deck. However, not every freeboard exceedance produces the same water elevation over the deck [44]. The water elevations measured outside the deck may be higher than those measured on it [45].

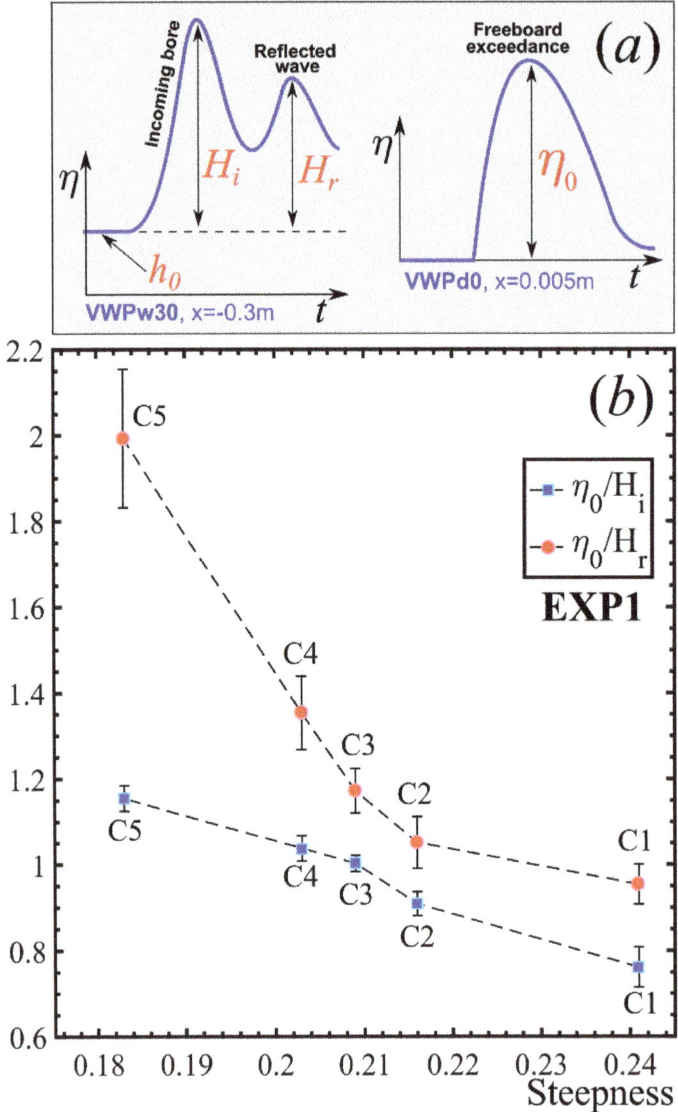

Figure 12. Relationship between the maximum freeboard exceedance, η_0, and the height of the incomming bore (H_i) and the reflected wave by the structure (H_r) for EXP1 (Cases C1–C5). (a) Sketch that illustrates the selection of the parameters. H_i and H_r were measured at $x = -0.3$ m, from the bow edge (VWPw30), whereas η_0 is the maximum freeboard exceedance measured at $x = 0.005$ m, over the deck (VWPd0). (b) Mean and standard deviation values of the ratios, η_0/H_i and η_0/H_r, for different bore steepnesses.

Experimentally, the freeboard exceedance has been measured over the deck of simplified structures at positions very close to the deck edge (e.g., [1]). However, in more complex arrangements, such as ship-type structures, it has been necessary to install the wave probes outside the hull to monitor the relative deck-wave motions (e.g., [42,45–48]). Details of the exact location of the wave probes with respect to the edge of the deck are scarce in most works. However, a typical distance can be inferred from the studies of [49,50], who used wave probes ~1 cm off the bow to measure relative ship-wave motions.

Figure 12 shows the relationship of the maximum freeboard exceedance (η_0) with the maximum elevations of the incoming bore (H_i) and of the wave reflected by the structure (H_r), considering EXP1. H_i and H_r were measured at the position of point B defined in Figure 6 (VWPw30, $x = -0.3$ m), whereas η_0 was measured at 5 mm from the bow edge (VWPd0), as illustrated in Figure 12a. Figure 12b shows the ratios, η_0/H_i and η_0/H_r, for different bore steepnesses. First of all, note that from C1 (higher freeboard, steeper bore) to C5 (lower freeboard, longer bore), the values of η_0 range from ~80% to ~120% of the values of H_i. For C3, η_0 is almost the same as H_i. Rearding the reflected wave, η_0 is very close to H_r for C1 and C2; however, for longer bores and shorter freeboards (from C3 to C5), η_0 overestimated H_r, reaching values of ~200% the values of H_r for C5.

In this study, we evaluated the differences obtained when selecting freeboard exceedance at the beginning of the deck and at some distances upstream (Figure 13). Although the present work does not apply to all green water problems found in real cases, this type of analysis may be extended to estimate correction factors in cases where green water studies were performed with freeboard exceedance data obtained outside the deck. Regarding the present application, we made a comparison between the resultant VWP time series obtained from VWPw0 ($x = -0.005$ m) subtracting the structure height (i.e., VWPw0-H_{st}, with H_{st} as structure height) and the time series measured on the deck (VWPd0), as shown in Figure 13a. The time series include the mean and standard deviation values estimated from all the repetitions. It is observed that the trend of both series is very similar, despite very small shifts in time between them and some differences in water elevations, mainly at the tail of the curves for $t > 1.7$ s, where the VWPd0 presents almost constant elevations. These elevations are due to a layer of water remaining at the edge of the deck during the event (see snapshots in Figure 7b). Considering the time at which the water elevations started rising, the time difference was about 0.03 s between both series.

In Figure 13a, the maximum freeboard exceedance of the time series taken upstream the deck edge (i.e., $\eta_{0,out}$) overestimated that of VWPd0 (η_0) by approximately 7%. These maximum values occurred at the time $t_{\eta_{0,out}}$ and t_{η_0}, respectively, which presented a very small time difference of about ~0.002 s.

Figure 13b,c show, respectively, the variation of the ratios $\eta_{0,out}/\eta_0$ and $t_{\eta_{0,out}}/t_{\eta_0}$ at several positions (x/L_{st}, where L_{st} is the structure length) upstream the bow edge for all the study cases. The parameter, $\eta_{0,out}$, was obtained from the mean values of time series provided by VWPs at these positions, substracting the structure height. In Figure 13b, it can be noted that for each case, there is a region from the structure at which $\eta_{0,out}$ overestimated between 10% and 25% of the freeboard exceedance over the deck (η_0), which avoids underestimation of values measured over the deck. However, there is a region in which $\eta_{0,out}$ starts underestimating η_0 (see [$\eta_{0,out}/\eta_0$] < 1 in the figure). The condition $\eta_{0,out} < \eta_0$ is accomplished at shorter distances from the structure for the cases with higher freeboards (steeper bores) and at longer distances for the lowest freeboards (longer bores).

It is also important to analyse the difference in time at which $\eta_{0,out}$ and η_0 occurred, as shown in Figure 13c. This difference will increase as the freeboard exceedance is considered farther from the structure. All cases presented this behaviour, showing differences between $t_{\eta_{0,out}}$ and t_{η_0} of ~3% tto 7% for a distance about 25% of the length of the structure.

Disregarding the differences observed in the time series of freeboard exceedance measured outside and over the deck, the former could be considered for practical applications if a suitable range is considered to avoid underestimation of η_0. For other applications, including oscillatory motions of the incident flow or motions of the structure, it may be relevant to verify suitable regions to estimate $\eta_{0,out}$ in order to avoid underestimation of the resultant water on deck and to reduce the shifts in time in simulations. Perhaps, it should be necessary to consider a correction factor in the estimation of the η_0 parameter.

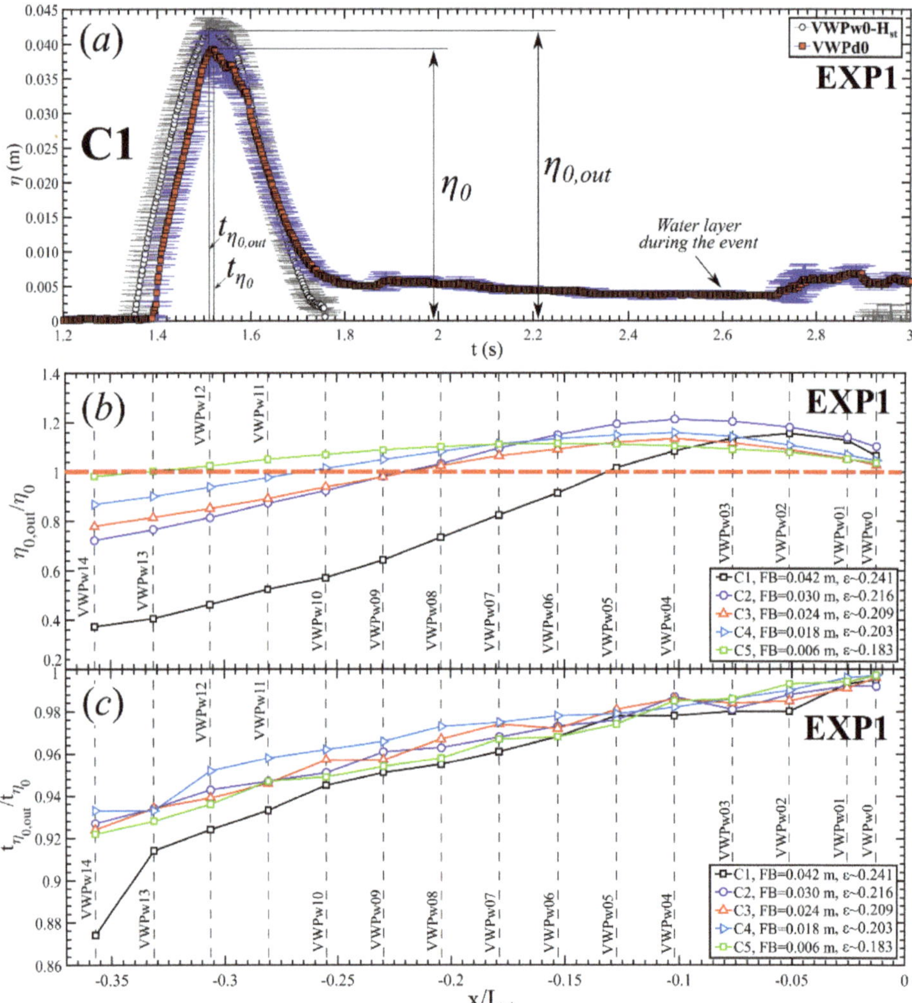

Figure 13. Maximum freeboard exceedance data. (**a**) Comparison of freeboard exceedance time series between VWPs measured before (VWPw0-H_{st}, $x = -0.005$ m) and after (VWPd0, $x = 0.005$ m) the bow edge of the deck (C1, EXP1, CAM1); see also [24]. η_0 and $\eta_{0,out}$ represent the maximum freeboard exceedance for VWPd0 and VWPw0, respectively. (**b**) Ratios of maximum freeboard exceedances $\eta_{0,out}/\eta_0$, considering $\eta_{0,out}$ obtained at different positions from the deck edge (x/L_{st}, where L_{st} is the structure length) for all the study cases. (**c**) Ratios of time of occurrence of the maximum freeboard exceedances ($t_{\eta_{0,out}}/t_{\eta_0}$), considering $\eta_{0,out}$ obtained at different positions from the deck edge for all the study cases.

6.3. Green Water Kinematics

The present methodology also includes the measurement of the wavefront velocity, U_{front}, which is another important parameter in green water research. It is used mainly to estimate the horizontal loading in structures located over the deck [2]. Information about this parameter has been extracted from visual inspection of the wavefront edge displacement onto the deck, as described in Section 3.4.2.

Mean U_{front} values for the five study cases were calculated considering different domains over the deck. These values were obtained by applying linear regression analyses to the wavefront velocity data for different domains in x, considering $x = 0$ at the bow edge. At the early stages of green water occurring at the beginning of the deck, U_{front} was between 0.29 and 0.31 m/s and 0.38 and 0.44 m/s for the domains $0 < x < 0.02$ m and $0 < x < 0.05$ m, respectively, for all cases. Next, when the wavefront developed further over the deck, all cases reached U_{front} values between 1 and 1.1 m/s ($0.1 < x < 0.35$ m). It was verified that the wavefront presented acceleration over the deck, from $U_{front} \approx 0.2$–0.3 m/s to $U_{front} \approx 1.1$ m/s for all cases. The velocities found for the well-developed wavefront are of similar orders of magnitude to those reported globally by [1], for green water experiments with regular waves in a wave flume.

Figure 14 presents the relationship between mean values of the well-developed U_{front} (measured between $0.1 < x < 0.35$ m) and kinematics of the incoming bores. To do this, ratios of U_{front} with the bore front velocity estimated teoretically (U_0, Stoker) and with the bore front velocity calculated experimentally (U_{AB}) for the cases with (EXP1) and without the structure (EXP2) are presented. U_{AB} was calculated from EXP1 and EXP2, considering the domain defined by $x = -0.4$ m (point A) and $x = -0.3$ m (point B) (Figure 6) as $U_{AB} = (x_B - x_A)/(t_B - t_A)$, where $x_B - x_A$ is the distance between points A and B, and t_A and t_B are the points in time at which the mean maximum elevation occurred in A and B, respectively.

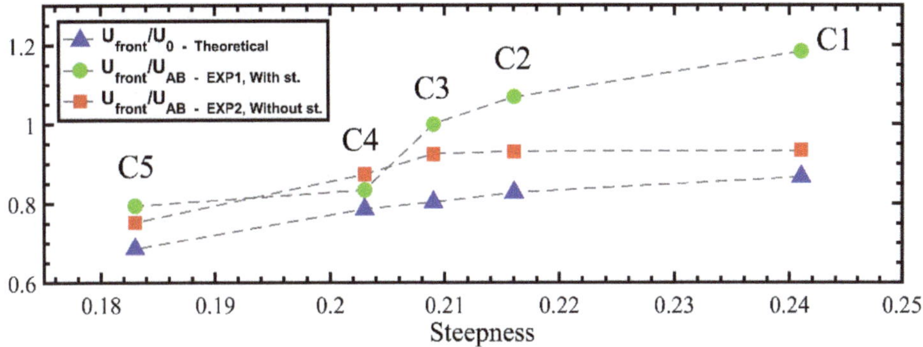

Figure 14. Ratios of mean values of wavefront velocity over the deck (U_{front}, $0.1 < x < 0.35$ m) with respect to the theoretical bore front velocity (U_0, [16]) and the experimental bore front velocities U_{AB} estimated for the cases with (EXP1) and without (EXP2) the structure, for different bore steepnesses (C1–C5).

For the velocities obtained with the experiments without a structure and calculated theoretically, U_{front} was always lower than the bore front velocity. However, for the case with the structure, U_{front} was closer to U_{AB}, particularly for C2 and C3. For C1 (steeper bore), U_{front} was ~20% higher than U_{AB}, whereas for C4 and C5, it was ~20% lower.

On the other hand, it can be noted that the theoretical values for the bore velocity (U_0) overestimate the experimental ones (U_{AB}, EXP1, and EXP2) for all cases, being closer to the ones obtained for EXP2 (case without the structure). Thus, it can be inferred that the presence of the structure reduced the velocity of the incoming bores. This suggests that it might be wise to consider a correction factor in green water analyses that employ kinematic information from the undisturbed incident wave.

7. Conclusions

The study of green water on structures requires detailed systematic experimental analyses that allow green water patterns to be identified and analytical or numerical models to be validated.

Considering the interaction of wet dam-break bores with a fixed structure, this paper presented an alternative image-based experimental study of green water. The experiments had a duration of ~3 s, which allowed the use of high sampling rates in cameras to capture details of wave propagation and green water flow, using two-dimensional open-source image-based methods.

The main conclusions of this work are summarized as follows:

- Five different study cases were performed, considering the same wet dam-break ratio $h_0/h_1 = 0.6$ and five different freeboards ($0.006 \leq FB \leq 0.042$). These conditions generated undular bores with similar heights (0.055–0.060 m for all cases) and theoretical steepnesses in the range 0.183–0.241. These bores generated four green water events with small cavities formed at the beginning of the deck (PDB-types of green water) for the steeper bores and a case where no cavity was observed (DB-type of green water) for the longest bore. Some concepts used to analyse the cavities formed in flip-through events in vertical walls were introduced in the present work to practically describe the evolution of the PDB cavity in a practical way.

- The consideration of the maximum freeboard exceedance (η_0) is important in the implementation of analytical and numerical models. In this study, it was verified that selecting η_0 at some distances outside the deck may yield differences with respect to the one measured at the bow. It is suggested that for applications in which the relative wave-deck motions were considered outside the deck, a correction factor should be estimated and included to approximate the real freeboard exceedance that occurs at the edge of the deck, considering also its time of occurrence, which is relevant to green water simulations.

- The proposed experimental setup allowed the use of image-based methods available in the literature to analyse the temporal and spatial evolution of the incident wave and green water on deck in a two-dimensional framework. This is an advantage over traditional techniques that employ obstructive wave probes over the deck to monitor green water elevations at a few positions. Regarding these results, a database of water elevations has been made available to allow model validations by other authors. These include time series of water elevations for the five repetitions of the five study cases, considering the experiments with and without the fixed structure. The database was made available in a Mendeley data repository: http://dx.doi.org/10.17632/zjrsmffh4d.1 as Supplementary Materials.

The physics of the wet dam-break bore used in the present method is different from that of regular waves, which are commonly used in green water analyses. However, it may be an alternative to acquire systematic and high-resolution local details of water propagating over the deck due to the small duration of the experiments. With tests of small duration, the sampling frequency of cameras could be significantly increased, and the high-speed visualization could be performed in a repeatable and reproducible way. The method can be extended to analyse more types of green water events, including the systematic analysis of other magnitudes, considering different types of wet dam-break waves (e.g., broken, unbroken) by changing the h_0/h_1 wet dam-break ratios. Moreover, different structure configurations can be tested, such as rounded or sloped 2D structures with and without substructures over the deck.

Supplementary Materials: A database of time series of water elevations for each repetition of the five study cases presented in this work has been made available in a Mendeley Data Repository: http://dx.doi.org/10.17632/zjrsmffh4d.1.

Author Contributions: Conceptualization, J.V.H.-F., P.d.T.T.E., and S.H.S.; methodology, J.V.H.-F., P.d.T.T.E., S.H.S., and R.S.; software, J.V.H.-F.; validation, J.V.H.-F., and J.F.B.G.; formal analysis, J.V.H.-F., P.d.T.T.E., and J.F.B.G.; investigation, J.V.H.-F., P.d.T.T.E., J.F.B.G., and R.S.; resources, P.d.T.T.E., S.H.S., and R.S.; data curation, J.V.H.-F.; writing-original draft preparation, J.V.H.-F., P.d.T.T.E., S.H.S., J.F.B.G., and R.S.; writing-review and editing, P.d.T.T.E., S.H.S., and R.S.; visualization, J.V.H.-F., and J.F.B.G.; supervision, P.d.T.T.E., S.H.S., and R.S.; project administration, S.H.S.; funding acquisition, P.d.T.T.E., S.H.S., J.F.B.G., and R.S.

Funding: This research was funded by the Brazilian National Petroleum Agency (ANP), the Mexican National Council of Science and Technology (CONACYT) and the Mexican Secretary of Energy (SENER/Hidrocarburos) grant number [407583]. The APC was funded by Universidad del Caribe.

Acknowledgments: J.V.H.-F. would like to thank the support given by the DGAPA-UNAM postdoctoral fellowship for the redaction and revision of the manuscript. S.H.S. would like to thank the Brazilian Agency for Research and Development (CNPq) for their support. Special thanks are given to Marcelo Vitola for his comments during the experiments and to Jill Taylor for her help in the revision of the manuscript.

Conflicts of Interest: The authors declare no conflict of interest.

References

1. Greco, M. *A Two-dimensional Study of Green-Water Loading*; Norwegian University of Science and Technology: Trondheim, Norway, 2001.
2. Buchner, B. *Green Water On Ship-Type Offshore Structures*; Delft University of Technology: Delft, The Netherlands, 2002.
3. Hernández-Fontes, J.V.; Vitola, M.A.; Esperança, P.T.T.; Sphaier, S.H. An alternative for estimating shipping water height distribution due to green water on a ship without forward speed. *Mar. Syst. Ocean Technol.* **2015**, *10*, 38–46. [CrossRef]
4. Ogawa, T.; Taguchi, H.; Ishida, S. Experimental study on shipping water volume and its load on deck. *J. Soc. Nav. Archit. Jpn.* **1997**, *182*, 177–185. [CrossRef]
5. Nielsen, K.B.; Mayer, S. Numerical prediction of green water incidents. *Ocean Eng.* **2004**, *31*, 363–399. [CrossRef]
6. Silva, D.F.C.; Esperança, P.T.T.; Coutinho, A.L.G.A. Green water loads on FPSOs exposed to beam and quartering seas, Part II: CFD simulations. *Ocean Eng.* **2017**, *140*, 434–452. [CrossRef]
7. Greco, M.; Colicchio, G.; Faltinsen, O.M. Shipping of water on a two-dimensional structure. Part 2. *J. Fluid Mech.* **2007**, *581*, 371–399. [CrossRef]
8. Barcellona, M.; Landrini, M.; Greco, M.; Faltinsen, O.M. An experimental investigation on bow water shipping. *J. Sh. Res.* **2003**, *47*, 327–346.
9. Lee, H.H.; Lim, H.J.; Rhee, S.H. Experimental investigation of green water on deck for a CFD validation database. *Ocean Eng.* **2012**, *42*, 47–60. [CrossRef]
10. Ryu, Y.; Chang, K.A.; Mercier, R. Application of dam-break flow to green water prediction. *Appl. Ocean Res.* **2007**, *29*, 128–136. [CrossRef]
11. Song, Y.K.; Chang, K.A.; Ariyarathne, K.; Mercier, R. Surface velocity and impact pressure of green water flow on a fixed model structure in a large wave basin. *Ocean Eng.* **2015**, *104*, 40–51. [CrossRef]
12. Van Veer, R.; Boorsma, A. Towards an improved understanding of green water exceedance. In Proceedings of the ASME 2016 35th International Conference on Ocean, Offshore and Arctic Engineering OMAE 2016, Busan, Korea, 19–24 June 2016; pp. 1–10.
13. Ryu, Y.; Chang, K.A.; Mercier, R. Runup and green water velocities due to breaking wave impinging and overtopping. *Exp. Fluids* **2007**, *43*, 555–567. [CrossRef]
14. Hernández-Fontes, J.V.; Vitola, M.A.; Silva, M.C.; Esperança, P.T.T.; Sphaier, S.H. Use of wet dam-break to study green water problem. In Proceedings of the ASME 2017 36th International Conference on Ocean, Offshore and Arctic Engineering OMAE 2016, Tondrheim, Norway, 25–30 June 2017.
15. Hernández-Fontes, J.V.; Vitola, M.A.; Silva, M.C.; Esperança, P.D.T.T.; Sphaier, S.H. On the generation of isolated green water events using wet dam-break. *J. Offshore Mech. Arct. Eng.* **2018**, *140*. [CrossRef]
16. Stoker, J.J. *Water Waves: Pure and Applied Mathematics*; Interscience Publishers: New York, NY, USA, 1957.
17. Binnie, A.M.; Davies, P.O.A.L.; Orkney, J.C. Experiments on the flow of water from a reservoir through an open horizontal channel I. The production of a uniform stream. *Proc. R. Soc. Lond. Ser. A Math. Phys. Sci.* **1955**, *230*, 225–236.
18. Ritter, A. Die fortpflanzung der wasserwellen. *Zeitschrift des Vereines Deutscher Ingenieure* **1892**, *36*, 947–954.
19. Nakagawa, H.; Nakamura, S.; Ichihashi, K. *Generation and Development of a Hydraulic Bore Due to the Breaking of a Dam*; Bulletin of the Disaster Prevention Research Institute, Kyoto University: Kyoto, Japan, 1969; Volume 19.

20. Stansby, P.K.; Chegini, A.; Barnes, T.C.D. The initial stages of dam-break flow. *J. Fluid Mech.* **1998**, *374*, 407–424. [CrossRef]

21. Ozmen-Cagatay, H.; Kocaman, S. Dam-break flows during initial stage using SWE and RANS approaches. *J. Hydraul. Res.* **2010**, *48*, 603–611. [CrossRef]

22. Oertel, M.; Bung, D.B. Initial stage of two-dimensional dam-break waves: Laboratory versus VOF. *J. Hydraul. Res.* **2012**, *50*, 89–97. [CrossRef]

23. Kocaman, S.; Ozmen-Cagatay, H. Investigation of dam-break induced shock waves impact on a vertical wall. *J. Hydrol.* **2015**, *525*, 1–12. [CrossRef]

24. Hernández-Fontes, J.V.; Vitola, M.A.; Esperança, P.D.T.T.; Sphaier, S.H. Analytical convolution model for shipping water evolution on a fixed structure. *Appl. Ocean Res.* **2019**, *82*, 415–429. [CrossRef]

25. Hernández-Fontes, J.V.; Vitola, M.A.; Esperança, P.D.T.T.; Sphaier, S.H. Assessing shipping water vertical loads on a fixed structure by convolution model and wet dam-break tests. *Appl. Ocean Res.* **2019**, *82*, 63–73. [CrossRef]

26. Kocaman, S.; Ozmen-Cagatay, H. The effect of lateral channel contraction on dam break flows: Laboratory experiment. *J. Hydrol.* **2012**, *432–433*, 145–153. [CrossRef]

27. Yeh, H.H.; Ghazali, A.; Marton, I. Experimental study of bore run-up. *J. Fluid Mech.* **1989**, *206*, 563–578. [CrossRef]

28. Hernández-Fontes, J.V. An Analytical and Experimental Study of Shipping Water Evolution and Related Vertical Loading. Ph.D. Thesis, COPPE, Federal University of Rio de Janeiro, Janeiro, Brazil, 2018.

29. Hernández, I.D.; Hernández-Fontes, J.V.; Vitola, M.A.; Silva, M.C.; Esperança, P.T.T. Water elevation measurements using binary image analysis for 2D hydrodynamic experiments. *Ocean Eng.* **2018**, *157*, 325–338. [CrossRef]

30. Hager, W.H.; Lauber, G. Hydraulische experimente zum talsperren bruchproblem (hydraulic experiments to dambreak problem). *Schweizer Ing. und Archit.* **1996**, *114*, 515–524.

31. Lauber, G.; Hager, W.H. Experiments to dambreak wave: Horizontal channel. *J. Hydraul. Res.* **1998**, *36*, 291–307. [CrossRef]

32. Shigematsu, T.; Liu, P.L.-F.; Oda, K. Numerical modeling of the initial stages of dam-break waves. *J. Hydraul. Res.* **2004**, *42*, 183–195. [CrossRef]

33. Liu, H.; Liu, H. Experimental study on dam-break hydrodynamic characteristics under different conditions. *J. Disaster Res.* **2017**, *12*, 198–207. [CrossRef]

34. Daily, J.W.; Stephan, S.C. The solitary wave. *Coast. Eng. Proc.* **1952**, *1*, 2.

35. Goring, D.G. *Tsunamis: The Propagation of Long Waves onto A Shelf*; California Institute of Technology: Pasadena, CA, USA, 1979; Volume 1979.

36. Silva, R.; Losada, I.J.; Losada, M.A. Reflection and transmission of tsunami waves by coastal structures. *Appl. Ocean Res.* **2000**, *22*, 215–223. [CrossRef]

37. Lugni, C.; Brocchini, M.; Faltinsen, O.M. Wave impact loads: The role of the flip-through. *Phys. Fluids* **2006**, *18*, 122101. [CrossRef]

38. Colicchio, G.; Greco, M.; Faltinsen, O.M. Domain-decomposition strategy for marine applications with cavity entrapments. *J. Fluids Struct.* **2011**, *27*, 567–585. [CrossRef]

39. Lugni, C.; Miozzi, M.; Brocchini, M.; Faltinsen, O.M. Evolution of the air cavity during a depressurized wave impact. I. The kinematic flow field. *Phys. Fluids* **2010**, *22*, 1–17. [CrossRef]

40. Lugni, C.; Brocchini, M.; Faltinsen, O.M. Evolution of the air cavity during a depressurized wave impact. II. The dynamic field. *Phys. Fluids* **2010**, *22*, 1–13. [CrossRef]

41. Peregrine, D.H.; Topliss, M.E. The pressure field due to steep water waves incident on a vertical wall. In Proceedings of the 24th International Conference on Coastal Engineering (ASCE, Kobe, 1994), Kobe, Japan, 23–28 October 1994; pp. 1496–1510.

42. Ogawa, Y. Long-term prediction method for the green water load and volume for an assessment of the load line. *J. Mar. Sci. Technol.* **2003**, *7*, 137–144. [CrossRef]

43. Hu, Z.; Xue, H.; Tang, W.; Zhang, X. A combined wave-dam-breaking model for rogue wave overtopping. *Ocean Eng.* **2015**, *104*, 77–88. [CrossRef]

44. Buchner, B. On the impact of green water loading. In Proceedings of the Sixth International Symposium on Practical Design of Ships and Mobile Units, PRADS 1995, Seoul, Korea, 17–22 September 1995.

45. Silva, D.F.C.; Coutinho, A.L.G.A.; Esperança, P.T.T. Green water loads on FPSOs exposed to beam and quartering seas, part I: Experimental tests. *Ocean Eng.* **2017**, *140*, 419–433. [CrossRef]
46. Buchner, B. Green water from the side of an weathervaning FPSO. In Proceedings of the OMAE99, 18th International Conference on Offshore Mechanics and Arctic Engineering, St. Johns, NL, Canada, 11–16 July 1999; pp. 1–11.
47. Buchner, B. The impact of green water on FPSO design. In Proceedings of the 27th Offshore Technology Conference, OTC, Houston, TX, USA, 1–4 May 1995; pp. 45–57.
48. Fonseca, N.; Pascoal, R.; Guedes Soares, C.; Clauss, G.; Schmittner, C. Numerical and experimental analysis of extreme wave induced vertical bending moments on a FPSO. *Appl. Ocean Res.* **2010**, *32*, 374–390. [CrossRef]
49. Xiao, L.; Tao, L.; Yang, J.; Li, X. An experimental investigation on wave runup along the broadside of a single point moored FPSO exposed to oblique waves. *Ocean Eng.* **2014**, *88*, 81–90. [CrossRef]
50. Xiao, L.; Yang, J.; Tao, L.; Li, X. Shallow water effects on high order statistics and probability distributions of wave run-ups along FPSO broadside. *Mar. Struct.* **2015**, *41*, 1–19. [CrossRef]

Article

Comparison of a Floating Cylinder with Solid and Water Ballast

Roman Gabl [1,2,]*, Thomas Davey [1], Edd Nixon [1], Jeffrey Steynor [1] and David M. Ingram [1]

[1] School of Engineering, Institute for Energy Systems, FloWave Ocean Energy Research Facility, The University of Edinburgh, Max Born Crescent, Edinburgh EH9 3BF, UK; tom.davey@flowave.ed.ac.uk (T.D.); Edward.Nixon@sintef.no (E.N.); Jeff.Steynor@ed.ac.uk (J.S.); david.ingram@ed.ac.uk (D.M.I.)

[2] Unit of Hydraulic Engineering, University of Innsbruck, Technikerstraße 13, 6020 Innsbruck, Austria

[*] Correspondence: roman.gabl@ed.ac.uk

Received: 30 September 2019; Accepted: 20 November 2019; Published: 26 November 2019

Abstract: Modelling and understanding the motion of water filled floating objects is important for a wide range of applications including the behaviour of ships and floating platforms. Previous studies either investigated only small movements or applied a very specific (ship) geometry. The presented experiments are conducted using the simplified geometry of an open topped hollow cylinder ballasted to different displacements. Regular waves are used to excite the floating structure, which exhibits rotation angles of over 20 degrees and a heave motion double that of the wave amplitude. Four different drafts are investigated, each with two different ballast options: with (water) and without (solid) a free surface. The comparison shows a small difference in the body's three translational motions as well as the rotation around the normal axis to the water surface. Significant differences are observed in the rotation about the wave direction comparable to parametric rolling as seen in ships. The three bigger drafts with free surface switch the dominant global rotation direction from pitch to roll, which can clearly be attributed to the sloshing of the internal water. The presented study provides a new dataset and comparison of varying ballast types on device motions, which may be used for future validation experiments.

Keywords: floating cylinder; water filled; motion capturing; wave tank; wave gauges; fluid–structure interaction; free surface; sloshing

---

1. Introduction

The presented study investigates the response of a floating structure with two different ballast options, namely liquid (water) and solid. This allows the effects of the water sloshing to be isolated and quantified. Regular waves of constant amplitude and varying frequency f_W are used to excite the motion of the simplified geometry. An additional background is presented in Gabl et al. [1] and the original data set is shared via the digital repository of the University of Edinburgh [2].

In general, water filled structures can be found in a wide range of applications including Wave Energy Converters (WEC) [3–5], Oscillating Water Columns (OWC) [6,7], and energy storage concepts [8,9]. The correct prediction of the motion and forces acting on a floating body are essential for the structural and mooring designs [10,11]. The response of Very Large Floating Structures (VLFS) [12–14], including the inner water level [15], have only been investigated for small motions. A wide range of studies investigate sloshing inside various containments and mainly focus on the resultant peak wall pressure [16–18], as well as being used for validation experiments [19–21]. The presented study aims to identify the influence of sloshing during large motions of a floating structure.

A comparable application of the investigated floating object can be found in ships filled with Liquefied Natural Gas (LNG) [22–26]. The influence of a rectangular tank on a rectangular barge is investigated by Su and Liu [27] based on a nonlinear Boussinesq-type, which failed in the case of steep sloshing phenomena. Huang et al. [28] also found significant limitations of the numerical approach in predicting large motions due to the influence of internal sloshing. Zhao and McPhail [29] compared frozen and liquid cargo in two spherical tanks on a vessel and identified a small difference in the rolling motion for half filled conditions. Bigger tanks are more susceptible to damage from pressure peaks caused by the internal sloshing [30,31]. Xu et al. [32] and Zhao et al. [33] investigated the side-by-side operation of two vessels through experiments at a scale of 1:60. A numerical simulation of such a combination is presented by Zhao et al. [34]. All these studies are focused on a very specific (ship) geometry and the presented work focuses on a more general approach by simplifying the floating structure to a hollow cylinder containing an internal water body and free surface.

Vortex-induced vibration on a cylinder is investigated for a wide range of geometries both experimentally and numerically [35–37] . Different commercial numerical products allow dynamic mesh zones in which the computational grid is reformed after every iteration. Zhu et al. [38] presented a recent study of an elliptic cylinder incorporating a two-way Fluid–Structure Interaction (FSI). The moving zone is in the range of three diameters for the comparable case of a cylinder mounted with fin-shaped strips [39]. Those studies, as well as the numerical sloshing experiment presented by Jamalabadi et al. [40], are only simulated as two-dimensional cases. Yang et al. [41] found a significant difference between the 2D- and 3D-approach for multi-bodies in a numerical wave tank. Different examples of a full coupling in 3D include propeller blades [42], offshore wind turbine towers [43], OWC [44], and point-absorbing WEC [45]. Alternatively, FLOW-3D leaves the grid constant and changes the discretisation of the geometry depending on the object's movement [46–48].

Meshless methods, such as Smoothed Particle Hydrodynamics (SPH), which are based on a purely Lagrangian technique, are a possible alternative option for large motions. Successful applications can be found from the simulation of the flow around cylinder incorporating an erosion model [49], the design of a floating tsunami shelter [50], the simulation of complex wave tanks [51], and the coupling with wave propagation models for wide areas [52]. Different studies apply this method for the simulation of a floating buoy [53] and WEC [54]. Some studies investigate damaged ships [55], which includes one large chamber with a hole [56] as well as simulations of one of three separate ones [57,58]. The comparison with experimental observations show good agreement for these cases, which makes SPH a valuable option to simulate the presented big motions of a floating structure.

This paper presents the experimental results of a simplified geometry floating in a wave tank under regular wave conditions. Consequently, the response of the floating object is influenced by the inner and the outer water bodies. This adds a further complexity to those sloshing experiments caused by excitation of the containment [19–21]. The aim of this study is to identify the influence of the internal liquid by comparing it to a solid ballast option. Four different drafts are investigated and the response in all six degrees of freedom are separately compared and presented in the following sections. The complete data-set is available according to Gabl et al. [1,2], and the results found are used to refine future measurement with the water filled cylinder. The presentation of this data set and experimental results will greatly benefit the numerical modelling community, who require a validation case for new solutions to capture this phenomenon.

2. Experimental Set-Up

2.1. General

The investigated geometry is a floating cylinder with an outer diameter D_{out} of 0.5 m and a total height H of 0.5 m. Figure 1 shows the transparent structure, which has wall thicknesses of 5 mm and a 7 mm floor plate. Wave gauges (WG) made of copper tape are applied on the inside wall. This instrumentation is introduced by Gabl et al. [59] and allows the measurement of internal run up at the walls. In the presented investigation, the movement of the internal water surface is not specifically monitored, hence the main goal is the comparison of the body motions between the ballast variations, namely water or solid, for a wide range of wave conditions. This allows the identification of the interesting cases where future additional investigations of the water ballast may be performed.

Figure 1. (**a**) overview of experimental set-up in the wave tank including wave gauges and main wave direction; (**b**) detail of the solid ballast; (**c**) cylinder filled with inner water on the raised tank floor with connected mooring lines.

Six markers are mounted at the top of the structure to use an optical motion capture system (Section 2.2). This in turn allows the 6 degree-of-freedom measurement of the motions of the floating cylinder under different wave conditions. The mass of all additional loads (measurement equipment) on the structure are reduced as far as possible but need to be located at the dry top. Consequently, the initial structure was observed to be unstable at smaller displacements (or low inner water depth h values). A reduction in the body's vertical centre of gravity was therefore required to allow a wider range of h values. A layer of lead ballast with a thickness of approximately 2.6 mm was located at the bottom inside of the cylinder. This had an added advantage in that it concealed a drain plug in the centre of the floor plate (R_{Plug} = 5 cm) and thus provided an uninterrupted plane on the internal floor.

The floating cylinder was tested in the FloWave test basin at the University of Edinburgh, which allows the combination current with non-directional waves [60–63]. The model was moored in the centre of the circular wave tank which has a diameter of 25 m and a depth of 2 m (upper volume of the tank). For the presented investigations, long crested sinusoidal wave inputs were applied with an initial wave amplitude of 0.05 m and a frequency range f_W covering 0.3 Hz to 1.1 Hz parallel to the gantry (Figure 1a). A set of 16 runs were predefined to cover the whole frequency band. Depending on a preliminary analysis, further cases were individually added to investigate specific observed behaviour and refine near resonances. The test length was 180 sec in total starting simultaneously with the wave makers. Between the tests, a settling time was allowed so that the motions of the floating cylinder were negligibly small ($\leq$0.5 mm).

The presented investigation focuses on the comparison of the solid ballast case for a range of water volumes inside of the cylinder. The solid ballast modules are located so as to maintain symmetry in the two main axes parallel to the still water surface (Figure 1b). A layer of Styropor (an expandable polystyrene material) was added on the top and bottom of the resulting cross arrangement so that the vertical centre of gravity is the same as that of the still water ballast case. In total, three different masses are combined (two crosses and one cylinder, the exact masses can be found in Gabl et al. [1]), which allowed comparison of a wide range of different inner water levels. The water filled case is investigated after the solid case and the exact displacement and external still water line is reproduced. Figure 1c shows the model on the raised tank floor with an indicative inner water level h.

Table 1 lists the four different cases, which are defined by the draft (vertical distance between the bottom of the cylinder and the external still water level). The distinction between the four draft options is made based on the comparison to the minimal draft $d = d_{Min}$ of 0.1822 m. The inner water height h is defined as the vertical distance between the internal bottom of the cylinder and the internal still water level. This value changes from nearly a quarter to more than a half of the total height of the cylinder H. The observed motions are comparably large but an exchange of water between inner and outer water body—spilling and over topping—did not occur.

Table 1. List of the four different investigated cases with the corresponding water depth h in relation to the total height H of the cylinder and the resulting draft d.

	$d \times 1.0 = d_{Min}$	$d \cdot 1.25$	$d \cdot 1.5$	$d \cdot 1.75$
Total Mass (kg)	35.70	44.65	53.50	62.45
h (m)	0.1232	0.1707	0.2178	0.2653
h/H (–)	0.25	0.34	0.44	0.53
draft (m)	0.1822	0.2279	0.2730	0.3187

2.2. Measurement

Two different measurement systems are used: (a) wave gauges and (b) an optical motion capture system. In total, seven wave gauges are aligned parallel to the direction of wave propagation through the centre of the tank. This allows the surface elevation in the tank to be measured at discrete points, five in front and two behind the floating cylinder. At the beginning of each test day, all probes are calibrated to maintain the high accuracy of the measurements. The accuracy of a resistive wave gauge is typically in the range of 1 mm [59,64,65].

The motion of the floating cylinder is measured with the motion capture system installed in the FloWave facility. This includes eight Qualisys cameras (Göteborg, Sweden) , which are distributed over one side of the test tank at different heights. The global coordinate system is orientated as part of the calibration process of the motion capturing system in the centre of the round wave tank. The positive x-axis is defined parallel to the gantry, which is identical to the wave direction used in these tests. Between each tested experimental set-up, the calibration is refined so that the overall accuracy of the localisation of the points could be kept smaller than 1 mm.

The 3D-locations of each point are assembled together to calculate the motion of the floating cylinder as part of the post-processing of the motion capturing software, Qualisys Track Manager (QTM, version 2019.3, Qualisys, Göteborg, Sweden) . Therefore, a body definition is needed, which connects the local body coordinate system to the measured markers and is dependent on each investigated experimental set-up (draft). Therewith, it is assured that the vertical position of the origin of the local body coordinate system is always level with the still water surface in the wave tank. All local axes are orientated similarly to the global coordinate system. The rotational symmetry of the cylinder is a big advantage over more specific ship geometries by providing independence to orientation of the incoming waves. Furthermore, a small rotation around the z-axis has nearly no influence on the behaviour of the floating structure in relation to the wave. The post-processing software, QTM, also allows for identify pitch and roll angles independently of yaw, thereby simplifying the analysis in this case. Consequently, the measured rotations for pitch and roll are always co-linear with the global tank coordinates, and therefore fixed relative to the wave direction, regardless of the cylinder rotation.

All instrumentation is triggered by a Transistor—transistor logic (TTL) pulse upon the start of wave generation and operate at the same measurement frequency of 128 Hz. Further information is given in Gabl et al. [1], which describes the available full data set of this investigation.

2.3. Mooring

A station keeping mooring system is necessary, but it should be soft enough not to introduce any effects on the wave frequency motion response. A horizontal approach near the water surface was chosen, which is comparable to Zhao et al. [66], who investigated the roll motion of a barged vessel with two cylindrical tanks. Four lines are connected to the quadrants of the round tank in 45° to the main axes (Figure 1c). Each one consists of a hollow elastic of 3 m long (diameter 3 mm) with a very high stretch factor and is expanded with standard rope (6.5 m) to the tank side. A small but balanced preload was introduced. Two of the mooring lines are joined together in one point with a symmetry along the x–z-plane. Both connection points are adjustable with the draft (Figure 1c) so that the rotation axis introduced by the mooring system is equal to the local y-axis (in the height of the still water surface). This minimises the influence for the main rotational motions (pitch; around the y-axis). The roll motion is affected more significantly, but the mooring set-up is generally very soft. The natural frequency of the mooring system is presented in Section 3.2.

3. Results

3.1. Overview

The analysis of the results is split into four parts. First, the different experimental set-ups are investigated in the still wave tank. In this case, the floating structure is manually excited in a series of free oscillation tests to verify the model's behaviour in connection with the mooring system (Section 3.2). All further analysis discusses the results of the response motion of the different drafts and ballast options. As an overview of all six degrees of freedom, Section 3.3 presents the minimum and maximum values after the initial start-up period. The following detailed analysis of the translation in the z-direction is shown in Section 3.4. Furthermore, it was observed that significant roll motion is connected to a reduction in pitch. Based on this, both rotations are presented together in Section 3.5. Section 3.6 summarises the results with the different drafts and allows a direct comparison between the liquid and solid ballasts.

3.2. Free Oscillation

For these tests, the cylinder is manually excited out of its static equilibrium state in the centre of a still water tank. The oscillation is captured for each degree of freedom and repeated multiple times. Figure 2 presents the response frequency f for the two different ballast configurations (solid and liquid) in relation to the draft d (normalised by the minimum draft d_{Min} of 18.22 cm).

In surge and sway, the response is dominated by the mooring lines and is very slow. Consequently the frequency is small and decreases as draft increases. A comparable behaviour—but at a higher frequency level—can be observed in the z-direction (heave) with a response of around 0.9 Hz for the minimal draft and 0.7 Hz for the heaviest ballast. For all translational motion, the difference between the two ballast configurations is not significant.

The tilting rotation around the x-axis is slightly more restricted than around the y-one, a deliberate feature of mooring setup intended to result in the smallest possible influence in pitch (Section 2.3). As expected, the water filled version of the floating cylinder shows a different response in roll and pitch. For the solid case, the changes due to a variation in the draft are relatively small and a typical value is around 0.9 Hz. In contrast to this, the sloshing inner water leads to a significantly smaller frequency and also a bigger dependence. The frequency is nearly doubled for the bigger draft in comparison to the version with the minimal draft d_{min} of 18.22 cm. A small difference can be observed for the rotation around the z-axis with the draft having nearly no influence.

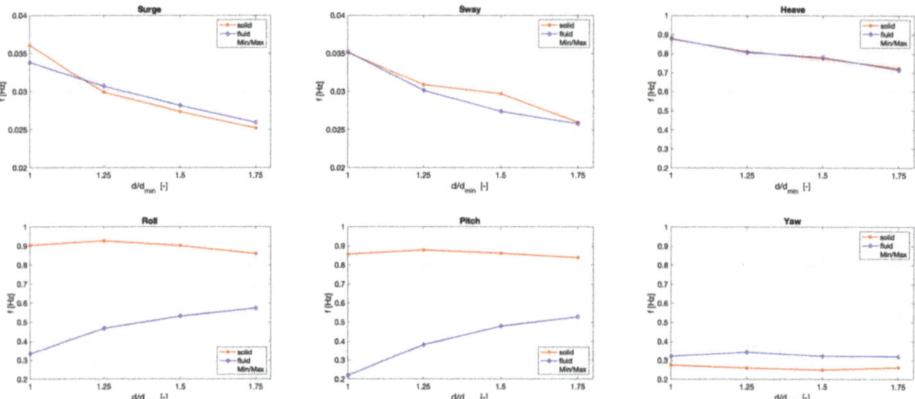

Figure 2. Response frequency f of the free oscillation experiments for each degree of freedom (upper row movements, lower row rotations)—comparison of solid and liquid ballast for four draft cases d normalised by d_{min} = 18.22 cm.

3.3. Minimum and Maximum Response Motion

The response motion of the floating cylinder under wave conditions is analysed for all six degrees of freedom separately. Figure 3 shows the results for the movement and the complementary rotations are presented in Figure 4. In all cases, the first 52 s from the starting of the wave makers are excluded to limit the analysis to the fully developed oscillation of the floating cylinder in the wave tank. This has the advantage that delayed effects caused by the sloshing of the inner water body are fully developed. On the other hand, due to the different final position of the cylinder in the tank (especially the x-direction, Figure 3), the reflections can have a different influence on the response of the floating structure [60]. The chosen approach is focused on an overall comparison. For a specific validation in the time-domain, the original measurement [1,2] should be checked for transient behaviour (especially switching between roll and pitch), and it is advisable to focus on the ramp up as well as the first stable oscillations [67].

In the following figures, the dashed line shows the minimum and uses the same colour as the maximum line. The upper row of graphics present the solid case and the lower one the ballast option with water. Generally, it can be noticed that the response is relatively small up to 0.6 Hz and peaks at around a range of 0.7 Hz to 0.9 Hz.

The biggest absolute movements can be found in x-direction (surge; wave direction), which is independent of the investigated option (Figure 3). The floating model drifts until the restricting forces of the mooring lines provide a new equilibrium. This drift reaches up to 2 m in surge, but the individual oscillations range up to around 200 mm. This behaviour is as expected due to the very soft mooring system. The model oscillates also in the approximate range of 200 mm in the orthogonal direction parallel to the y-axis. In heave, the movement can reach up to the double the wave amplitude of 50 mm (the presented results are limited to this main value). The observed response in the z-direction is analysed in detail in Section 3.4.

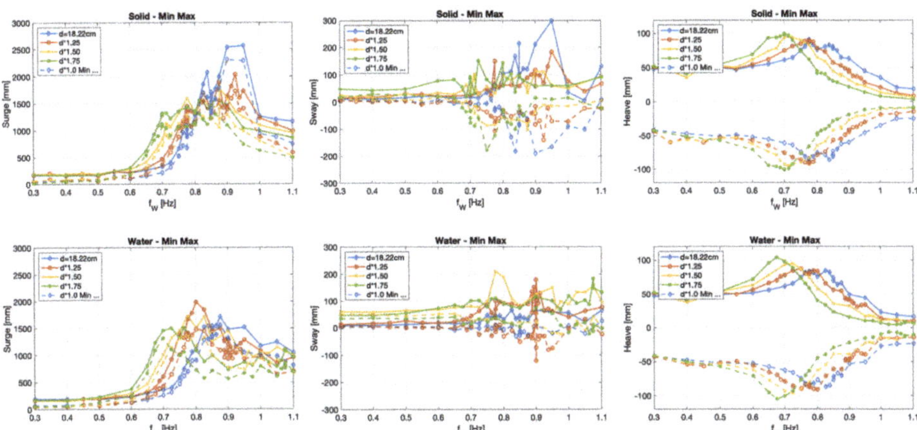

Figure 3. Summary of the minimum and maximum value of the movement in x-(Surge), y (Sway) and z-direction (heave)—values in mm—Solid (upper) and liquid (lower) case in relation to the initial wave frequency f_W.

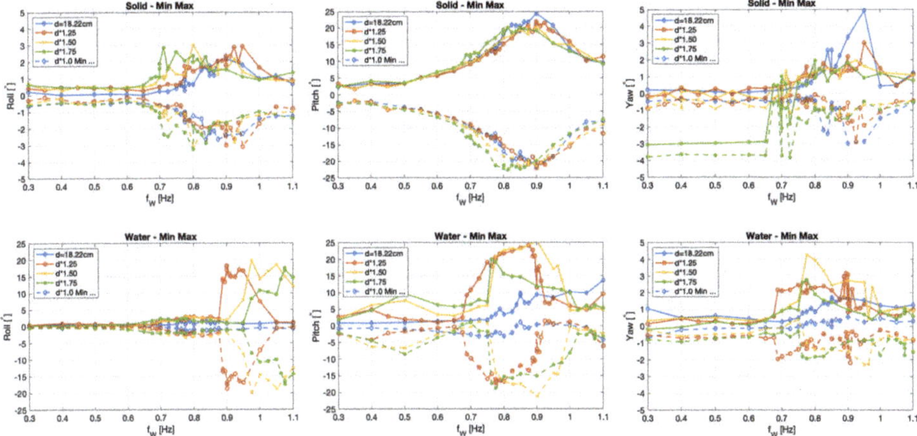

Figure 4. Summary of the minimum and maximum value of the rotation around the x- (Roll), y- (Pitch) and z-axis (Yaw)—values in °—Solid (upper) and fluid (lower) case in relation to the initial wave frequency f_W.

The solid ballast cases, in addition to the smallest water draft, have similarly small rotations around the x-axis (Roll). The range is approximately 3° to 4°. In contrast to this, the three cases with more water inside the cylinder have a very clear response in roll, which will be discussed in Section 3.5. The response in pitch peaks around a wave frequency f_W of 0.8 to 0.9 Hz. With up to 25°, the rotations are large, but nevertheless no spilling out or overtopping could be observed.

As mentioned in Section 2.2, the post-processing resolves the two main rotations around the x- and y-axis independently of a rotation around the vertical axis. Figure 4 shows that the angles are nevertheless very small as it is expected for a rotational symmetric geometry. For the solid case with the draft of $d \cdot 1.75$, the model was turned approximately 3° for the first tests with a lower wave frequency f_W, which was corrected (Figure 4; right graph in the upper row—Yaw Solid—green lines). A repetition was not conducted; hence, the analyses are not sensitive to this rotation.

For all three translational degrees of freedom (Figure 3), the difference between the two investigated ballasts are comparably small. A significantly different response could be identified for the three deeper drafts in the roll motion (Figure 4), which could not be observed in the solid case. Further analyses are limited to the vertical movement (Section 3.4) of the floating cylinder as well as the two main rotations (Section 3.5).

3.4. Heave Response

Based on the overview of the three different motions in Figure 3, the movement in the vertical direction is chosen to be presented in detail. For all following analysis, it is assumed that the response of the floating structure can be described as a (sum of) sine oscillation(s) similar to the incoming sine wave. The responding amplitude as well as the frequency f_R is evaluated based on the initial wave frequency f_W produced by the wave makers.

Similar to Section 3.3, the first 52 s of the total measured data (180 s) is excluded to ensure the system has reached steady state. The remaining period is split in eight equal sub-datasets for the following evaluation. A sine function is fitted to each subset separately. The analysis code (developed in MATLAB (version R2019a, MathWorks, Natick, Massachusetts, USA) considers a least-squares cost function as part of the fitting process. Incorrect results for the amplitudes are automatically sorted out based on the relation to the range of the local minimum/maximum of the measured data as well as the Root Mean Square Error (RMSE). The method was repeated for 4, 16 and 32 interval cases, which showed either a worse fitting rate for specific frequencies or only insignificant differences to the chosen approach with 8. The gained amplitudes of all sub-data sets are average for each initial wave frequency f_W. The results are verified with a single Fast Fourier transform (FFT) analysis [67], which showed a very good agreement.

In case of heave, those values for the amplitude are normalised by the measured wave amplitude a_W in front of the floating cylinder (details are given in Gabl et al. [1]) and therewith presented in Figure 5. The upper row shows the results for the solid ballast option and the lower row the complimentary ones for the cylinder filled with water. It is evident that all investigated cases indicate a comparable behaviour for the heave response. In the lower frequency band, the floating cylinder moves equal to the incoming waves and the amplitudes exceed those value with a peak in the range of 0.7 to 0.9 Hz. In this case the response frequency observed in the free oscillation experiments (Section 3.2, Figure 2) is reproduced, which leads to resonance. The observed movements reach up to double the wave amplitude. Higher frequencies lead to a near decoupling of the response and the movement of the floating cylinder decreases significantly. Consequently, the investigated wave frequency covers the full spectrum.

The response of the floating cylinder shows in some frequencies significant additional oscillations. This can also be observed based on the Fast Fourier transformation of the measurement. To isolate those additional oscillations, the fitted sine function (first step presented as a blue line) is subtracted from the original data, and a further fitting is conducted for the modified dataset. This process is repeated until no significant fitting is possible. The results are presented as 2nd and 3rd peaks in the following graphics. In the lower frequencies as well as in the range around the response frequency of the free oscillation (Section 3.2), significant secondary phenomena are detectable.

Figure 6 presents the response frequency f_R and includes indicative lines for the wave frequency multiplied with different constants. For the first sine fitting, the frequency f_R^{1st} is identical to the incoming f_W. Only in the highest frequencies is a small deviation found. This is caused by reflection in the wave tank, hence those high frequency waves cannot be absorbed by the receiving wave-maker paddles. For the solid ballast option (Figure 6, upper row), the second successful sine fitting has a doubled response frequency relative to the first one (equal to f_W). This is also true for the lower frequencies for the water filled ballast option. A factor of 0.5 and 1.5 could be identified, which is assumed to be caused by the sloshing of the inner water stored in the cylindrical tank.

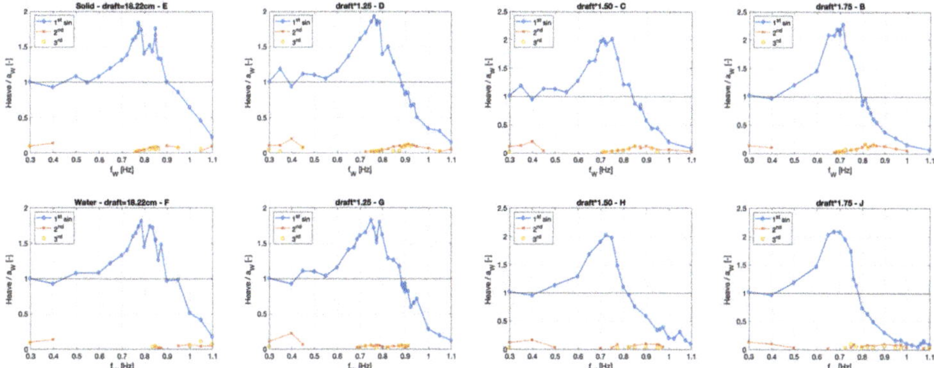

Figure 5. Amplitudes of the sine fitting for the movement in the z-direction (heave) normalised by the measured wave amplitude a_W in relation to the wave frequency f_W—upper row solid ballast option of the four different drafts and the corresponding water filling in the lower row.

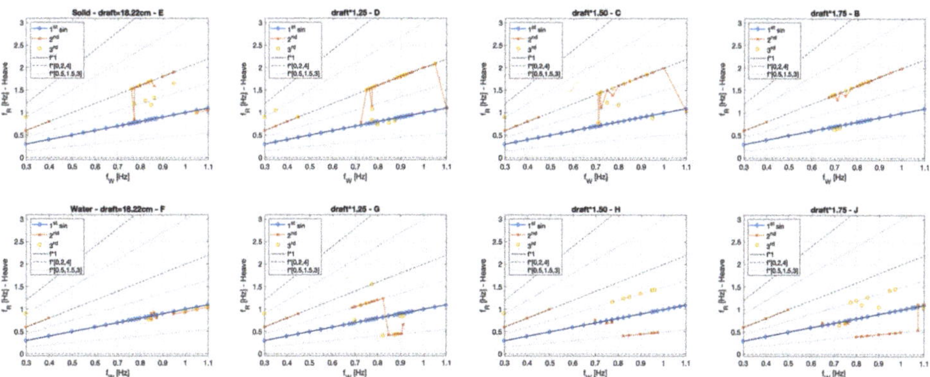

Figure 6. Response frequency f_R of the sine fitting for the movement in the z-direction (heave) in relation to the wave frequency f_W—upper row solid ballast option of the four different drafts and the corresponding water filling in the lower row.

Generally, the differences between the to ballast options, namely the solid case and the cylinder filled with water, are very small for the movements in the z-direction (heave) and more significant differences can be found in the rotations presented in Section 3.5.

3.5. Roll and Pitch Response

As shown in Section 3.3, the rotation around the z-axis is very small and is not significant for the analysis of the other degrees of freedom based on the rotationally symmetrical design of the floating cylinder. Figure 7 presents the results of the different stages of the sine fitting for the rotation around the x-axis (roll) and those around the y-axis (pitch) in Figure 8. The evaluation process is similar to the analysis of heave, which is described in Section 3.4. In these particular graphs, the measured unit [°] is chosen and frequency bands with a response smaller than 0.5° are marked grey.

Two specific characteristics can be observed for the pitch rotation in Figure 8. The water filled cylinder has a first response peak in the range of the natural frequency (Section 3.2, Figure 2), which has to be investigated in further tests. A second one is in the range of the double response frequency f_R based on the free oscillation test, which is in the same range as the f_R of the solid ballast option. This analysis is supported by Figure 10. The first sine fitting for the solid case shows a direct connection between the wave frequency and the response, but, for the water filled variation, the primary oscillation is connected with half of the f_W.

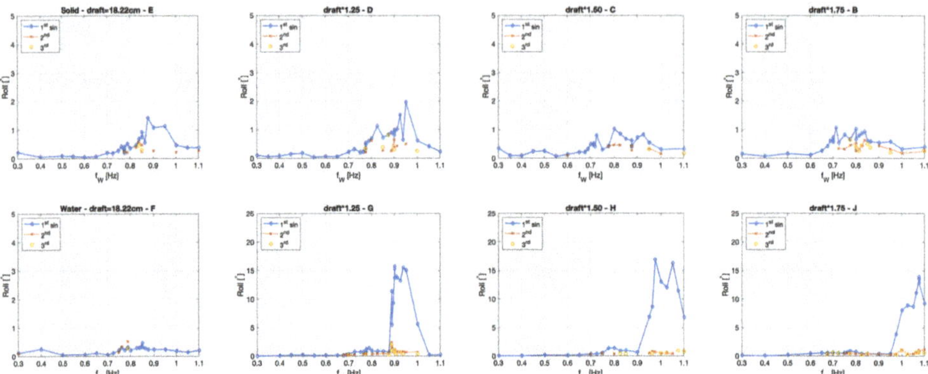

Figure 7. Amplitudes of the sine fitting for the rotation around x-direction (roll) in relation to the wave frequency f_W—upper row solid ballast option of the four different drafts and the corresponding water ballast in the lower row.

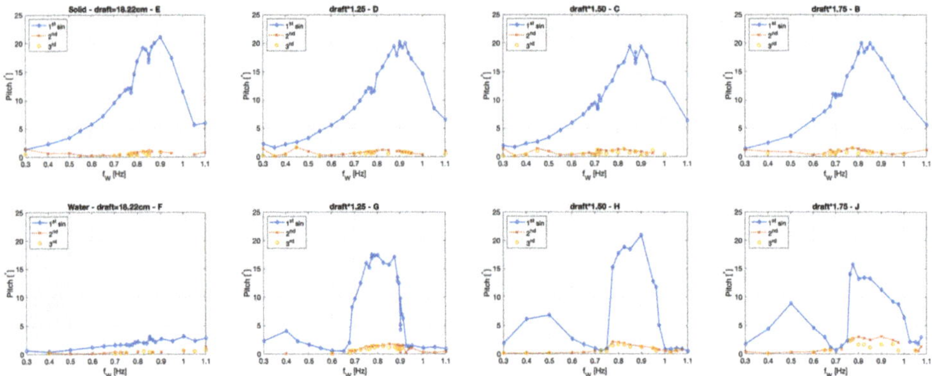

Figure 8. Amplitudes of the sine fitting for the rotation around y-direction (pitch) in relation to the wave frequency f_W—upper row solid ballast option of the four different drafts and the corresponding water filling in the lower row.

In most of the cases, the roll motion is comparably small (Figure 7), except for the three deepest draft water ballast cases. In these cases, the pitch motion of the water filled cylinder decreases significantly as the roll motion develops. Roll angles over 20° are observed. In the transient frequency zone, multiple switches between the two rotations direction can be observed. This indicates that the response caused by the incoming waves is not stable, clearly caused by the sloshing inside of the floating cylinder. This leads to an unstable phenomenon, which can be compared with parametric rolling of ships. Such a stability variation occurs based on the governing parameter of wave height and length as well as the ship geometry [68–70]. It can be reduced by changes of speed and direction to limit the loads in the securing system of container ships [71,72].

The analysis of the corresponding response frequency for roll (Figure 9) and pitch (Figure 10) indicates that the cylinder oscillates in both directions at half of the wave frequency f_W. France et al. [68] show that the parametric rolling in ships occurs in the ratio 2:1 of the period of pitch and roll, which is not the case for the tested floating cylinder. Consequently, the observed phenomenon is only similar to parametric rolling and more likely connected to a resonance to the natural response frequency found in the free oscillation (Section 3.2, Figure 2). Further investigations are needed to study this in detail.

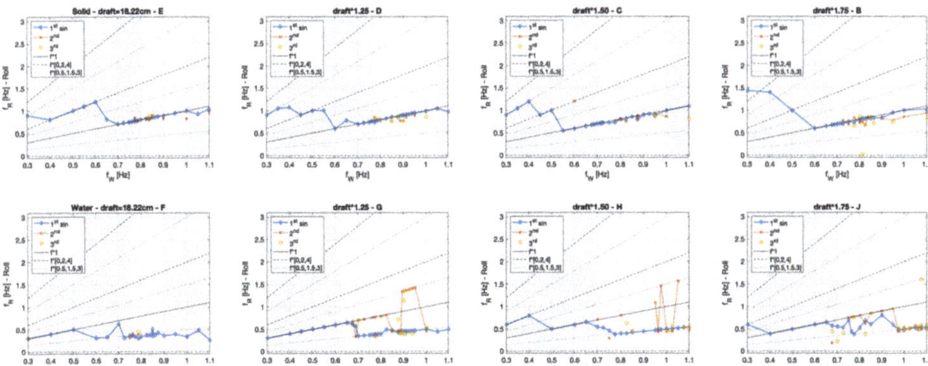

Figure 9. Response frequency f_R of the sine fitting for the rotation around the *x*-direction (roll) in relation to wave frequency f_W—upper row solid ballast option of the four different drafts and the corresponding water filling in the lower row.

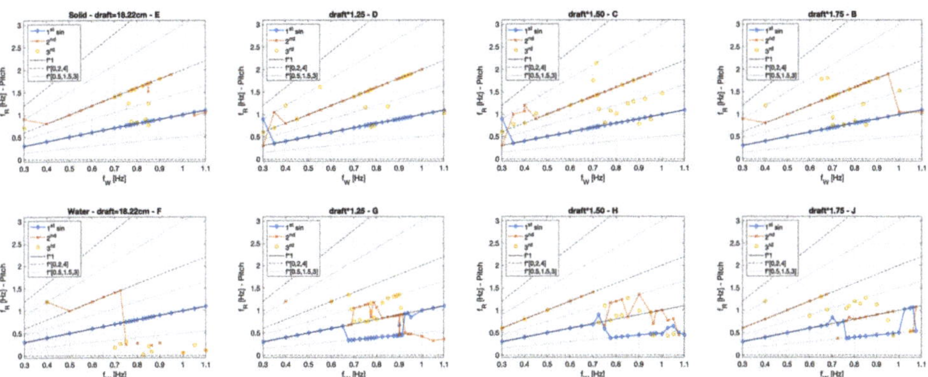

Figure 10. Response frequency f_R of the sine fitting for the rotation around the *y*-direction (pitch) in relation to wave frequency f_W—upper row solid ballast option of the four different drafts and the corresponding water filling in the lower row.

For the experimental set-up filled with the deepest inner water level h, and hence the biggest draft, this response is observed in a f_W-range near the absorption capacity of the wave tank. Consequently, the response is not as clear as it would be without this disturbance. Nevertheless, it is important to mention that this effect also occurs for all three deeper draft cases, but, for further investigations, smaller inside water levels are better suited for investigations in FloWave. In contrast to the small roll angles, the pitch results are significantly larger. Only the lowest water ballast draft configuration reaches a nearly stable condition in the incoming waves and shows nearly no motion. For all other investigated cases, the goal to investigate big motions are reached.

3.6. Summary and Comparison of the Two Different Ballast Options

In Sections 3.4 and 3.5, each of the four investigated drafts is presented separately. Figure 11 summarises the amplitudes of the first sine fitting for heave (Figure 5), roll (Figure 7) and pitch

(Figure 8). The splitting between the two ballast options, solid (upper row) and water, is maintained as well as the line colour of the Figures in Section 3.3.

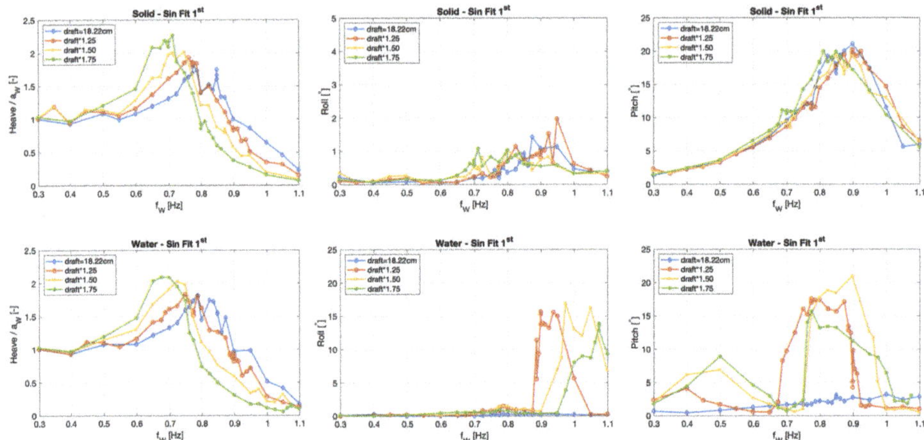

Figure 11. Summary of the fitting with sinusoidal functions for the movement in z-direction (heave normalised by the measured wave amplitude a_W) and rotation around x-(Roll) and y-axis (Pitch)—Solid (upper) and fluid (lower) case in relation to the initial wave frequency f_W.

The responding movement of the floating cylinder in the z-direction (heave) is for both ballast options directly connected to the wave frequency f_W of the tank. The peak values of the normalised response is in the range of more than double the incoming wave amplitude and increases with a bigger draft (distance between the outside bottom of the floating cylinder to the still water surface in the wave tank). The shift of the peak frequency is similar to the findings of the free oscillation experiments (a bigger draft results in smaller response frequency; Section 3.2 and Figure 2), which also shows no significant difference between the two ballast options. Comparable results are found by Xu et al. [32] for the side-by-side arrangement of a floating LNG tanker.

A different response to the incoming waves can be observed in the roll motion (rotation around the x-axis). Nearly no rolling motion occurs for the solid cases as well as the smallest draft for the liquid ballast configuration. In contrast to this, the three set-ups filled with water oscillate orthogonal to the incoming wave front instead of in the wave direction. This behaviour indicates the occurring of an instability phenomenon comparable to parametric rolling in ships [68]. The shift in the peaks of this roll motion also correspond with the change of the response frequency of the free oscillation (a deeper draft leads to a bigger frequency), but occurs at a wave frequency, which is approximately the doubled response frequency. In a transient phase, both rotations can be alternately observed. This effect is clearly initiated by the sloshing inner water body, which is stored inside of the floating cylinder. A further investigation of this phenomena has to be conducted, which includes a detailed monitoring of the inner water surface.

The set-up with the shallowest inner water level is the only one that doesn't respond with a significant pitch angle. For all other cases, values of over $20°$ are observed and therewith the requested big motions reached. Nearly no difference between the different draft variations are found for the solid ballast option. The three corresponding water filled set-ups reach nearly the same maximum level, but, in the higher frequency band, the rotation starts to switch to orthogonal direction. These three water filled cylinders also respond in an additional peak when the wave frequency is equal to the response frequency found in the free oscillation. The deeper draft corresponds with a higher frequency. This frequency band should be further investigated to clarify this effect.

4. Discussion

The main goal of this investigation is to gain an overview of the behaviour of water filled floating cylinders under wave conditions and large motions. Four different solid cases with a similar mass distribution are investigated and compared to the water filled variation.

A cylinder is chosen as a simplified geometry for this investigation. Keeping typical space discretisations in mind, this is not ideal for further comparison with numerical simulations. On the other hand, this approach reduces the influence of the orientation of the model to the incoming waves to a minimum as well as vortexes caused by sharp edges. From a hydrodynamic point of view, an addition of a hemisphere on the bottom of the cylinder would be a good addition, thereby avoiding the sharp edge between the side wall and the bottom surface. This option can be investigated in a further measurement campaign along with different cylinder diameters.

In nearly all cases, a stable oscillation could be observed for the response of the floating object. The single exception is the transient phase for the water filled variation, when the motion switches between pitch and roll. The rotational symmetry also allows the use of the post-processing features of the Qualisys system to evaluate pitch and roll motion independently of a rotation around the vertical z-axis. Nevertheless, the yaw angles are small.

An important point for a good comparison is the solid representation of the water. This approach should have the mass distribution but no influence caused by sloshing. The first approach considered was to fill the cylinder with a gravel mixture of an equal density. This would have the big advantage that a wider range of inner water levels could be investigated. On the downside, it is also possible that such a loose material could start to move inside the cylinder (even if this would be significantly smaller than the liquid case). An accidental pollution of the water in the wave tank is also possible, which has the potential for major disruption to FloWave's complex wave and current generation systems. Furthermore, it was considered that a mixture of the correct density would be hard to reproduce and guarantee that it stays homogeneous. Therefore, the option of building fixed ballast options based on iron blocks is chosen. The two crosses only allow a symmetrical distribution according the two main axes, but can be exchanged very quickly and repeatably. In addition, a cylindrical weight is used to combine the four different draft variations. The investigation showed no significant influence based on this assumption.

The positioning of the marker point on the top of the cylinder is not ideal considering the stability of the floating cylinder but allows for using the standard motion capturing system of FloWave. This measurement system further allows for reach the exact same inner water level, so that the draft is equal to the solid case. A difference smaller than 1 mm could be reached. A change of the global z-coordinate would also indicate that water flows outside or inside the cylinder. Despite very big rotations, no water exchange could be observed.

In the higher range of the wave frequency, the absorption capacity of the tank is limited and during the test period the reflections can reach the floating object. For the model set-up with the deepest draft filled with water, this has a negative effect on the peak roll motions, but, in all other cases, the influence is relatively small. In general, the chosen frequency band width triggers different responses including all the response frequencies found in the still water experiments.

The soft mooring system is in principal a good choice and with minimal influence on the response of the floating cylinder. The intention was to allow an undisturbed behaviour as long as possible to have a good data set to compare against numerical simulations, which could be reached by focusing on the ramp up time as well as the first oscillations [1]. The downside of the soft approach is that the final position of the model can vary over a wider range in the tank. This can potentially cause different influences due to the reflections as well as expands the needed ramp up time to reach a stable response of the floating structure. In hindsight, the pre-load of all mooring lines should have been increased to minimise the movement in the wave direction. This would further increase the comparability between the different ballast options.

As part of further experiments, the influence of the mooring system will be investigated using the full 360 wave production capability of FloWave. The measurement of the mooring forces would be an option as well as the expansion for irregular wave conditions. Furthermore, the documentation of the behaviour of the inner water body will be improved to allow measuring slamming load depending on the tank geometry [73,74], run-up heights on the walls [59] and detection of the occurrence of breaking waves inside the cylinder.

In the presented investigation, the phase shift between the waves and the response is not the focus of the analysis. In principle, a theoretical wave gauge could be calculated at the position of the floating cylinder and compared with the response. This will be done as part of the further investigations as well as additional wave gauges in a parallel line to the main axis, so that this calculation can be verified.

5. Conclusions

The paper presents the results of an investigation of a cylindrical floating object in a wave tank. Regular waves of 0.3 Hz to 1.1 Hz are applied to initiate the response oscillation of the moored floating cylinder. Four different drafts (distance between the bottom point and the still water surface) are tested with two different ballast options, solid and water. Based on the comparison between the liquid and solid, the influence of the sloshing of the stored water can be identified. For all three motions, the differences are relatively small. Due to symmetry, the rotation around the z-axis (vertical to the still water surface) is very small. The pitch motion is directly caused by the impacting wave, and large motions over $20°$ are reached. The biggest difference could be identified in the roll motion (rotation around the x-axis, which is parallel to the wave direction). These rotational motions are small except for the three deepest drafts for the water ballasted cylinders. In these cases, a change from pitch to roll occurs at a wave frequency in the range of the double response frequency found in the still water test. This is clearly caused be the sloshing of the inside water, and this unstable phenomenon will be further investigated. The full dataset is available based on Gabl et al. [1,2] and can be used as a validation experiment for numerical models to simulate such interactions as well as moving boundary conditions under large motions.

Author Contributions: R.G., T.D., E.N., J.S., and D.M.I. are responsible for the conceptualisation of the experimental investigation. R.G., T.D., and E.N. measured the data and analysed the data. R.G. and T.D. wrote the initial draft and E.N., J.S., and D.M.I. reviewed and edited the paper.

Funding: This work was supported by the Austrian Science Fund (FWF) under Grant J3918.

Acknowledgments: Open Access Funding by the Austrian Science Fund (FWF).

Abbreviations

d	cylinder draft (m)
D	cylinder diameter (m)
f	frequency (Hz)
f_R	frequency response (Hz)
f_W	frequency wave (Hz)
g	gravity acceleration (ms^{-2})
h	water depth inside the cylinder (m)
H	height of the cylinder (m)
x	main wave direction (m)
y	orthogonal to the main wave direction (m)
FFT	Fast Fourier Transformation
QTM	Qualisys Track Manager
RMSE	Root Mean Square Error
SPH	Smoothed Particle Hydrodynamics
VLFS	Very Large Floating Structures
WEC	Wave Energy Converter

References

1. Gabl, R.; Davey, T.; Nixon, E.; Steynor, J.; Ingram, D.M. Experimental Data of a Floating Cylinder in a Wave Tank: Comparison Solid and Water Ballast. *Data* **2019**, *4*, 146. [CrossRef]
2. Gabl, R.; Davey, T.; Nixon, E.; Steynor, J.; Ingram, D.M. *Experimental Data of a Floating Cylinder in a Wave Tank—Comparison Solid and Water Ballast*; DataShare Edinburgh [Dataset]; University of Edinburgh: Edinburgh, UK, 2019. [CrossRef]
3. Zheng, S.; Zhang, Y. Theoretical modelling of a new hybrid wave energy converter in regular waves. *Renew. Energy* **2018**, *128*, 125–141. [CrossRef]
4. Zeng, X.; Shi, M.; Huang, S. Hydrodynamic interactions of water waves with a group of independently oscillating truncated circular cylinders. *Acta Mech. Sin.* **2016**, *32*, 773–791. [CrossRef]
5. Chatjigeorgiou, I.K. Water wave trapping in a long array of bottomless circular cylinders. *Wave Motion* **2018**, *83*, 25–49. [CrossRef]
6. Chaplin, J.R.; Heller, V.; Farley, F.J.M.; Hearn, G.E.; Rainey, R.C.T. Laboratory testing the Anaconda. *Philos. Trans. R. Soc. A Math. Phys. Eng. Sci. R. Soc.* **2012**, *370*, 403–424.
7. Farley, F.J.M.; Rainey, R.C.T.; Chaplin, J.R. Rubber tubes in the sea. *Philos. Trans. R. Soc. A Math. Phys. Eng. Sci. R. Soc.* **2012**, *370*, 381–402. [CrossRef]
8. Klar, R.; Steidl, B.; Sant, T.; Aufleger, M.; Farrugia, R.N. Buoyant Energy—Balancing wind power and other renewables in Europe's oceans. *J. Energy Storage* **2017**, *14 Pt 2*, 246–255. [CrossRef]
9. Klar, R.; Steidl, B.; Aufleger, M. A floating energy storage system based on fabric. *Ocean Eng.* **2018**, *165*, 328–335. [CrossRef]
10. Ai, C.; Ma, Y.; Yuan, C.; Dong, G. Semi-implicit non-hydrostatic model for 2D nonlinear wave interaction with a floating/suspended structure. *Eur. J. Mech. B/Fluids* **2018**, *72*, 545–560. [CrossRef]
11. Cheng, L.; Lin, P. The num. modeling of coupled motions of a moored floating body in waves. *Water* **2018**, *10*, 1748. [CrossRef]
12. Kim, K.-T.; Lee, P.-S.; Park, K.C. A direct coupling method for 3D hydroelastic analysis of floating structures. *Int. J. Numer. Methods Eng.* **2013**, *96*, 842–866. [CrossRef]
13. Wei, W.; Fu, S.; Moan, T.; Lu, Z.; Deng, S. A discrete-modules-based frequency domain hydroelasticity method for floating structures in inhomogeneous sea conditions. *J. Fluids Struct.* **2017**, *74*, 321–339. [CrossRef]
14. Yoon, J.S.; Cho, S.P.; Jiwinangun, R.G.; Lee, P.S. Hydroelastic analysis of floating plates with multiple hinge connections in regular waves. *Mar. Struct.* **2014**, *36*, 65–87. [CrossRef]
15. Lee, K.H.; Cho, S.; Kim, K.T.; Kim, J.G.; Lee, P.S. Hydroelastic analysis of floating structures with liquid tanks and comparison with experimental tests. *Appl. Ocean Res.* **2015**, *52*, 167–187. [CrossRef]
16. Ariyarathne, K.; Chang, K.-A.; Mercier, R. Green water impact pressure on a three-dimensional model structure. *Exp. Fluids* **2012**, *53*, 1879–1894. [CrossRef]
17. Chuang, W.L.; Chang, K.A.; Mercier, R. Impact pressure and void fraction due to plunging breaking wave impact on a 2D TLP structure. *Exp. Fluids* **2017**, *58*, 68. [CrossRef]
18. Kim, S.Y.; Kim, K.H.; Kim, Y. Comparative study on pressure sensors for sloshing experiment. *Ocean Eng.* **2015**, *94*, 199–212. [CrossRef]
19. Chen, Y.; Xue, M.A. Numerical Simulation of Liquid Sloshing with Different Filling Levels Using OpenFOAM and Experimental Validation. *Water* **2018**, *10*, 1752. [CrossRef]
20. Cruchaga, M.A.; Reinoso, R.S.; Storti, M.A.; Celentano, D.J.; Tezduyar, T.E. Finite element computation and experimental validation of sloshing in rectangular tanks. *Comput. Mech.* **2013**, *52*, 1301–131. [CrossRef]
21. Caron, P.A.; Cruchaga, M.A.; Larreteguy, A.E. Study of 3D sloshing in a vertical cylindrical tank. *Phys. Fluid* **2018**, *30*, 082112. [CrossRef]
22. Hirdaris, S.E.; Bai, W.; Dessi, D.; Ergin, A.; Gu, X.; Hermundstad, O.A.; Huijsmans, R.; Iijima, K.; Nielsen, U.D.; Parunov, J.; et al. Loads for use in the design of ships and offshore structures. *Ocean Eng.* **2014**, *78*, 131–174. [CrossRef]
23. Jiao, J.; Ren, H.; Chen, C. Model testing for ship hydroelasticity: A review and future trends. *J. Shanghai Jiaotong Univ. (Sci.)* **2017**, *22*, 641–650. [CrossRef]
24. Sharma, R.; Kim, T.W.; Storch, R.L.; Hopman, H.; Erikstad, S.O. Challenges in computer applications for ship and floating structure design and analysis. *CAD Comput. Aided Des.* **2012**, *44*, 166–185. [CrossRef]

25. Bureau Veritas (BV). *Design Sloshing Loads for LNG Membrane Tanks—Guidance Note NI 554 DT R00 E*; Bureau Veritas: Paris, France, 2010.

26. Malenica, S.; Diebold, L; Kwon, S.H.; Cho, D.-S. Sloshing assessment of the LNG floating units with membrane type containment system where we are? *Mar. Struct.* **2017**, *56*, 99–116. [CrossRef]

27. Su, Y.; Liu, Z.Y. Coupling effects of barge motion and sloshing. *Ocean Eng.* **2017**, *140*, 352–360. [CrossRef]

28. Huang, S.; Duan, W.; Han, X.; Nicoll, R.; You, Y.G.; Sheng, S.W. Nonlinear analysis of sloshing and floating body coupled motion in the time-domain. *Ocean Eng.* **2018**, *164*, 350–366. [CrossRef]

29. Zhao, W.; McPhail, F. Roll response of an LNG carrier considering the liquid cargo flow. *Ocean Eng.* **2017**, *129*, 83–91. [CrossRef]

30. Zhao, W.; Taylor, P.H.; Wolgamot, H.A.; Taylor, R.E. Identifying linear and nonlinear coupling between fluid sloshing in tanks, roll of a barge and external free-surface waves. *J. Fluid Mech.* **2018**, *884*, 403–434. [CrossRef]

31. Zhao, W.; Milne, I. A.; Efthymiou, M.; Wolgamot, H.A.; Draper, S.; Taylor, P.H; Taylor, R.E. Current practice and research directions in hydrodynamics for FLNG-side-by-side offloading. *Ocean Eng.* **2018**, *158*, 99–110. [CrossRef]

32. Xu, Q.; Hu, Z.; Jiang, Z. Experimental investigation of sloshing effect on the hydrodynamic responses of an FLNG system during side-by-side operation. *Ships Offshore Struct.* **2018**, *12*, 804–817. [CrossRef]

33. Zhao, D.; Hu, Z.; Chen, G. Experimental investigation on dynamic responses of FLNG connection system during side-by-side offloading operation *Ocean Eng.* **2017**, *136*, 283–293. [CrossRef]

34. Zhao, D.; Hu, Z.; Zhou, K.; Chen, G.; Chen, X.; Feng, X. Coupled analysis of integrated dynamic responses of side-by-side offloading FLNG system. *Ocean Eng.* **2018**, *168*, 60–82. [CrossRef]

35. Zhao, M.; Cheng, L.; An, H.; Lu, L. Three-dimensional numerical simulation of vortex-induced vibration of an elastically mounted rigid circular cylinder in steady current. *J. Fluids Struct.* **2014**, *50*, 292–311. [CrossRef]

36. Kang, Z.; Ni, W.; Zhang, L.; Ma, G. An experimental study on vortex induced motion of a tethered cylinder in uniform flow. *Ocean Eng.* **2017**, *142*, 259–267. [CrossRef]

37. Assi, G.; Bearman, P. Vortex-induced vibration of a wavy elliptic cylinder. *J. Fluids Struct.* **2018**, *80*, 1–21. [CrossRef]

38. Zhu, H.; Zhao, Y.; Zhou, T. Numerical investigation of the vortex-induced vibration of an elliptic cylinder free-to-rotate about its center. *J. Fluids Struct.* **2018**, *83*, 133–155. [CrossRef]

39. Zhu, H.; Gao, Y. Hydrokinetic energy harvesting from flow-induced vibration of a circular cylinder with two symmetrical fin-shaped strips. *Energy* **2018**, *165*, 1259–1281. [CrossRef]

40. Jamalabadi, M.Y.A.; Ho-Huu, V.; Nguyen, T.K. Optimal Design of Circular Baffles on Sloshing in a Rectangular Tank Horizontally Coupled by Structure. *Water* **2018**, *10*, 1504. [CrossRef]

41. Yang, Y.; Zhu, R.; Chen, X.; Jiang, Y. Numerical study on fluid resonance of 3D multi-bodies by a non-reflection numerical wave tank. *Appl. Ocean Res.* **2018**, *80*, 166–180. [CrossRef]

42. Sodja, J.; De Breuker, R.; Nozak, D.; Drazumeric, R.; Marzocca, P. Assessment of low-fidelity fluid-structure interaction model for flexible propeller blades. *Aerosp. Sci. Technol.* **2018**, *71*, 71–88. [CrossRef]

43. Dagli, B.Y.; Tuskan, Y.; Gokkus, U. Evaluation of Offshore Wind Turbine Tower Dynamics with Numerical Analysis. *Adv. Civ. Eng.* **2018**, *2018*, 3054851. [CrossRef]

44. O'Connell, K.; Thiebaut, F.; Kelly, G.; Cashman, A. Development of a free heaving OWC model with nonlinear PTO interaction. *Renew. Energy* **2018**, *117*, 108–115. [CrossRef]

45. Sjökvist, L.; Wu, J.; Ransley, E.; Engström, J.; Eriksson, M.; Göteman, M. Numerical models for the motion and forces of point-absorbing wave energy converters in extreme waves. *Ocean Eng.* **2017**, *145*, 1–14. [CrossRef]

46. Li, X.; Xu, L.; Yang, J. Study of fluid resonance between two side-by-side floating barges. *J. Hydrodyn.* **2016**, *28*, 767–777. [CrossRef]

47. Tsai, C.P.; Chen, Y.-C.; Sihombing, T. O.; Lin, C. Simulations of moving effect of coastal vegetation on tsunami damping. *Nat. Hazards Earth Syst. Sci.* **2017**, *17*, 693–702. [CrossRef]

48. Gabl, R.; Gems, B.; De Cesare, G.; Aufleger, M. Contribution to Quality Standards for 3D-Numerical Simulations with FLOW-3D (Anregungen zur Qualitätssicherung in der 3D-numerischen Modellierung mit FLOW-3D). *WasserWirtschaft* **2014**, *3*, 15–20. [CrossRef]

49. Wang, D.; Shao, S.; Li, S.; Shi, Y.; Arikawa, T. 3D ISPH erosion model for flow passing a vertical cylinder. *J. Fluids Struct.* **2018**, *78*, 374–399. [CrossRef]

50. Ardianti, A.; Mutsuda, H.; Kawawaki, K.; Doi, Y. Fluid structure interactions between floating debris and tsunami shelter with elastic mooring caused by run-up tsunami. *Coast. Eng.* **2018**, *137*, 120–132. [CrossRef]

51. Kanehira, T.; Mutsuda, H.; Doi, Y.; Taniguchi, N.; Draycott, S; Ingram, D. Development and experimental validation of a multidirectional circular wave basin using smoothed particle hydrodynamics. *Coast. Eng. J.* **2019**, *61*, 109–120. [CrossRef]

52. Verbrugghe, T.; Manuel Dominguez, J.; Crespo, A.J. C.; Altomare, C.; Stratigaki, V.; Troch, P.; Kortenhaus, A. Coupling methodology for smoothed particle hydrodynamics modelling of nonlinear wave-structure interactions. *Coast. Eng.* **2018**, *138*, 184–198. [CrossRef]

53. Gunn, D.F.; Rudman, M.; Cohen, R.C.Z. Wave interaction with a tethered buoy: SPH simulation and experimental validation. *Ocean Eng.* **2018**, *156*, 306–317. [CrossRef]

54. Madhi, F.; Yeung, R.W. On survivability of asymmetric wave-energy converters in extremewaves. *Renew. Energy* **2018**, *119*, 891–909. [CrossRef]

55. Cao, X.Y.; Ming, F.R.; Zhang, A.M.; Tao, L. Multi-phase SPH modelling of air effect on the dynamic flooding of a damaged cabin. *Comput. Fluids* **2018**, *163*, 7–19. [CrossRef]

56. Guo, K.; Sun, P.-N.; Cao, X.-Y.; Huang, X. A 3D SPH model for simulating water flooding of a damaged floating structure. *J. Hydrodyn.* **2017**, *29*, 831–844. [CrossRef]

57. Ming, F.R.; Zhang, A.M.; Cheng, H.; Sun, P.N. Numerical simulation of a damaged ship cabin flooding in transversal waves T with Smoothed Particle Hydrodynamics method. *Ocean Eng.* **2018**, *165*, 336–352. [CrossRef]

58. Cheng, H.; Zhang, A.M.; Ming, F.R. Study on coupled dynamics of ship and flooding water based on experimental and SPH methods. *Phys. Fluids* **2017**, *29*, 107101. [CrossRef]

59. Gabl, R.; Steynor, J.; Forehand, D.I.M.; Davey, T.; Bruce, T.; Ingram, D.M. Capturing the motion of the free Surface of a fluid stored within a floating structure. *Water* **2019**, *11*, 50. [CrossRef]

60. Draycott, S.; Noble, D.; Davey, T.; Bruce, T.; Ingram, D.; Johanning, L.; Smith, H.; Day, A.; Kaklis, P. Re-creation of site-specific multi-directional waves with non-collinear current. *Ocean Eng.* **2018**, *152*, 391–403. [CrossRef]

61. Draycott, S.; Davey, T.; Ingram, D. Simulating extreme directional wave conditions. *Energies* **2017**, *10*, 1731. [CrossRef]

62. Sutherland, D.R.J.; Noble, D.R.; Steynor, J.; Davey, T.; Bruce, T. Characterisation of current and turbulence in the FloWave Ocean Energy Research Facility. *Ocean Eng.* **2017**,*139*, 103–115. [CrossRef]

63. Draycott, S.; Sellar, B.; Davey, T.; Noble, D.R.; Venugopal, V.; Ingram, D. Capture and Simulation of the Ocean Environment for Offshore Renewable Energy. *Renew. Sust. Energ. Rev.* **2019**, *104*, 15–29. [CrossRef]

64. Gomit, G.; Chatellier, L.; Calluaud, D.; David, L. Free surface measurement by stereo-refraction. *Exp. Fluids* **2013**, *54*, 1540. [CrossRef]

65. MARINET Work Package 2: Standards and Best Practice–D2.1 Wave Instrumentation Database. Revision: 05. 2012. Available online: http://www.marinet2.eu/wp-content/uploads/2017/04/D2.01-Wave-Instrumentation-Database.pdf (accessed on 22 November 2019).

66. Zhao, W.; McPhail, F.; Efthymiou, M. Effect of partially filled spherical cargo tanks on the roll response of a bargelike vessel. *J. Offshore Mech. Arctic Engi.* **2016**, *138*, 031601. [CrossRef]

67. Draycott, S.; Payne, G.; Steynor, J.; Nambiar, A.; Sellar, B.; Venugopal, V. An experimental investigation into nonlinear wave loading on horizontal axis tidal turbines. *J. Fluids Struct.* **2019**, *84*, 199–217. [CrossRef]

68. France, W.; Levadou, M.; Treakle, T.; Randolph-Paulling, J.; Keith-Michel, R.; Moore, C. An investigation of head-sea parametric rolling and its influence on container lashing systems. *Mar. Technol.* **2003**, *40*, 1–19.

69. Ma, S.; Ge, W.; Ertekin, R.C.; He, Q.; Duan, W.Y. Experimental and Numerical Investigations of Ship Parametric Rolling in Regular Head Waves. *China Ocean Eng.* **2018**, *32*, 431–442. [CrossRef]

70. Kianejad, S.S.; Enshaei, H.; Duffy, J.; Ansarifard, N. Prediction of a ship roll added mass moment of inertia using numerical simulation. *Ocean Eng.* **2019**, *173*, 77–89. [CrossRef]

71. Schumacher, A.; Ribeiro e Silva, S.; Guedes Soares, C. Experimental and numerical study of a containership under parametric rolling conditions in waves. *Ocean Eng.* **2016**, *124*, 385–403. [CrossRef]

72. Lin, Y.; Ma, N.; Gu, X. Approximation Method for Flare Slamming Analysis of Large Container Ships in Parametric Rolling Conditions. *J. Mar. Sci. Appl.* **2018**, *17*, 406–413. [CrossRef]

73. Xue, M. A.; Zheng, J.; Lin, P.; Yuan, X. Experimental study on vertical baffles of different configurations in suppressing sloshing pressure. *Ocean Eng.* **2017**, *136*, 178–189. [CrossRef]
74. Xue, M. A.; Chen, Y.; Zheng, J.; Qian, L.; Yuan, X. Fluid dynamics analysis of sloshing pressure distribution in storage vessels of different shapes. *Ocean Eng.* **2019**, *192*, 106582. [CrossRef]

Article

Flow Depths and Velocities across a Smooth Dike Crest

Sara Mizar Formentin *, Maria Gabriella Gaeta, Giuseppina Palma, Barbara Zanuttigh and Massimo Guerrero

Department of Civil, Chemical, Environmental and Materials Engineering, University of Bologna, Viale del Risorgimento 2, 40136 Bologna, Italy; g.gaeta@unibo.it (M.G.G.); giuseppina.palma2@unibo.it (G.P.); barbara.zanuttigh@unibo.it (B.Z.); massimo.guerrero@unibo.it (M.G.)
* Correspondence: saramizar.formentin2@unibo.it; Tel.: +39-051-2093661

Received: 19 July 2019; Accepted: 17 October 2019; Published: 22 October 2019

Abstract: This contribution proposes a systematic analysis of the overtopping process at dikes, focused on the statistical description of the extreme flow characteristics across the dike crest. The specific objective of the analysis is the investigation of structures subjected to high run-up levels and low freeboards, under severe or extreme conditions that are likely to occur in the future due to climate change. The adopted methodology is based on the collection of new experimental and numerical tests of wave overtopping at smooth dikes at various crest levels. The reliability of the new data is checked in terms of average overtopping discharge and wave reflection coefficient, against consolidated predicting methods from the literature. An update and refitting of the existing formulae for the prediction of the extreme flow depths and velocities at the dike off-shore edge is proposed based on the experimental and numerical outcomes. The dynamics of the overtopping flow propagation along the dike crest under breaking and non-breaking waves, in emerged and submerged conditions, is investigated. Guidelines to update the state-of-the-art formulae for a more cautious estimation of the water depths and the velocities of propagation of the flow in the landward area are provided.

Keywords: wave overtopping; flow velocity; flow depth; dike; wave breaking; experiments; numerical modelling

---

1. Introduction

For design purposes, the characterization of the overtopping flow for a dike crest represents a key information for the assessment of the hydraulic and geotechnical vulnerability of coastal structures and of landward areas. Specifically, the accurate estimation of the extreme depths, velocities and volumes during overtopping events is essential for the assessment: (i) of the critical velocity for the infiltration rate at the crest and the erosion threshold at the landward slope [1]; (ii) of the seaward hydraulic loads to model the wave overtopping propagation; e.g., in case of the wave overtopping simulator [2] and flooding scenarios in the landward areas; and (iii) of the local scour due to the impinging of the breaking, overtopping waves at the toe of the landward slope [3].

The first conceptual model of the wave run-up and overtopping of an emerged smooth dike was developed by Schüttrumpf [4], who delivered the expressions of the flow depths (h) and velocities (u) at the seaward slope, along the structure crest and at the landward slope. The proposed expressions were then fitted by other authors based on physical model tests that highlighted the effects of the seaward slope [1,5–8]. According to that model, the seaward and landward values of h and u are functions of the wave run-up, of the crest freeboard and of the seaward slope. Both h and u exponentially decay along the dike crest due to frictional losses and overtopping volume deformation [9]. Recently, an analytical model to predict h and u along the dike crest and the landward slope, offering two new decay functions for h and u over the crest, was proposed by Van Bergeijk et al. [10]. Both the studies [9,10] indicate that the entity of the decay increases with increasing the roughness and the width of the dike crest.

The analysis of the overtopping flow characteristics was recently extended to rubble-mound breakwaters [11]. This recently published study focused on extreme events and provided new formulae for h and u based on small-scale physical tests and punctual measurements of the overtopping flow depths at the middle of the dike crest and of the overtopping flow velocities from micro-propellers at the edges and in the middle of the dike crest.

The research on h and u in the case of over-washed dikes and levees started with Hughes and Nadal [12], who performed a series of small-scale tests on a smooth over-washed dike with a single seaward slope. They refined the representation of the overtopping and overflow components of the average wave overtopping discharge q after EurOtop [13] and delivered two modified equations for the landward mean flow thickness and velocity as functions of q.

References [14] and [15] described the trends of h and u along the crest of low-crested and submerged dikes by means of numerical simulations with the 2DV RANS-VOF code by the University of Cantabria [16]. The authors provided a sensitivity analysis to selected key structural (seaward slope and crest freeboard) and hydraulic (wave steepness) parameters. For overtopped structures in emerged conditions, a slightly increasing trend of the flow velocities along the crest was found, in contrast with the existing formulae.

Guo et al. [17] numerically investigated the effects of the wave breaking on the overtopping flow describing a relationship among the values of q and the maximum u over the dike crest. They derived a decreasing and an increasing trend of the layer thickness and of the overtopping velocities, respectively, over the crest. Their analyses were limited to a single geometry (i.e., a trapezoidal dike with 1:3 off-shore slope) in emerged conditions under regular waves.

The goal of this paper is to systematically describe the wave overtopping process in terms of depths and velocities across a smooth dike at any crest level, providing a verification, and where required, a revision or an update of the existing formulae. Specific attention is paid to the analysis of the overtopping at low-crest freeboards to extend the existing literature—which is targeted to emerged structures only—to dikes working in over-washed or breached conditions, consequently to the effects of sea-level rise and of increased storm frequency and intensity in a climate-change scenario.

Toward that purpose, new laboratory experiments of wave run-up and overtopping against dikes were carried out in the wave flume of the University of Bologna. The overtopping values of h and u values were obtained from the installation of ultrasonic velocity profilers (UVPs) over the dike crest, providing for the first time, the vertical profiles of h and u instead of punctual measurements from micro-propellers. The use of a full-HD camera was also introduced to record the overtopping processes during the experiments and derive estimates of the flow depths' envelopes over the dike crest.

The laboratory experience was integrated with new numerical modelling. The original numerical database collected by the authors [15] was extended with a series of new simulations to include a wider variety of wave attacks and structure crest heights representing overtopped, over-washed and breached dikes. The new simulations were carried out with a modified version [18] of the original RANS-VOF code [16], which includes the possibility to model the over-washed conditions and the propagation of the overtopping flow over the landward slope. Overall, the new data collected consists of 60 experiments and 94 numerical simulations. For each test, the following quantities were measured or evaluated: q, the wave reflection coefficient; K_r, the instantaneous values of the overtopping flow depths h; and velocities u at different positions from the off-shore to the in-shore edge of the dike crest. The new data of h and u were used to derive a statistical representation of the overtopping flow characteristics over the structure crest.

The contribution is organized as follows. Section 2 includes first a short description of the new experimental and numerical data and of the setups adopted to carry out the experiments and the numerical modelling. The new data are then checked against literature formulae in Section 3. The validation of the numerical code is presented in Section 4 by reproducing a subset of the experimental tests. Section 5 focuses on the analysis of h and u at the beginning of the dike crest. The new data are herein compared with the commonly adopted formulae available from the literature [1,5,8] and a new, more cautious fitting for the prediction of the extreme flow depths and velocities at the dike off-shore edge is proposed. Section 6 is dedicated to the evolution of the flow characteristics along the dike crest. The punctual values of h and u were compared to the theoretical trends proposed by Schüttrumpf and Oumeraci [7], leading to an update and extension of the existing approach. Some conclusions on the relevance of these results for design purposes are drawn in Section 7.

2. Characterization of the Database

The database consists of new experimental and numerical data on wave run-up and wave overtopping at dikes. Sections 2.1 and 2.2 describe the main characteristics and the setup of the numerical and experimental tests, respectively. Section 2.3 illustrates the adopted methodologies to elaborate the new data and derive the statistics of the overtopping flow depths and velocities occurring at the dike off-shore and in-shore edges and along the dike crest. A few remarks and warnings about model and scale effects related to the experimental modelling are given in Section 2.4.

2.1. Numerical Setup and Tested Conditions

The numerical database employed in this contribution was collected by Formentin et al. (2014) [15] and updated by Formentin et al. (2018) [18]. It consists of 94 simulations of wave overtopping against smooth trapezoidal structures in a 2D numerical flume, 43 m long and 2.0 m deep. Irregular waves were generated in all the simulations, characterized by Jonswap spectrum, with a peak enhancement factor $\gamma = 3.3$. The variety of the tested conditions (reported in Table 1) included two significant wave heights H_s (0.1 or 0.2 m) and various peak periods T_p (ranging approximately from 1.3 to 6.5 s) specifically identified to test values of the wave steepness $H_s/L_{m-1,0}$ between 0.02 and 0.05, where $L_{m-1,0}$ is the wave length computed from the spectral wave period $T_{m-1,0}$.

Table 1. Synthesis of the range of configurations tested in the numerical model. The "dry" landward conditions were carried out with $H_s = 0.2$ m and $H_s/L_{m-1,0,t} = 0.02$ and 0.03 only.

R_c/H_s	−1.5	−1	−0.5	−0.2	0	+0.5	+1	+1.5
$H_s/L_{m-1,0,t}$	0.02; 0.03	0.02; 0.03; 0.04	0.02; 0.03; 0.05	0.02	0.02; 0.03; 0.04; 0.05	0.02; 0.03; 0.04	0.02; 0.03; 0.05	0.03
H_s (m)	0.1; 0.2	0.1; 0.2	0.1; 0.2	0.2	0.1; 0.2	0.1; 0.2	0.2	0.2
$\cot(\alpha_{off})$	4; 6	4; 6	4; 6	4; 6	4; 6	4; 6	4; 6	4; 6
$\cot(\alpha_{in})$	3	3	3	3	3	3	3	3
Landward	wet; dry	wet; dry	wet; dry	dry	wet; dry	wet; dry	dry	dry

Figure 1 displays the layout of the numerical flume, with reference to the adopted symbols. The structures consisted of smooth dikes with varying seaward slope ($\cot(\alpha_{off}) = 4$ or 6) and fixed landward slope ($\cot(\alpha_{off}) = 3$). The crest width (G_c) and the structure height (h_c) were kept constant and respectively equal to 0.3 m and 0.85 m. The seven crest freeboard values (R_c), varying with the still water depth (wd), were selected to test positive, zero and negative freeboard conditions. The geometrical features and the wave characteristics were defined starting from the experiments on wave overtopping at dikes carried out in the small wave flume of the Leichtweiss Institute for Hydraulic Engineering of the Technical University of Braunschweig in Germany [4] and in the flume of WL|Delft Hydraulics in The Netherlands [5]. Those experiments represent, indeed, the basis of most of the literature studies aimed at characterizing the overtopping flow over the dike slopes and crest.

The structure off-shore edge was placed at a distance of 39 m from the numerical wave generator (see Figure 1). Following Van Gent [5], a 1:100 sloping foreshore was located in front of the structure to allow wave generation in deep water. A free outflow boundary condition was set at the right boundary of the numerical flume. The water depth landward the structure was either set to zero, "dry" conditions, or equal to the seaward water depth *wd*, "wet" conditions. The dry condition was used to represent the case of over-washed dikes or levees, while the wet condition represents the case of fully breached or submerged dikes. The first case (dry) represents, therefore, ordinary conditions in a climate change scenario, while the second one (wet) represents extreme, catastrophic conditions. The numerical modelling of the dry condition was achieved [18] modifying the right boundary condition of the original IH-2VOF numerical code [16], which was conceived to work with a constant water depth across the whole channel. The two different landward conditions, wet and dry, determine completely different processes of wave overtopping, and in turn, different values and evolution trends of the overtopping flow characteristics along the structure crest.

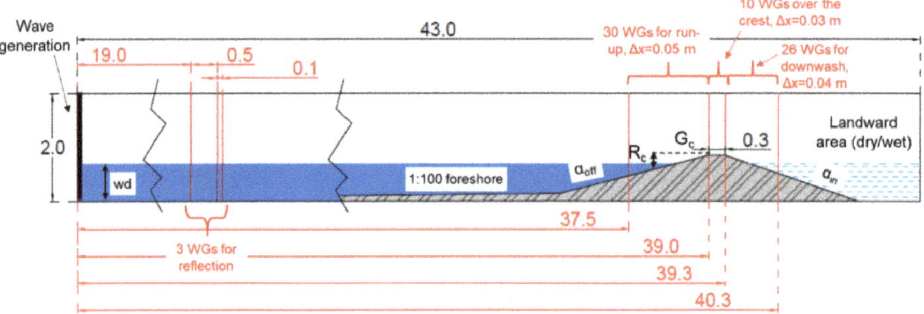

Figure 1. Layout of the numerical flume, including the dike's cross-section and the position of the wage gauges (wgs), and references to the adopted symbols. Measures in m.

A constant mesh of $\Delta z = 0.01$ m was adopted to discretize the whole computational domain in the vertical direction (*z*, hereinafter), while a variable mesh was chosen for the cross-shore direction (*x*, hereinafter). The *x*-mesh was set up by keeping the it fine and constant $\Delta x = 0.01$ m in the area around the structure and by gradually increasing the grid size towards the right and the left boundary sections up to $\Delta x \approx 0.02$ and 0.04 m, respectively.

For all the simulations, 69 wave gauges (wgs, hereinafter) were installed in the numerical flume to record the time-series of the free surface elevations *h*, the pressures *p* and the cross-shore directed flow velocities *u* in front of, over and behind the structure. In particular:

- Three wgs were placed at approximately 19 m from the wave generator to estimate the wave reflection coefficient K_r based on the methodology by Zedlt and Skjelbreia [19];
- Thirty and 26 wgs were placed, respectively, on the off-shore slope (between 37.5 and 39 m from the wave generator, interspaced with a uniform interval of $\Delta x = 0.05$ m) and on the in-shore slope (from 39.3 to 40.3 m, with $\Delta x = 0.04$ m); these gauges were installed for analyzing the wave run-up and the wave overtopping and calculating the wave transmission coefficient;
- Eleven wgs ($\Delta x = 0.03$ m) were placed across the dike crest, from 39 to 39.3 m, to characterize the flow over the crest.

2.2. Experimental Setup and Tested Conditions

Sixty new experiments on wave overtopping at dikes were carried out in the wave flume of the Hydraulics Laboratory of the University of Bologna. The wave flume—displayed schematically in Figure 2a—is 12 m long, 0.5 m wide and 1.0 m deep and the waves are generated by the vertical movement of a cuneiform-shaped piston-type wave-maker under the control of the mass conservation law [20,21]. The maximum wave height and length are, respectively, $H_s = 0.06$ m and $L_{m-1,0} = 3$ m, while the water depth *wd* before the wave-maker may be at maximum of about 0.4 m.

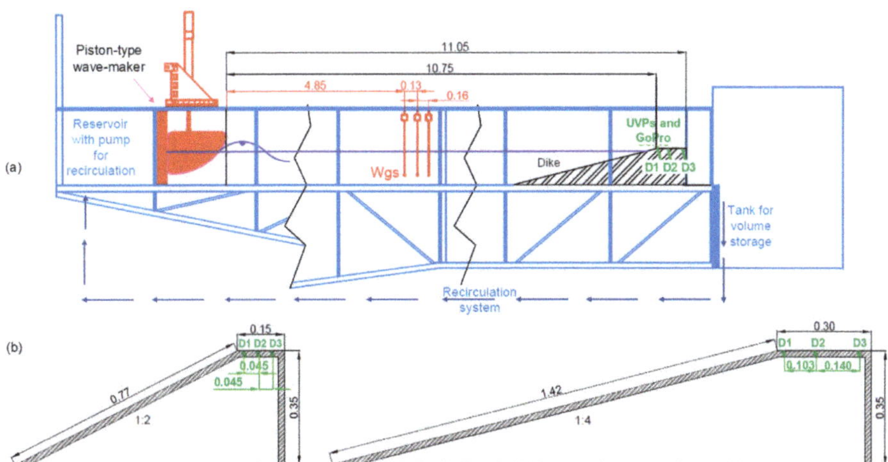

Figure 2. Panel (**a**): scheme of the wave flume and layout of the equipment. Panel (**b**): cross-sections of the dikes (slopes $\cot(\alpha_{off}) = 2$ and 4, $G_c = 0.15$ and 0.3 m) with reference to the position of the ultrasonic velocity profilers (UVPs) along the dike crest (D1, D2 and D3). Measures in m.

The setup of the lab experiments was based on the range of configurations and wave attacks tested with the numerical code, aiming at verifying and extending the numerical experience. Subject to the size of the wave flume of the laboratory and the constraints of the wave-maker, the experiments were conducted in 1:2 scale with respect to the numerical model scale. They consisted of irregular waves characterized by Jonswap spectrum with a peak enhancement factor $\gamma = 3.3$, wave heights H_s in the range 0.04–0.06 m and wave periods $T_p \approx 0.85$–1.51 s, realizing values of $H_s/L_{m-1,0} \approx 0.03$–0.04. The wave attacks were performed against four dikes, whose cross-sections differed for the crest widths, $G_c = 0.15$ and 0.3 m, and/or the off-shore slope, $\cot(\alpha_{off}) = 2$ and 4. Figure 2b shows two out of the four dike configurations, where $\cot(\alpha_{off}) = 2$ is combined with $G_c = 0.15$ m (left) and $\cot(\alpha_{off}) = 4$ is combined with $G_c = 0.30$ m (right). The remaining two configurations (not shown for sake of brevity) were realized by combining $\cot(\alpha_{off}) = 2$ with $G_c = 0.30$ m and $\cot(\alpha_{off}) = 4$ with $G_c = 0.15$ m. All the structures were 0.35 m high (see Figure 2b) and were positioned in the wave-flume so that the spot of the dike's off-shore crest edge was always at the same distance of 10.75 m from the wave-maker, as indicated in Figure 2a. Each dike configuration was tested with three crest-freeboard conditions, $R_c/H_s = 0$, 0.5 and 1, obtained by varying the water depth wd in the channel from 0.29 to 0.35 m. For each freeboard, several wave attacks were performed, by varying the values of H_s and T_p, resulting in a total of 60 tests. All the experiments were performed in dry landward conditions.

A specific label (ID) was defined to identify each test performed. This ID consists of 12 alphanumeric symbols forming five groups of chars, e.g., "R00-H04-s4-G15-c2," where:

- The 1st group includes three symbols, which may be "R00," "R05" or "R10," and refers to the target value of R_c/H_s (R00 stands for $R_c/H_s = 0$; R05 stands for $R_c/H_s = 0.5$ and R10 stands for $R_c/H_s = 1.0$);
- The 2nd group may be "H04," "H05" or "H06" and each represents the target H_s value (respectively 0.04, 0.05 and 0.06 m);
- The 3rd group includes "s3" or "s4" and refers to the target wave steepness $s_{m-1,0} = H_s/L_{m-1,0} = 0.03$ or 0.04;
- The 4th group is "G15" or "G30," which refer respectively to $G_c = 0.15$ or 0.3 m;
- The 5th group is "c2" or "c4" and refers to $\cot(\alpha_{off}) = 2$ or 4, respectively.

The summary of the tested configurations in the laboratory is given in Table 2.

Table 2. Summary of the target conditions of the 60 experiments of wave overtopping at dikes performed in the Laboratory of Bologna.

R_c/H_s	0	+0.5	+1
$s_{m-1,0}$ (%)	3; 4	3; 4	3; 4
H_s (m)	0.04; 0.05; 0.06	0.04; 0.05; 0.06	0.05; 0.06
wd (m)	0.35	[0.32; 0.325]	[0.29; 0.30]
$\cot(\alpha_{off})$	2; 4	2; 4	2; 4
G_c (m)	0.15; 0.30	0.15; 0.30	0.15; 0.30
Landward	dry	dry	dry
Tot. #	24	18	18

As shown in Figure 2a, the wave-flume was equipped with the following instruments:

- Three wgs, placed at approximately 1.5 times the maximum $L_{m-1,0}$ from the wave-maker ($\approx$5 m) to record the free-surface elevation with a sampling frequency of 100 Hz and separate the incident and reflected waves; the positions of the wgs are displayed in Figure 2a in red color.
- Three ultrasonic Doppler velocity profilers (UVPs), which were installed along the structure crest and were used to record the time series of the vertical profiles of the horizontal flow velocities u and track the free surface elevation h. The positions of the three UVPs, shown in Figure 2b in green color and referenced as D1, D2 and D3, were selected to reconstruct the statistics of h and u in proximity of the dike crest off-shore edge (D1), in the middle of the crest (D2) and close to the in-shore edge (D3).
- A tank for the storage and the measurement of the overtopping volumes, placed at the end of the wave flume and below the channel and connected to recirculation system, regulated by a flowmeter (precision $q = 1 \times 10^{-5}$ m^3/s), which collects the overtopped water from the tank and brings it back to the reservoir placed upstream the channel.
- A 30 Hz full-HD camera to film the wave run-up and overtopping process; the camera was installed in front of the channel and corresponding to the upper part of the dike slope and the dike crest.
- All the structures were realized in a very smooth plexiglas material, which can be characterized by a roughness factor of $\gamma_f = 1$ [13].

2.3. Methodology for the Reconstruction of the Overtopping Flow Characteristics

For the numerical tests, the instantaneous values of h and u were directly derived from the time series recorded at the 10 numerical wgs placed along the structure crest (see Figure 1), while for the laboratory experiments, they were derived from the three UVPs measurements of the particle velocities and echoes. The monostatic sonar emits and receives short bursts composed by acoustic pulses to measure profiles of flow velocity at the probe location [22] by exploiting the Doppler effect. The following procedure was adopted.

- u. At each time step, the UVPs recorded a vertical profile of the radial velocity along the acoustic beam that was forming a 15°-angle (i.e., the Doppler angle) with the horizontal crest of the dike. The actual velocity was then reconstructed by assuming horizontal vectors (i.e., aligned flow with the dike crest). The range of the measured velocities depended on the settings of the probes; i.e., emitting frequency, pulse repetition period, the beam width and the Doppler angle. In the present experiments, the UVP settings yielded ~10 cm as the maximum layer thickness (as expected to occur above the dike crest during the largest wave attacks with $H_s = 0.06$ m), with a profile resolution of 1.01 mm, and 4.2 m/s as the maximum velocity, with a nominal accuracy of 0.1% depending on the developing turbulence patterns in the test.
- h. At each time step and with the same spatial resolution of the flow velocities, the UVPs also recorded the vertical profiles of echo (dB) of the acoustic impulse reflected by the particles

transported by the water. In accordance with the free surface, the acoustic impulse undergoes a strong reflection induced by the density variation, which determines, in turns, a sharp peak of the echo value. The time series of the free surface elevations, and of the layer thicknesses h, were reconstructed based on the peak position in the instantaneous vertical profiles of the echo.

For both the numerical simulations and the laboratory experiments, the instantaneous depth-averaged values of u were calculated from the vertical profiles. The time series of h and u were treated as stochastic variables, calculating the following statistics: mean, standard deviation, median, minimum, maximum and upper 2% percentiles (namely, $h_{2\%}$ and $u_{2\%}$). Based on the literature [5], these latter quantities were, respectively, the values of h and u exceeded by 2% of the incoming waves and were used by several authors [1,4,5,8] to characterize the extreme overtopping flow.

The records of the tests obtained with the camera were also elaborated to derive the envelopes of the free-surface elevations through the implementation of ad hoc image processing algorithms. In this study, a few results of the envelopes are presented as a validation of the numerical code. More details about this technique and the results obtained are given in reference [23].

2.4. Scale and Model Effects

Neither the experimental nor the numerical tests refer to real prototype structures, but the setup of the hydraulic and structural parameters aimed at represent "realistic" conditions. In particular, the whole numerical and experimental campaign was inspired to the experiments on dikes carried out in 1:10 scale by Schüttrumpf and Oumeraci [7]. Since the same structural dimensions and the same wave conditions tested by Schüttrumpf and Oumeraci were adopted for the numerical modelling, it can be assumed that the numerical tests were conducted at 1:10 scale by the Froude similarity law. Since the experiments were conducted in 1:2 scale with respect to the numerical modelling, a scale factor of 1:20 can be considered for the laboratory tests.

The adoption of the Froude similarity law necessarily involves the distortion of the ratios of other forces, such as inertia to surface tension (Weber number) and inertia to viscosity (Reynolds number).

An incorrect scaling of the surface tension can affect the dynamics and the type of the wave breaking [24]. Specifically, the surface tension increases with decreasing length scale, and for shorter waves (wave lengths $L < 4$ m) the increased surface tension tends to inhibit the wave plunging [24]. A plunging breaker type at prototype scale might, thus, transform into a spilling breaker at model scale, with a reduction of the wave energy dissipation. Despite all the experimental tests present values of $L < 4$ m ($0.9 < L < 3.30$ m), the occurrence of fully-plunging was systematically observed and filmed, and, globally, no relevant effect of a reduced wave energy dissipation was observed (see Section 3).

A distorted representation of the fluid viscosity leads to lower Reynolds numbers and larger viscous forces [25–27] in the model tests. Indeed, the experiments were characterized by Reynolds numbers in the range 7×10^3–5×10^4, which, rescaled to prototype conditions, correspond to Reynolds numbers in the range 5×10^5–2×10^6. Overall, reduced Reynolds numbers might determine higher drag coefficients, and consequently, smaller run-up heights and less overtopping at small scale [26]. However, the effects of increased drag forces on the overtopping due to scale effects are considered to be relevant for rubble mound structures, while they tend to be less effective or negligible for smooth structures [28,29].

The results of the experimental tests confirm that, on average, the effects of the surface tension and of the viscosity on the overtopping are modest or negligible: the average values of q are indeed in line with the predictions by literature methods (see Section 3.1), and no systematic overestimation or underestimation was observed. This analysis cannot assure, though, that viscosity and surface tension may not induce local and/or instantaneous effects, which are not quantifiable with the available instrumentation.

The small scale adopted for the experiments may also result in a smaller amount of air bubbles entrapped in the overtopping flow [30]. The reduced presence of air bubbles (with respect to larger scales or prototype conditions) is expected to induce smaller wave energy dissipation, and overall, higher run-up, higher overtopping rates and higher values of the overtopping flow characteristic. Though the effective rate of the air entrainment has not been quantified yet, the presence of air bubbles was systematically observed during all the experiments. Research is ongoing to get precise estimations based on the analysis of the videos and the image processing.

The experimental tests might be also affected by "side effects," due to the relatively small width of the wave flume (50 cm). These side effects might determine distortions or asymmetries of the wave shape in the longshore direction, which were not measured but were not visually observed. To limit the influence of these effects, all the measurements (free surface elevations, flow depths, flow velocities) were taken in the middle section of channel.

3. Verification of the Data

This Section is dedicated to the characterization and verification of the new experimental and numerical data and to the model validation. The data are compared to existing prediction methods in terms of q and K_r in Sections 3.1 and 3.2, respectively. For q, the EurOtop equations [28] and the artificial neural network (ANN) developed by the authors [31–33] were considered. The data of K_r were instead compared to the formula by Zanuttigh and van der Meer [34] and to the ANN.

3.1. Wave Overtopping Discharge

The most commonly-adopted method for the prediction of q in case of $R_c/H_s \geq 0$ and dry landward conditions is represented by the formulae of EurOtop (2018). For a probabilistic design and following the mean value approach, these formulae give:

$$\frac{q}{\sqrt{gH_{m0}^3}} = \begin{cases} \frac{0.023}{\sqrt{tan\alpha_{off}}} \cdot \gamma_b \cdot \xi_{m-1,0} \cdot exp\left(-\left(2.7 \cdot \frac{R_c}{\xi_{m-1,0} \cdot H_{m0} \cdot \gamma_b \cdot \gamma_f \cdot \gamma_v \cdot \gamma_\beta}\right)^{1.3}\right) & \text{(1a)} \\[3mm] \text{with a maximum of} \quad 0.09 \cdot exp\left(-\left(1.5 \cdot \frac{R_c}{H_{m0} \cdot \gamma_f \cdot \gamma_\beta}\right)^{1.3}\right) & \text{(1b)} \end{cases}$$

where $\xi_{m-1,0}$ is the Iribarren–Battjes breaker parameter and the factors γ_b, γ_f, γ_v and γ_b are the reducing coefficients for a berm, the roughness of the armor layer, the presence of a crown wall and oblique wave attacks. In the new numerical and experimental tests presented in this paper, all these factors are equal to 1. The two formulae Equations (1a) and (1b) are respectively valid for breaking (i.e., approximately for values of $\xi_{m-1,0} \leq 2$) and non-breaking ($\xi_{m-1,0} > 2$) wave conditions, where the wave breaking is supposed to occur for the interaction between the wave and the structure slope (α_{off}).

The new data of q are compared to the curves representing Equations (1a) and (1b) in Figure 3a,b, respectively. In both charts, the data are distinguished by numerical (triangles) and experimental (circles) tests and are plotted against the relative freeboard. The values of R_c/H_s were computed from the incident wave heights H_s, measured at the structural toe instead of the target values reported in Tables 1 and 2. In both charts, only the tests at $R_c/H_s \geq 0$ and dry landward conditions are plotted. On average, the data are straightly and symmetrically distributed around the [28] curves. Most of the data fall within the 90% confidence bands associated to the predicting formulae (dashed lines in Figure 3). The greatest scatter is observed around the zero freeboard, where the formulae seem to underestimate part of the numerical and experimental data. Quantitatively, the agreement among the data and Equations (1a) and (1b) is provided for both the numerical and the experimental data in Table 3 in terms of the error indexes R^2 (coefficient of determination) and of $\sigma_{\%}$ (relative standard deviation or coefficient of variation). Numerical data are significantly better predicted by Equations (1a) and (1b) than the experimental data, being $\sigma_{\%} = 16\%$ and $R^2 = 0.95$ in case of the numerical dataset and $\sigma_{\%} = 56\%$ and $R^2 = 0.87$ in case of the laboratory dataset. The lower agreement found for the experimental data is principally caused by the underestimation bias at $R_c/H_s = 0$, as it can be appreciated by Figure 3.

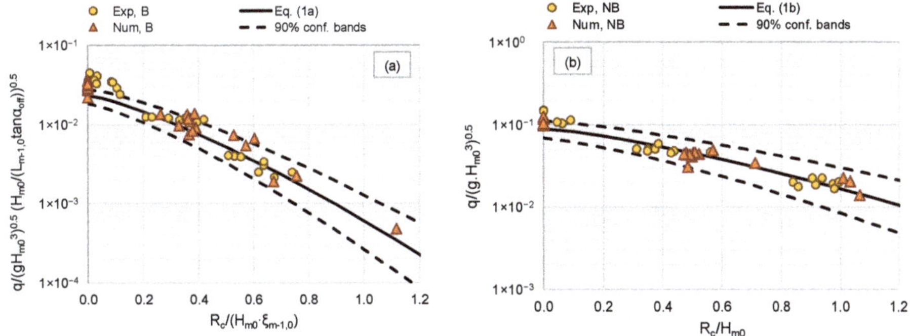

Figure 3. Experimental and numerical dimensionless values of q as functions of R_c/H_s compared to the curves representing the formulae by EurOtop (2018) for the prediction of q, Equations (1a) and (1b), in panels (**a**) and (**b**), respectively. The dashed lines represent the 90% confidence bands associated to the formulae. Data at $R_c/H_s \geq 0$ and dry landward conditions.

The data were further checked against the predictions of q obtained with the ANN predicting tool [30,32,33]. Differently from the EurOtop equations, the ANN can be applied to structures at negative freeboard as well, but still in dry landward conditions exclusively. The comparison among experimental/numerical values of q and the corresponding ANN predictions is given in Figure 4, while the error indexes $\sigma_\%$ and R^2 are reported in Table 3. The agreement between the predictions and the measurements is remarkable in case of the laboratory dataset ($\sigma_\% = 12\%$ and $R^2 = 0.91$), as all the data are concentrated around the bisector line in Figure 4. The slight underestimation bias ($\mu(q_{ANN}/q_{lab}) = 0.89$) could be explained with the particular smoothness of the dike material (plexiglas), which has been quantified with $\gamma_f = 1$, whereas in the ANN training database the value of $\gamma_f = 1$ is generally associated to concrete, asphalt, plywood, grass, etc. The values of q associated to the numerical dataset ($\sigma_\% = 27\%$ and $R^2 = 0.81$) are more scattered but still fairly represented by the ANN, as almost all the data fall within the 90% confidence bands (dashed lines in Figure 4) and no significant bias is observable.

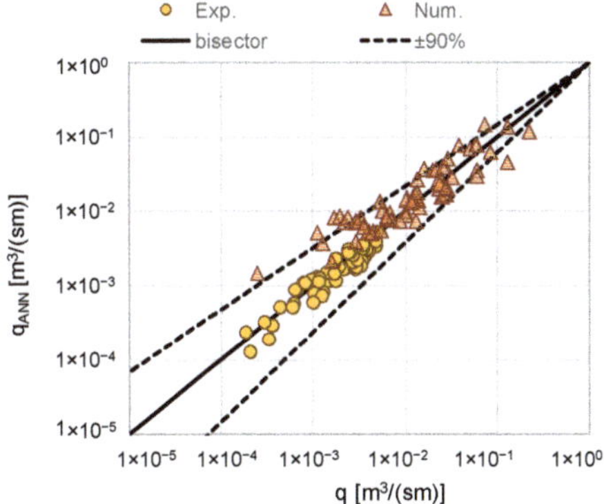

Figure 4. Comparison among experimental and numerical values of q and corresponding predictions from the artificial neural network (ANN). The dashed lines represent the 90% confidence bands associated to the formulae. Data in dry landward conditions.

The analysis of the data at negative freeboard and wet landward conditions is provided in [18], where a specific method for the prediction of q in such conditions is presented. As this method is the only available in the literature so far, the data are not reported here again, as they cannot be used for validation. Anyway, the main focus of this paper is on dry landward conditions. The dry landward conditions represent, indeed, the ordinary conditions, while the wet landward conditions represent the case of fully breached dikes; i.e., extreme conditions.

Table 3. Error indexes R^2 and $\sigma_\%$ characterizing the agreement between the new laboratory/numerical data of q and K_r and corresponding predictions from literature methods.

	q				K_r			
Dataset	**Equations (1a)–(1b)**		**ANN**		**Equation (2)**		**ANN**	
	$\sigma_\%$	R^2	$\sigma_\%$	R^2	$\sigma_\%$	R^2	$\sigma_\%$	R^2
Laboratory	56%	0.87	12%	0.91	45%	0.77	40%	0.89
Numerical	16%	0.95	27%	0.81	28%	0.90	6.0%	0.93

3.2. Wave Reflection Coefficient

The data of K_r resulting from the numerical simulations and the laboratory tests were compared to the predicting formula by Zanuttigh and van der Meer [34], which gave:

$$K_r = \tan h(a \times \xi_{m-1,0}^b) \cdot \begin{cases} 1, & \text{if } \frac{R_c}{H_s} \geq 0.5 \\ 0.67 + 0.37 \times \frac{R_c}{H_s}, & \text{if } -1 \leq \frac{R_c}{H_s} < 0.5 \end{cases} \tag{2}$$

where $a = 0.16$ and $b = 1.43$ when $\gamma_f = 1$. Equation (2) is valid for values of the wave steepness $s_{m-1,0h} \geq 0.01$.

The qualitative comparison among the new data of K_r and the curve representing Equation (2) is provided in Figure 5a. In this chart, the quantity K_r^* is shown as a function of $\xi_{m-1,0}$, where $K_r^* = K_r$ if $R_c/H_s \geq 0.5$, and $K_r^* = K_r/(0.67 + 0.37 \cdot R_c/H_s)$ if $-1 \leq R_c/H_s < 0.5$. All the data beyond the range of validity of Equation (2) have been removed from the plot.

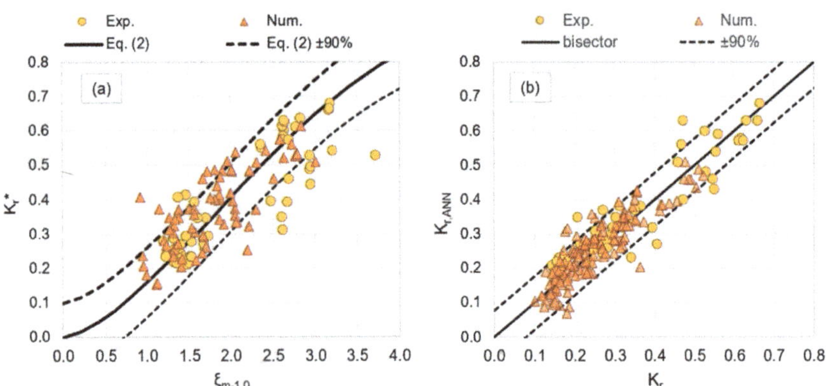

Figure 5. (**a**) Experimental and numerical values of K_r as functions of $\xi_{m-1,0}$ compared to the curve representing Equation (2) for the prediction of K_r. (**b**) Comparison among experimental and numerical values of K_r and corresponding predictions from the ANN. The dashed lines represent the 90% confidence bands associated to the formula and the ANN predictions.

Figure 5a shows that almost all the numerical tests (triangles) are included within the 90% confidence bands (dashed lines), while a slightly higher scatter was observed for the laboratory tests (circles), especially when $\xi_{m-1,0} > 2.5$. The presence of a few "outliers," i.e., data that fall outside the 90% confidence bands, can be explained by considering that Equation (2) represents an average trend and was fitted mostly on rubble mound data. Quantitatively, the agreement among the data and the formula is represented by $\sigma_\% = 45\%$ and $R^2 = 0.77$ in case of the laboratory tests and $\sigma_\% = 28\%$ and R^2

= 0.90 in case of the numerical tests, as reported in Table 3. These error indexes are in line with the uncertainty associated to Equation (2) (see [34]).

The comparison with the predictions of K_r obtained with the ANN [33] is qualitatively given in Figure 5b and quantitatively characterized by the error indexes of Table 3. All the numerical and experimental tests fall within the range of applicability of the ANN. Both Figure 5b and Table 3 indicate that the data are slightly better represented by the ANN than Equation (2), with $\sigma_\% = 40\%$ and 6.0% and $R^2 = 0.89$ and 0.93 for the laboratory and numerical datasets, respectively. This can be explained with the higher number of parameters involved in the ANN with respect to the formula.

4. Validation of the Numerical Code

The numerical code was validated against a dataset of 12 tests selected among the 60 new experiments carried out in the wave flume (see Section 2.2). The 12 tests were reproduced with the modified IH-2VOF code and the resulting values of the average wave overtopping discharge q, of the wave reflection coefficient K_r (Section 4.1) and of the flow depth h and velocities u at the dike crest (Sections 4.2 and 4.3) were compared to the corresponding laboratory measurements.

A specific numerical channel and an ad hoc mesh grid were set up to exactly reproduce the wave flume and the position of the wgs in the lab. The mesh resolution Δz was set constant across the numerical channel and equal to 0.004 m, while the horizontal mesh was made variable by keeping the finest grid of $\Delta x = 0.004$ m in the area around the structures. Six numerical wgs were set in accordance with the laboratory instruments; i.e., three of the resistive gauges for the separation of the incident and reflected wave heights (at around 5 m from the wave-maker) and three of the UVPs along the dike crests. The layout of the numerical channel and the displacement of the numerical gauges are shown in Figure 6. The reproduced dike configurations with G30 and G15 are respectively reported at the top and at the bottom of Figure 6.

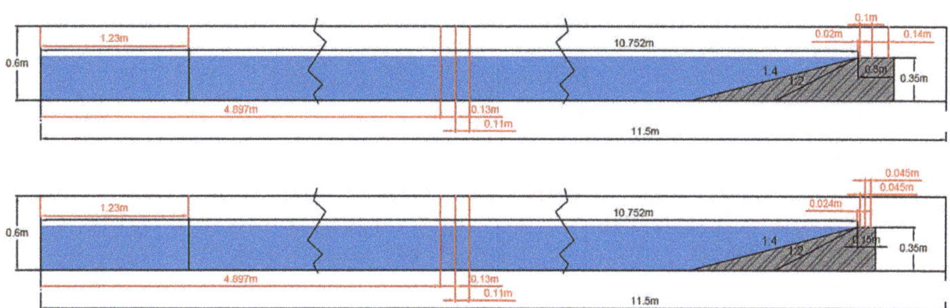

Figure 6. Layouts of the numerical channel set up for the validation of the code. Structure configuration with $G_c = 0.30$ and 0.15 m to the top and to the bottom, respectively.

The 12 selected tests, whose characteristics are reported in Table 4, cover the variety of the geometrical configurations tested in the laboratory, precisely: two dike slopes, c2 and c4; three relative crest freeboards R00, R05 and R10; and two crest widths, G15 and G30. The simulated wave conditions were the same for all the 12 tests and refer to the target wave height H05 and the target wave steepness s3. Each simulation was made lasting 480 s (like in the lab) and a sampling frequency of 20 Hz was adopted for recording the values of h and u at the numerical wave gauges.

Table 4. Summary of the 12 experimental tests carried out at the Laboratory of Bologna and reproduced in the numerical code.

Test ID	H_{m0} (m)		$T_{m-1,0}$ (s)		R_c/H_s (-)	$cot\alpha$ (-)	G_c (m)	$\xi_{m-1,0}$ (-)		$q/(gH_sT_{m-1,0})$ (-)		K_r (-)	
	Lab	*Num*	*Lab*	*Num*	-	-	-	*Lab*	*Num*	*Lab*	*Num*	*Lab*	*Num*
R00H05s3G15c4	0.047	0.042	1.048	1.083	0.00	4	0.15	1.52	1.68	7.80×10^{-3}	7.6×10^{-3}	0.16	0.20
R00H05s3G15c2	0.049	0.049	1.036	1.040	0.00	2	0.15	2.93	2.93	7.09×10^{-3}	1.8×10^{-3}	0.37	0.41
R00H05s3G30c4	0.046	0.042	1.048	1.190	0.00	4	0.30	1.52	1.81	7.02×10^{-3}	7.5×10^{-3}	0.17	0.18
R00H05s3G30c2	0.048	0.048	1.036	1.190	0.00	2	0.3	2.94	3.40	7.48×10^{-3}	8.0×10^{-3}	0.32	0.36
R05H05s3G15c4	0.053	0.044	1.076	1.105	0.50	4	0.15	1.46	1.64	2.54×10^{-3}	2.5×10^{-3}	0.34	0.26
R05H05s3G15c2	0.053	0.046	1.374	1.105	0.50	2	0.15	3.72	3.22	2.49×10^{-3}	2.7×10^{-3}	0.53	0.48
R05H05s3G30c4	0.048	0.044	1.076	1.105	0.50	4	0.30	1.53	1.64	2.49×10^{-3}	2.4×10^{-3}	0.28	0.26
R05H05s3G30c2	0.054	0.048	1.052	1.105	0.50	2	0.3	2.55	3.15	3.22×10^{-3}	2.4×10^{-3}	0.53	0.51
R10H05s3G15c4	0.056	0.042	1.	1.040	1.00	4	0.15	1.37	1.59	8.02×10^{-4}	9.3×10^{-4}	0.41	0.27
R10H05s3G15c2	0.057	0.045	1.008	1.040	1.00	2	0.15	2.64	3.06	1.43×10^{-3}	1.1×10^{-3}	0.62	0.57
R10H05s3G30c4	0.049	0.042	1.036	1.040	1.00	4	0.30	1.46	1.59	6.87×10^{-4}	9.7×10^{-4}	0.29	0.27
R10H05s3G30c2	0.057	0.046	1.008	1.040	1.00	2	0.3	2.63	3.04	1.50×10^{-3}	1.1×10^{-3}	0.63	0.58
μ (Num/Lab)	0.87		1.02		-	-	-	1.11		0.94		0.96	
$\sigma_\%$ (Num/Lab)	8.0%		8.4%		-	-	-	9.7%		28%		16%	

4.1. Wave Overtopping and Wave Reflection

For each test, Table 4 reports the values of the hydraulic parameters (H_{m0}, $T_{m-1,0}$ and $\xi_{m-1,0}$) measured in the laboratory (column "Lab") and derived from the numerical model (column "Num"), including the dimensionless average values of q, $q/(gH_{m0}T_{m-1,0})$ and the values of K_r. For each parameter, the mean μ and the standard deviation σ representative of the distribution of the ratio between numerical and experimental values (num/lab) are provided. Overall, the values of μ(num/lab) related to H_{m0}, $T_{m-1,0}$ and $\xi_{m-1,0}$, which are respectively equal to 0.87, 1.02 and 1.11, suggest that the code well reproduces the target wave conditions. The agreement between numerical and laboratory wave periods is remarkable ($\mu = 1.02$, with $\sigma = 8.4\%$), while the wave heights were slightly underestimated in the numerical code ($\mu = 0.87$, $\sigma = 8.0\%$). This result is line with the tendency of the IH-2VOF code to reduce the wave steepness by generating lower wave heights with respect to the target values, especially when the waves are generated in intermediate depth water (see, all of [15,16]), as in the present case. The lower values of $s_{m-1,0}$ obtained with the numerical code determined, in turn, slightly higher values of $\xi_{m-1,0}$ ($\mu = 1.11$, $\sigma = 9.75\%$); i.e., wave conditions slightly more distant from the wave breaking.

As for q and K_r, the values of μ respectively equal to 0.94 and 0.96 indicate that on average the code can very well represent both the overtopping and the reflection processes, giving a very modest underestimation of both the quantities ($\mu < 1$). The underestimations can be explained considering the different methodologies adopted in the numerical code and in the laboratory to calculate q and the different sampling frequencies used to record the free-surface elevations at the three wgs to reconstruct K_r (100 Hz in the lab and 20 Hz in the numerical code). The lab values of q correspond to the average quantities measured from the tank (see Section 2.2), while the numerical values were derived from the integration of the values of u by the corresponding values of h recorded at the dike off-shore edge. Nevertheless, the differences between numerical and experimental values are on average 5%, with standard deviations of 16% (K_r) and 28% (q), which are very limited and significantly lower than the average uncertainty associated to the common predicting methods (see, e.g., [34] or [35] for K_r, and [28] for q).

4.2. Water Depth Envelopes

The validation of the flow depths is given by comparing the maximum and mean envelopes of the h-values obtained with the numerical code to the corresponding envelopes derived from the image processing of the video of the tests recorded with the camera installed in the laboratory. The reconstruction of the free surface elevation from the image processing was based on automatic algorithms of pattern recognition, which were implemented after the phases of calibration of the camera, background removal and the application of filters to remove the noise from the signals. The details about the procedure and the algorithms applied to reconstruct the free surface elevation from the videos are given in [23].

The comparison among the envelopes is qualitatively given in Figure 7 for two tests (R00H05s3G30c4 and R10H05s3G30c2) selected from the dataset of validation (see Table 4) and presenting different crest freeboards (R00 and R10) and slopes (c4 and c2). In both cases, the numerical envelopes follow the trends from the videography, showing the same decrease rates and very similar values of the maximum and mean h at the extremes of the dike crest. To assess the level of agreement between the corresponding envelopes, the maximum reciprocal distances can be considered: ~2 mm for the maximum envelopes of both the tests, and respectively ~0.5 mm and ~1 mm for the mean envelopes of test R00H05s3G30c4 and test R10H05s3G30c2. These tests are representative of the whole dataset, and the same accuracy was achieved for all the 12 tests.

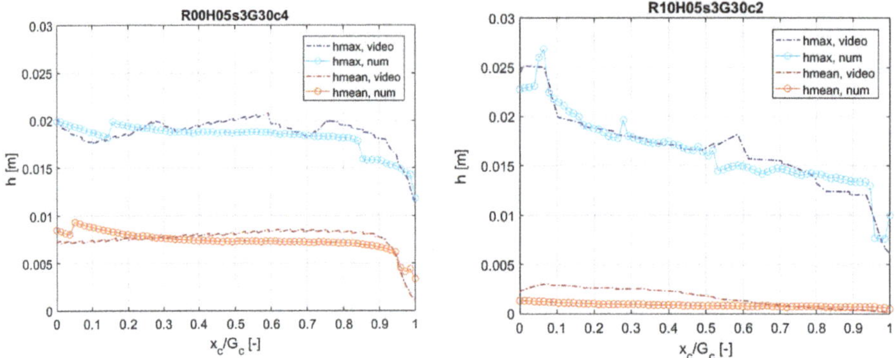

Figure 7. Maximum and mean depth (h) envelopes along the dike crest (x_c/G_c) as obtained from the image processing (dotted-dashed lines, "video") and the numerical code (dots, "num"). (**Left**) Test R00H05s3G30c4; (**right**) test R10H05s3G30c2.

4.3. Extreme Flow Depths and Velocities

The numerical values of $h_{2\%}$ and $u_{2\%}$ calculated at the edges of the dike crest (precisely, at D1 and D3, see Figure 2b) are compared here to the corresponding values extracted from the elaboration of the echo and velocity profiles recorded from the UVPs installed at the same positions on the dikes in the laboratory. This analysis proposes a punctual verification of the numerical data, as it is not possible to draw the envelopes of the h and u-values from the UVPs measurements, which are available in accordance with the three UVPs only.

The comparison is qualitatively shown for all the tests of the dataset of validation (Table 4) in Figure 8. The data of h and u are made dimensionless through the terms H_s and $(gH_s)^{0.5}$, respectively, to account for the different wave conditions generated in the numerical and laboratory flumes. The numerical values of $h_{2\%}$ (Figure 8a) at both D1 and D3 are symmetrically distributed around the bisector line, representing the perfect association with the corresponding laboratory data. The agreement is represented by values of the standard deviation σ respectively equal to 0.006 m and 0.005 m for D1 and D3; the corresponding relative standard deviations (or coefficient sof variation) $\sigma_\%$ are 26% and 30%.

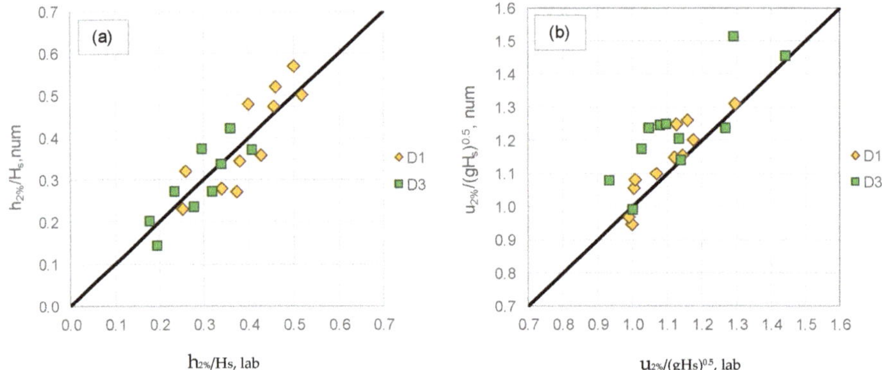

Figure 8. Laboratory (abscissa) and numerical (ordinate) values of $h_{2\%}$ (panel (**a**)) and $u_{2\%}$ (panel (**b**)) calculated at D1 (off-shore edge, diamonds) and D3 (in-shore edge, squares).

Most of the numerical values of $u_{2\%}$ in Figure 8b are aligned directly with or above the bisector line, suggesting that, on average, the numerical code gives slightly overestimations of $u_{2\%}$. This tendency is observed at both D1 and D3 and can be quantified with a value of $\mu(u_{2\%,num}/u_{2\%,lab}) = 1.03$ and 1.09 for D1 and D3, respectively. Overall, the standard deviation associated to the two distributions is $\sigma = 0.09$ m/s and 0.12 m/s at D1 and D3, respectively, with $\sigma_\% = 9.8\%$ at D1 and 13.7% at D3. The slight tendency to overestimate the u-values by the numerical code is probably due to the different representation of the friction with the dike surface and of the air entrainment. The IH2VOF code, indeed, is monophasic and cannot account for the air bubbles entrapped in the overtopping flow within the experiments (see Section 2.4). The air entrainment might induce a larger wave energy dissipation in the laboratory tests with respect to the numerical modelling and reduce, in turn, the experimental flow velocities.

5. Extreme Flow Depths and Velocities at the Dike's Off-Shore Edge

This Section proposes a systematic and detailed analysis of the overtopping flow characteristics at the off-shore edge of the dike crest. In Section 5.1, the extreme percentiles of $h_{2\%}$ and $u_{2\%}$ resulting from the numerical and experimental modelling are compared to the literature formulae. After discussing the performance of the existing methods and investigating the reasons of the discrepancies between data and formulae (Section 5.2), a new fitting is proposed to update the state-of-the-art formulae for the prediction of $h_{2\%}$ and $u_{2\%}$ at the off-shore edge.

5.1. Comparison with Literature Formulae

All the formulae available in the literature for the evaluation of h and u at the extremes of the structure crest [1,5,7,8] draw on the following overarching scheme originally [4]:

$$h_{2\%}(x_c = 0) = c_h \times (R_{u,2\%} - R_C), R_C \geq 0 \tag{3}$$

$$u_{2\%}(x_c = 0) = c_u \times [(R_{u,2\%} - R_C)]^{0.5}, R_C \geq 0 \tag{4}$$

where the values $h_{2\%}$ and $u_{2\%}$ are, respectively, the flow thicknesses and the flow velocities at the off-shore edge ($x_c = 0$) of the dike crest that are exceeded by the 2% of the incident waves. The estimations of $h_{2\%}$ and $u_{2\%}$ by Equations (3) and (4) depend on $R_{u,2\%}$, which is the value of the wave run-up exceeded by the 2% of the waves and can be calculated based on EurOtop [28]. The values or the formulations proposed by the various authors for the coefficients c_h and c_u are summarized in Table 5. In the earliest formulae [4,5], the fitting coefficients c_h and c_u are constant values, while in more recent methods [1,8], c_h and c_u are functions of the off-shore slope α_{off}. Moreover, while the formulations for $u_{2\%}$ by Schüttrumpf [4] and Van Gent [5] were targeted to represent the flow velocities, the formulations by Bosman et al. [8] and by van der Meer et al. [1] aim at predicting the front velocities, or wave celerities, giving higher estimations of $u_{2\%}$.

Figures 9 and 10 compare, respectively, the values of $h_{2\%}(x_c = 0)$ and $u_{2\%}(x_c = 0)$ gathered with the new numerical and experimental modelling to the corresponding predicting formulae; i.e., Equations (3) and (4). In both Figures 9 and 10, the results are grouped by values of $\cot(\alpha_{off})$ and refer to tests in dry landward conditions $R_c/H_s \geq 0$. The data are further distinguished between experimental (filled-in symbols) and numerical (void symbols). The two charts of Figure 10 show the tests at $R_c/H_s > 0$ (panel a) and $R_c/H_s = 0$ (panel b). Following the structure of Equations (3) and (4), the results are shown as functions of $(R_{u,2\%} - R_c)$ in Figure 9 and of $(g \cdot (R_{u,2\%} - R_c))^{0.5}$ in Figure 10.

To ease the comparison, only the formulations by Van Gent [5] and Bosman et al. [8] are shown. The curves representing the formulae by Schüttrumpf [4] are very close to the curves by Van Gent, while the curves by van der Meer et al. [1] for $u_{2\%}$ (Figure 10) are close to the curves by Bosman et al. [8]. It should be noted that Bosman et al. [8] provided fittings for $\cot(\alpha_{off}) = 4$ and 6, exclusively (see Table 5). The curves for the prediction of $h_{2\%}$ and $u_{2\%}$ in case of $\cot(\alpha_{off}) = 2$ have been extrapolated from the original formulae by Bosman et al. [8]. For brevity, the notations "c2," "c4" and "c6" will be used in the following to refer to $\cot(\alpha_{off}) = 2$, 4 and 6, respectively (see Section 2.2).

Table 5. Values of the coefficients c_h and c_u adopted by the several authors in the formulae for the evaluation of $h_{2\%}$ and $u_{2\%}$ based on the scheme of Equations (4) and (5).

Author(s)	Flow Characteristic	Coefficient	Adopted Value/Formulation	Validity Range
Schüttrumpf (2001) [4]	$h_{2\%}(x_c = 0)$	c_h	0.33	$\cot(\alpha_{off}) = 3, 4$ and $6; R_c \geq 0$
	$u_{2\%}(x_c = 0)$ (flow velocity)	c_u	1.37	
Van Gent (2002) [5]	$h_{2\%}(x_c = 0)$	c_h	0.15	$\cot(\alpha_{off}) = 4; R_c > 0$
	$u_{2\%}(x_c = 0)$ (flow velocity)	c_u	1.33	
Bosman et al. (2008) [8]	$h_{2\%}(x_c = 0)$	c_h	$0.01/\sin^2(\alpha_{off})$, for $\cot(\alpha_{off}) = 4$ or 6	$\cot(\alpha_{off}) = 4$ and $6; R_c > 0$
	$u_{2\%}(x_c = 0)$ (wave celerity)	c_u	$0.30/\sin(\alpha_{off})$, for $\cot(\alpha_{off}) = 4$ or 6	
Van der Meer et al. (2010) [1]	$u_{2\%}(x_c = 0)$ (wave celerity)	c_u	$0.35 \cdot \cot(\alpha_{off})$, for $\cot(\alpha_{off}) = 4$ or 6	$\cot(\alpha_{off}) = 3, 4$ and $6; R_c \geq 0$

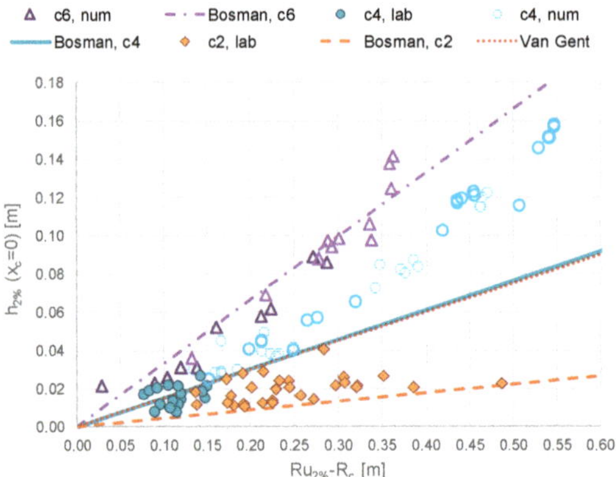

Figure 9. Comparison among the values of $h_{2\%}$ recorded at $x_c = 0$ within the new experimental and numerical tests and literature formulae. The data refer to dry landward conditions and $R_c/H_s \geq 0$, are grouped by values of $\cot(\alpha_{off})$ and are distinguished between experimental (filled in) and numerical (void).

The analysis of Figure 9 suggests that:

- All the values of $h_{2\%}$ fall within the straight lines representing the theoretical formulae, with the exception of one test at c6 that slightly exceeds the upper line by Bosman et al. [8] for c6 (dot-dashed line).

- In agreement with Bosman et al., all the data show a non-negligible effect of the structure slope: the milder the slope, the higher $h_{2\%}$.

- The formulation by Van Gent [5]—which does not account for the slope effect—can be used to get an "average" estimation of the values of $h_{2\%}$.

- For modest values of the wave run-up, i.e., for $(R_{u,2\%} - R_c) < 0.10$–0.15, the formulae by Bosman et al. (2008), give an accurate representation of the data, while for $(R_{u,2\%} - R_c) > 0.15$, the formulae underestimate the values of $h_{2\%}$ in case of c2 and c4. The underestimation increases when increasing $(R_{u,2\%} - R_c)$, and reaches ~100% when $(R_{u,2\%} - R_c) \approx 0.45$ (data c4, circles in Figure 7). This might be in part explained by considering that the experimental tests used to calibrate the formulae were characterized by values of $(R_{u,2\%} - R_c)$ ranging between 0 and 0.3.

- Overall, the data seem to follow a non-linear trend with $(R_{u,2\%} - R_c)$ and a refitting of Equation (3) to extend its validity to the cases of c2 and $(R_{u,2\%} - R_c) > 0.3$ will be discussed in Section 4.2.

A separate analysis was performed to investigate further potential effects of R_c/H_s and $s_{m-1,0}$ on $h_{2\%}$, without leading to any relevant result.

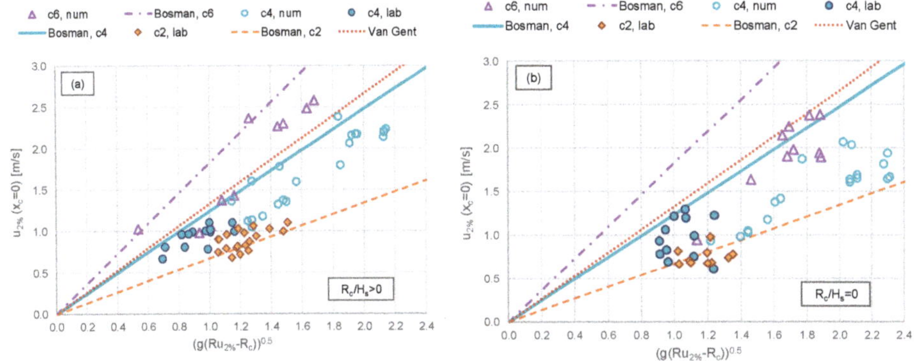

Figure 10. Comparison among the values of $u_{2\%}$ recorded at $x_c = 0$ within the new experimental and numerical tests and literature formulae. The data refer to dry landward conditions and are grouped by values of $\cot(\alpha_{off})$ and distinguished between experimental (filled in) and numerical (void). (**a**) Tests at $R_c/H_s > 0$; (**b**) tests at $R_c/H_s = 0$.

As for the values of $u_{2\%}(x_c = 0)$, the analysis of Figure 10 leads to the following considerations.

- Similarly to $h_{2\%}$, the formula by Van Gent [5] (dotted line) gave an average estimation of the $u_{2\%}$-values, representing, respectively, an upper and a lower envelope for the data at c4 and c6. The data at $R_c = 0$ (panel b)—which are out of the range of validity of the formula—were significantly over-estimated by Van Gent.
- The effect of $\cot(\alpha_{off})$ is still evident: the milder the slope, the higher $u_{2\%}(x_c = 0)$—but slightly smoothed, with respect to $h_{2\%}(x_c = 0)$. With the exception of the data at c2, most of the data of $u_{2\%}$ were over-predicted by the formulae by [8], by 30%–50% in case of $R_c/H_s > 0$ (filled-in points) and of 40%–80% in case of $R_c/H_s = 0$ (void points).
- On average, the data at $R_c/H_s = 0$, tend to be lower than the data at $R_c/H_s > 0$ for the same value of the abscissa; i.e., $(g\cdot(R_{u,2\%} - R_c))^{0.5}$.

The discussion about these results, specifically regarding the discrepancy among formulae and $u_{2\%}$ values and the effects of $\cot(\alpha_{off})$ and R_c/H_s, is given in the next Section 5.2.

5.2. Discussion of the Results

The overestimation given by the formulae by Bosman et al. [8] in case of $u_{2\%}(x_c = 0)$ can be explained considering the following three aspects.

First, Bosman et al. calibrated the coefficients c_u based on the results of the previous experiments [4] and [5]. In those experiments, the measurements of u were obtained from micro-propellers placed at a fixed height over the structures. Therefore, the formulae were calibrated against punctual measures of u, while the values of $u_{2\%}$ obtained from the new experimental and numerical modelling (and shown in Figure 10) were based on the average values of the velocities along the vertical above the structure crest. To further investigate this aspect, Figure 11 shows as example 2 instantaneous vertical profiles of u measured at D1 (i.e., approximately at $x_c = 0$) during the test R00H05s3G15c4 conducted in the laboratory (Figure 11a) and reproduced with the numerical code (Figure 11b). Both the profiles refer to the instant of occurrence of the maximum flow depth h at D1 (i.e., around 0.38 m). Both the charts show a high variability of the u along the vertical: the vertically-averaged values of the u are 0.34 and 0.39 m/s for the laboratory and numerical test, respectively, with standard deviations of ±0.10 and ±0.05 m/s, and maximum values of u which are higher than the average of approximately 30% and 20%. This example indicates how much the vertical position of the instrument may affect the measured value of u, especially in case of laboratory tests, where the uncertainty related to the measurements is higher (as evident by comparing Figure 11a to Figure 11b).

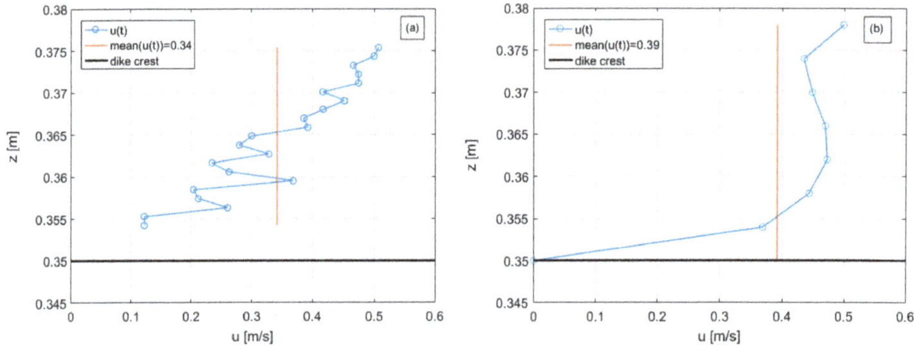

Figure 11. Instantaneous vertical profiles (z) of the flow velocity (u) measured at D1 ($\approx x_c = 0$) for the test R00H05s3G15c4 in the laboratory (panel (**a**)) and reproduced with the numerical code (panel (**b**)). The profiles refer to the instant of occurrence of the maximum flow depth at D1.

Second, the discrepancy between flow velocities and wave celerities, c. Bosman et al. refitted the original values of the coefficients c_u—which were originally calibrated by Schüttrumpf [4] on the measurements of u—to account for the higher values of the c observed for the tests. Based on the observation that, for the same test, c tended to be greater than u, Bosman et al. increased the values of c_u from 1.37 to 1.64 to achieve a more cautious approach. The formulations of c_u as functions of $\cot(\alpha_{off})$ further prompted by Bosman et al. were fitted again on the values of c instead of u.

Figure 12 provides the comparison among c and $u_{2\%}$-values for the new experimental dataset. The wave celerities have been computed using the procedure for the identification and coupling of the overtopping waves developed by references [36,37]. The procedure was applied to the time series of the h-signals registered at D1 and D3, providing as outputs, the time lags for the waves to propagate from D1 and D3. The c-values were computed from the time lags, known the distance D1–D3. Figure 12 shows that the c-values are indeed higher than the corresponding $u_{2\%}$, and the discrepancy among c and u tends to increase with increasing u.

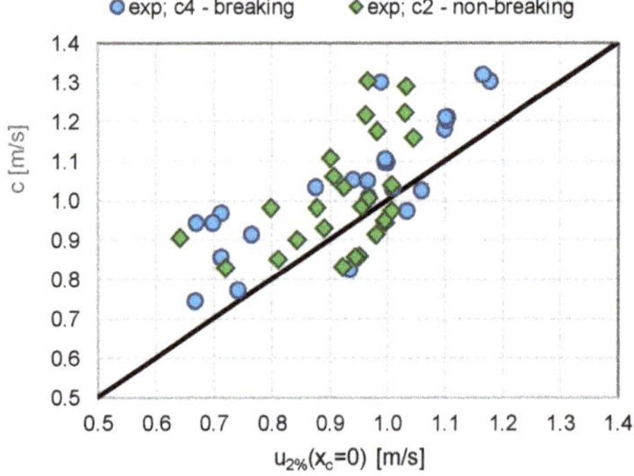

Figure 12. Comparison among values of $u_{2\%}$ at the dike off-shore edge ($x_c = 0$, D1) and average wave celerities c. The data belong to the experimental dataset and are grouped by values of $\cot(\alpha_{off})$.

These results are in line with the findings of Bosman et al. [8]. However, the present study followed the original work by Schüttrumpf [4] and the analyses of the overtopping flow characteristics were conducted based on the flow velocities.

Finally, the effect of R_c/H_s. The on-average lower $u_{2\%}$-values observed at $R_c/H_s = 0$ (Figure 10b) than at $R_c/H_s > 0$ (Figure 10a) are due to the different characteristics of the flow in the two freeboard conditions. When $R_c/H_s > 0$, the dike crest is located above the mean water level in the wave run-up area and the waves go overtop the dike crest during the crest phase exclusively, resulting, thus, in positive—i.e., in-shore directed—values of u at $x_c = 0$. On the contrary, when $R_c/H_s = 0$, the dike crest level is situated exactly in line with the mean water level; i.e., in the middle between the wave run-up and run-down area. In such conditions, the waves can reach and overtop the dike crest during the through phase also, determining a flow in regard to the dike edge which is intermittently in-shore and off-shore directed, with values of u that area alternatively >0 and <0. This phenomenon is clearly evident in the example of Figure 13, which reports the frequency distribution of all the u-values measured with UVP at D1 for the same experimental test (H05s3G30c4) conducted at $R_c/H_s = 0$ (R00) and $R_c/H_s = 1$ (R10). In this example, there was a predominance of negative u-values in case of R00, while for R10, the distribution clearly shifted towards positive u-values. The different overtopping flow conditions and the resulting different distributions of the u-values affected, in turn, the values of $u_{2\%}$.

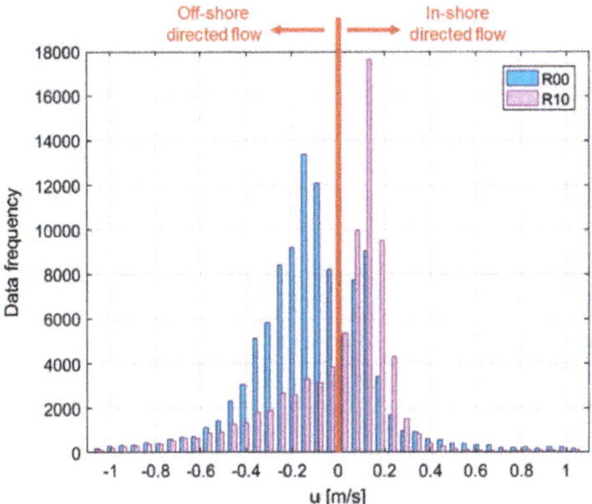

Figure 13. Frequency distribution of the values of u measured at D1 ($\approx x_c = 0$) for the same experimental test H05s3G30c4, conducted at zero (R00) and in emerged freeboard (R10) conditions. The flow is in-shore and off-shore directed when the values of u are, respectively, >0 and <0.

5.3. Refitting of the Formulae

Based on the analyses of the results, the following refitting of the literature formulae for the prediction of $h_{2\%}$ and $u_{2\%}$ at the dike off-shore edge ($x_c = 0$) are proposed:

$$\begin{cases} h_{2\%}(x_c = 0) = a_h \cdot (Ru_{2\%} - R_c)^b \\ a_h = 0.085 \cdot \cot \alpha_{off}; \ b = 1.35 \end{cases}, \ \text{for} \ \frac{R_c}{H_s} \geq 0 \tag{5}$$

$$\begin{cases} u_{2\%}(x_c = 0) = a_u \cdot [(g(Ru_{2\%} - R_c))^{0.5}]^b \\ a_u = 0.12 \cdot \cot \alpha_{off} + 0.41; \ b = 1.35 \end{cases}, \ \text{for} \ \frac{R_c}{H_s} > 0 \tag{6}$$

Equations (5) and (6) follow the same form of Equations (3) and (4), respectively, but present a non-linear relationship between $h_{2\%}$ and $(R_{u,2\%} - R_c)$, and $u_{2\%}$ and $[g(R_{u,2\%} - R_c)]^{0.5}$ through the introduction of the power coefficient $b = 1.35$. The coefficients c_h and c_u of Equations (3) and (4) are, respectively, replaced by the new coefficients a_h and a_u, whose formulations vary linearly with

$\cot(\alpha_{off})$, as indicated in Equations (5) and (6). The dimensions of a_h and a_u are, respectively, $[\text{m}^{0.65}]$ and $[\text{m}^{0.65}\text{s}^{-0.65}]$.

Note that, while Equation (5) applies to $R_c \geq 0$, Equation (6) is valid for $R_c > 0$ only. As discussed in Section 5.2, the values of $u_{2\%}$ follow different trends with $[g(R_{u,2\%} - R_c)]^{0.5}$ at positive and at zero freeboard. Since no further data were available from the literature for check, no fitting was proposed for $u_{2\%}(x_c = 0)$ at $R_c = 0$.

The agreement between the new fitting and the experimental and numerical data is qualitatively shown in Figures 14a and 15a for $h_{2\%}$ and $u_{2\%}$, respectively, and quantitatively represented by the error indexes R^2 and $\sigma_\%$ reported in Table 6. This Table includes also the values of R^2 and $\sigma_\%$ associated to the application of Equations (3) and (4) to the new data obtained by considering the formulations of c_h and c_u [8].

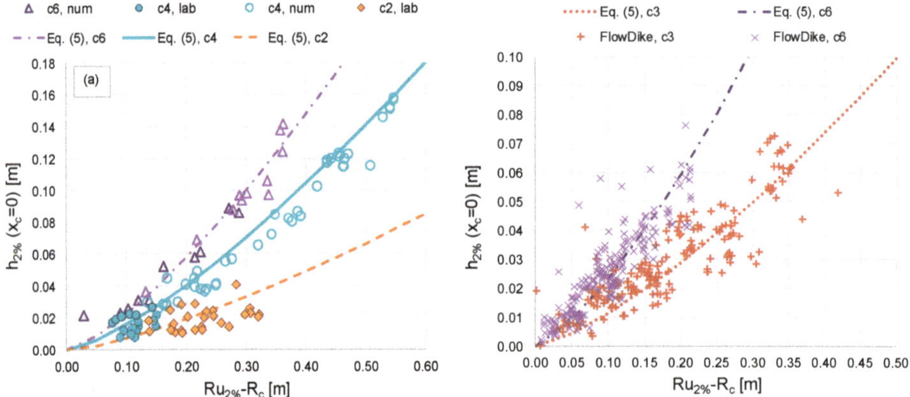

Figure 14. Comparison among the values of $h_{2\%}$ recorded at $x_c = 0$ during the new tests (**a**) and the *FlowDike* data experiments (**b**) and the new fitting by Equation (5). The data are grouped by values of $\cot(\alpha_{off})$.

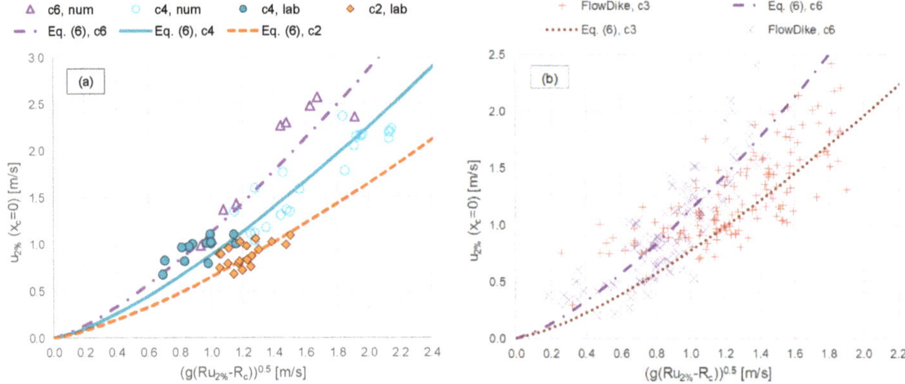

Figure 15. Comparison among the values of $u_{2\%}$ recorded at $x_c = 0$ during the new tests (**a**) and the *FlowDike* data experiments (**b**) and the new fitting by Equation (6). The data are grouped by values of $\cot(\alpha_{off})$ and relate to $R_c > 0$ only.

In order to check the validity of the new fitting, Equations (5) and (6) were also applied to the prediction of $h_{2\%}(x_c = 0)$ and $u_{2\%}(x_c = 0)$ for some tests of the datasets *FlowDike*1 and *FlowDike*2 [38,39]. These tests involved 2D and 3D wave attacks against smooth dikes characterized by different values of

G_c (0.6 and 0.7 m, model scale values) and of $\cot(\alpha_{off})$ (3 and 6, respectively in *FlowDike*1 and *FlowDike*2). The following tests from *FlowDike*1 and *FlowDike*2 were not considered for the analyses:

- All the tests with missing records of either $R_{u2\%}$, $h_{2\%}(x_c = 0)$ or $u_{2\%}(x_c = 0)$;
- The tests giving zero or negative overtopping discharge (considered unreliable);
- All the tests with wind velocity >10 m/s, as this fitting does not include the wind effect.

The values of $h_{2\%}(x_c = 0)$ and $u_{2\%}(x_c = 0)$ from the remaining tests are compared to Equations (5) and (6) in Figures 14b and 15b, respectively. The quantitative assessment of the agreement between the *FlowDike*1 and *FlowDike*2 data and the new formulae are given in Table 6, that also includes the performance of Equations (3) and (4) for comparison.

Overall, both new and existing data seem to follow the same trend of the curves representing the new fitting, either in case of $h_{2\%}$ (Figure 14) or of $u_{2\%}$ (Figure 15). The best agreement is found between Equation (5) and the values of $h_{2\%}$ from the numerical dataset, being that $R^2 = 0.996$ and $\sigma_\% = 4.1\%$ (see Table 6). A larger scatter was observed for the other datasets, as indicated by the values of $\sigma_\%$ ranging between 15% (new laboratory data of $u_{2\%}$ and Equation (6)) and 51% (*FlowDike*2 data of $u_{2\%}$ and Equation (6)). The relatively high values of $\sigma_\%$ are caused by the intrinsic scatter associated to the data, especially to the *FlowDike*1 and *FlowDike*2 experiments, and do not necessarily indicate a poor agreement with the new fitting. This is confirmed by the values of $\sigma_\%$ associated to the predictions by Equations (3) and (4), which are comparable or even higher to the $\sigma_\%$ associated to the new fitting. On the contrary, Figures 14 and 15 reveal that the new formulae provide a good, "average" representation of most of the data, which are randomly but symmetrically distributed around the curves following their trends. This qualitative analysis is confirmed and reinforced by the values of R^2 of Table 6, which are always >0.72 and in six cases out of eight, even >0.80. Equations (3) and (4) provide a similar—but lower—performance on *FlowDike*1 and *FlowDike*2 data, while they give a worse representation of the new experimental and numerical data, as indicated by Figures 9 and 10.

Table 6. Error indexes R^2 and $\sigma_\%$ characterizing the agreement between the data of $h_{2\%}(x_c = 0)$ and $u_{2\%}(x_c = 0)$ and the new fitting by Equations (5) and (6).

Dataset	$h_{2\%}(x_c = 0)$ -Equation (5)		$u_{2\%}(x_c = 0)$ -Equation (6)		$h_{2\%}(x_c = 0)$ -Equation (3), [8]		$u_{2\%}(x_c = 0)$ -Equation (4), [8]	
	$\sigma_\%$	R^2	$\sigma_\%$	R^2	$\sigma_\%$	R^2	$\sigma_\%$	R^2
New laboratory data	41%	0.82	15.2%	0.82	37.2%	-	24.8%	-
New numerical data	4.1%	0.99	22.1%	0.73	41.9%	-	71.9%	-
*FlowDike*1 data—c3	43.4%	0.85	41.8%	0.79	42.7%	0.85	43.2%	0.77
*FlowDike*2 data—c6	42.3%	0.84	50.7%	0.88	43.8%	0.67	69.9%	0.85

In conclusion, the new fitting by Equations (5) and (6) represents a more cautious alternative to that by Equations (3) and (4), respectively, also applying to steep slopes ($\cot(\alpha_{off}) = 2, 3$) and low freeboards. Its use is suggested especially in case of relatively high run-up levels, i.e., approximately for values of $(R_{u,2\%} - R_c)/H_s > 2$–2.5, which represent very severe surge conditions to catastrophic flooding scenarios.

The new formulae were applied in the following ranges: $0 \leq R_c/H_s \leq 4.0$ and $0 < R_c/H_s \leq 4.0$ in case of Equation (5) and Equation (6), respectively; $\cot(\alpha_{off}) = 2; 3; 4; 6; 0.72 \leq \xi_{m-1,0} \leq 6.13; 0.035 \leq H_s \leq 0.22$ m.

6. Evolution of Flow Characteristics along the Dike Crest

This Section focuses on the evolution of the flow depths and velocities between the dike off-shore and in-shore edges. Section 6.1 provides the comparison among the data of $h_{2\%}$ and $u_{2\%}$ and the literature formulae [7]. Sections 6.2 and 6.3 are dedicated to an in-depth analysis of the values and trends of the flow velocities, with specific attention to the effects of the crest freeboard and wave breaking. Section 6.4 draws some conclusions about the analyses of the flow velocities, providing a few guidelines for practical use.

6.1. Comparison with Literature Formulae

The existing methods for the prediction of the evolution of flow depths and velocities along the crest of a smooth dike were developed starting from the physical and theoretical analysis by Schüttrumpf [4], which was based on experimental evidence and on these assumptions:

- The dike crest is horizontal;
- The vertical velocities can be neglected;
- The pressure term is almost constant over the dike crest;
- The viscous effects along the flow direction are small;
- The bottom friction is constant over the dike crest.

Articles [4,7] describe the flow evolution of both h and u in terms of exponential decays along the dike crest, resulting in the following formulations:

$$\frac{h(x_c)}{h(x_c=0)} = \exp\left(-c_3 \frac{x_c}{G_c}\right), \ R_c \geq 0 \tag{7}$$

$$\frac{u(x_c)}{u(x_c=0)} = \exp\left(-\frac{f}{2}\cdot\frac{x_c}{h(x_c)}\right), \ R_c \geq 0 \tag{8}$$

where x_c is the horizontal coordinate along the crest, $h(x_c)$ is the overtopping flow depth on the dike crest at the coordinate x_c; $u(x_c)$ is the overtopping flow velocity on the dike crest at the coordinate x_c; c_3 is a dimensionless coefficient, varying according to the quantile used for h and u (50%, 10%, 2%); and f is the bottom friction coefficient.

The decay of h (Equation (7)) is due to the energy loss and to the deformation of the overtopping volume induced by the change of flow direction from the off-shore slope to the dike crest. The decay of u (Equation (8)) should be induced by the friction over the crest. However, as remarked by [7], the contemporary decay of u and h contradicts the continuity equation. In fact, the authors observed significant reductions of h (up to the 50%) but very modest changes of u, yielding to the conclusion that u was almost constant for the tested conditions; i.e., for a relatively short and smooth crest at zero freeboard and emerged conditions, $G_c = 0.3$ m and $R_c \geq 0$. Therefore, for a smooth dike, the authors suggested the use of the (very low) value of $f = 0.0058$, which reduces u approximately 8%.

Later, small and large-scale model tests were carried out to check the validity of Equations (7) and (8) and to update the formulations [1,8,40]. Based on these results, the EurOtop manual [28] provides the following indications.

The flow thickness h decreases of approximately 1/3 with respect to the value at the off-shore edge; no new formulation is proposed in place of Equation (7).

The flow velocity u decays along the crest as a function of the wave length $L_{m-1,0}$, according to the following relationship:

$$\frac{u_{2\%}(x_c)}{u_{2\%}(x_c=0)} = \exp\left(-1.4\cdot\frac{x_c}{L_{m-1,0}}\right), \ R_c \geq 0 \tag{9}$$

Reference [28] warns about the potential inadequacy of Equation (9) in case of large crests (such as promenades).

Recently, a new analytical model [10] was proposed to describe the evolution of the maximum flow velocities (U) on the dike crest and landward slope. The model consists of two coupled formulae to be applied in sequence for the dike crest and the landward slope, respectively. The first formula still predicts the decay of U along the dike crest due to the bottom friction f:

$$U(x_c) = \cfrac{1}{\cfrac{fx_c}{2Q} + \cfrac{1}{U(x_c=0)}}, \; R_c > 0 \tag{10}$$

where based on the "momentary discharge Q;" i.e., the discharge calculated at the instant of occurrence of the maximum flow velocity U. The approach by article [10] gives also a formulation for the evolution of the "momentary layer thickness", h_U, derived from the continuity equation of Q, but no validation is given for this formula.

So far, no predicting method exists for the evolution of h and u in case of $R_c < 0$. Moreover, the formulae for u are verified against the off-shore and the in-shore values $u(x_c = 0)$ and $u(x_c = G_c)$ only. The numerical simulations (see Table 1) provide instead, continuous records of u and h between $x_c = 0$ and $x_c = G_c$ for any crest emergence and submergence. Due to the limits of the laboratory equipment, the following analyses and discussion are principally based on numerical results: experimental evidence is available for the values of u and h corresponding to the UVPs; i.e., at D1, D2 and D3, for $R_c/H_s \geq 0$ and dry landward conditions.

Following EurOtop [13,28], the quantiles 2% were considered in this study. Figure 16 illustrates the trends of $\frac{h_{2\%}\,(x_c)}{h_{2\%}(x_c=0)}$ and $\frac{u_{2\%}\,(x_c)}{u_{2\%}(x_c=0)}$ derived from the numerical results for different values of R_c/H_s. In case of $R_c/H_s < 0$, only tests in dry landward conditions were considered.

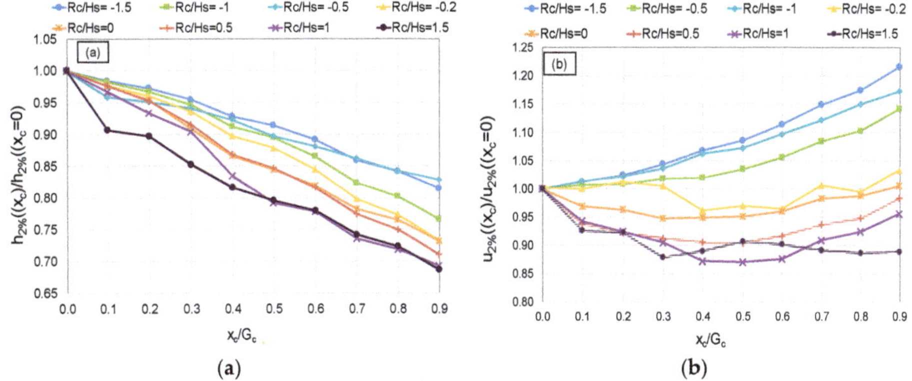

Figure 16. Evolution of the flow thickness ($h_{2\%}/h_{2\%,xc} = 0$, panel (**a**)) and of the flow velocity ($u_{2\%}/u_{2\%,xc} = 0$, panel (**b**)) along the dike crest (x_c/G_c). Average values from the numerical tests grouped by R_c/H_s.

The flow thickness $h_{2\%}$ (panel a) decays almost linearly with x_c, showing a certain dependency with R_c/H_s. The decrease of h, ranging between 15 and 20% for $R_c < 0$, and being 35% for $R_c/H_s = 1.5$, is in agreement with EurOtop [28] but it is lower than the estimations of Equation (7), which would predict a decrease of approximately the 50%. The same decays were approximately found when comparing the values of $h_{2\%}$(D1) with the corresponding values of $h_{2\%}$(D3) both in experiments and in simulations.

Based on the numerical and experimental results and on the indications of EurOtop [28], it is, therefore, suggested to use Equation (7) for $h_{2\%}$ by adopting the following values of the coefficient c_3:

$$\begin{cases} c_3 = 0.35, \; \text{if } R_c \geq 0 \\ c_3 = 0.18, \; \text{if } R_c < 0 \end{cases}, \; \text{for dry landward conditions only} \tag{11}$$

The comparison among the predictions of $h_{2\%}(x_c = G_c)$ derived from the application of Equation (7) with the coefficients given in Equation (11) and the corresponding numerical values led to standard deviations σ and coefficients of determination R^2, respectively equal to 0.06 and 0.93 for $R_c \geq 0$ and to 0.09 and 0.96 for $R_c < 0$. For $R_c < 0$ in wet landward conditions, i.e., fully submerged or breached dikes, no significant changes of the values of h were detected along the crest.

The trend of $u_{2\%}$ along the dike crest (Figure 16b) is significantly different for tests at positive or negative freeboard. In case of $R_c \geq 0$, a slight decay was observable until $x_c/G_c \approx 0.4$, while the trend seems to invert around $x_c/G_c = 0.6$ and u increases up to approximately the same value of the crest beginning; i.e., $u(x_c = G_c) \approx u(x_c = 0)$. In case of $R_c < 0$, u monotonically increases with an apparent quadratic function of x_c from the beginning to the end of the dike crest. The different trends of u were, therefore, investigated separately for positive and negative freeboards in Sections 6.2 and 6.3, respectively.

6.2. Flow Velocities at Zero and Positive Crest Freeboard

The initial decay and subsequent increasing trend of u for $x_c/G_c > 0.4$–0.5 was already observed by Guo et al. [17] in case of numerical tests with $R_c \geq 0$. The combination of decreasing flow thicknesses (Figure 14a) and increasing flow velocities (Figure 14b) along the dike crest fulfills the continuity and the momentum balance equations, accounting also for the (small) effect of the friction. However, it is clearly in opposition with the existing approach by Schüttrumpf and Oumeraci [7] that is based on the approximation of the Navier–Stokes equations. Schüttrumpf and Oumeraci highlighted indeed, that their approach contradicts the instantaneous continuity equation, but they found experimental evidence of the decay of u and pointed out that the continuity equation is globally fulfilled by considering the time-integral of the product of u by h. Guo et al. [17] argued that the discrepancy between the results of their numerical modelling and the equations by Schüttrumpf and Oumeraci can be explained by a number of elements, such as (i) the dynamics of the overtopping flow over the dike crest and the shape of the water front; (ii) the effect of the air entrainment; and (iii) the limitations imposed by the assumptions by Schüttrumpf and Oumeraci; specifically, the approximation of the boundary layer and the adoption of flow-depth integrated velocities. These aspects are analyzed in detail in the following Sections 6.2.1 and 6.2.2.

6.2.1. Effect of the Wave Breaking

One of the main points raised by Guo et al. [17] to justify the different trends of u along the dike crest with respect to Equation (8) regards the different dynamics of propagation of the breaking and non-breaking waves over the dike crest. When the wave reaches the dike crest in breaking or fully-broken conditions (see Figure 15), it has already dissipated most of its energy, and the flow over the dike crest is highly turbulent and is characterized by a large amount of air pockets entrapped in the water tongue. The air entrainment may induce significant differences between the results of physical, numerical and theoretical modelling due to the different approaches for the assessment of the flow velocity. In the laboratory, the UVPs provide measurements of the instantaneous vertical profiles of u, which can be severely affected by the presence of air pockets in the water column as the sonic impulse of the UVP is interrupted when travelling across air. The numerical code can calculate the u-values of the water phase exclusively, neglecting the air-phase, leading, therefore, to potential over-estimation of the u-values. Finally, Schüttrumpf and Oumeraci [7] considered a steady-flow boundary layer and flow-depth averaged velocities, hence evaluating the maximum u-values at the free surface. It is evident that these assumptions hold no more in case of broken waves in turbulent flow, where the maximum u is found close to the dike surface.

As an example, Figure 17 show two frames of the same overtopping event derived from the physical (left) and numerical (right) modelling of the test R00H05s3G30c4. The color map of Figure 17b illustrates the flow velocity field from computations. In both panels of Figure 17a,b, the flow propagates on the dike crest after the wave breaking has already occurred along the dike slope. In such conditions, the overtopping flow is similar to a weir-like stream governed by the outfall in-shore condition. Due to the modest or null friction, the value of $u_{2\%}$ hardly decays while propagating along the dike crest. Similarly to the results of Guo et al., the numerical model suggests that the flow velocity tends on the contrary to accelerate. The shape of the overtopping tongue is correctly represented by the numerical code (Figure 17b). However, the large amount of air pockets characterizing the flow in the laboratory (Figure 17a) cannot be captured by the mono-phase numerical code.

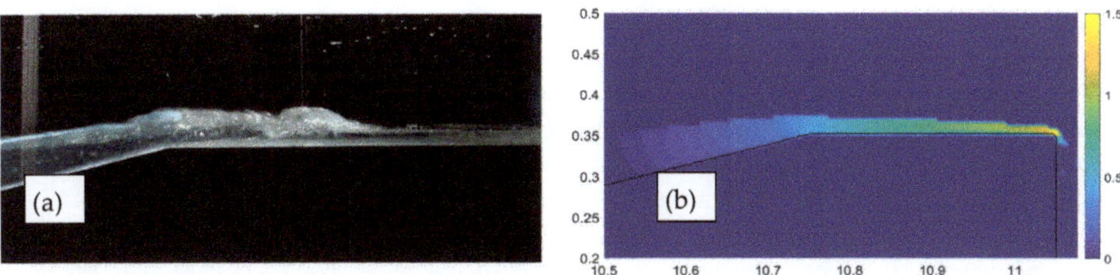

Figure 17. Frames of an overtopping event propagating over the dike crest during the same test R00H05s3G30c4, relative to a breaking wave, and carried out in the laboratory (**a**) and by the numerical code (**b**). The color map displays the computational field of the u-velocity (m/s).

A completely different situation occurs in case the wave reaches the dike crest before breaking. Guo et al. [17] observed that, in such conditions, the overtopping tongue overtops and jumps above the dike off-shore edge, hitting the crest around or just before its middle section (see Figure 18). The section of the impinging jet determines, thus, a sharp modification of the velocity field along the dike crest:

- From the off-shore edge to the impinging jet section, $x_c \approx [0; 0.4 \cdot G_c]$, the overtopping tongue dissipates its energy in the change of direction from the up-rush along the seaward slope to the horizontal stream over the crest;
- Corresponding to the section of the impinging jet ($x_c \approx 0.4 \cdot G_c$), the wave front hits violently against the dike crest surface and breaks; this section is subjected to the maximum impact (wave pressure), and as a consequence of the momentum balance equation, to the minimum velocity; the section is, therefore, associated to the maximum stress and possibly to the maximum scour risk;
- In the second half of the dike crest, $x_c \approx [0.5 \cdot G_c; G_c]$, the overtopping flow velocity tends to accelerate into a supercritical stream for the free-outfall boundary condition at the landward edge, while the potential energy accumulated at the hit turns into kinetic.

In the present work, we found numerical and experimental confirmation of the results by Guo et al. [17]. Figure 18a,b provides two consecutive frames of an impinging jet observed within the same numerical (18a) and experimental (18b) test R00H05s3G30c2. Both the numerical and the experimental frames show the wave overtopping the dike off-shore edge before breaking and impinging on the crest around its middle section, $x_c \approx [0.35; 0.4 \cdot G_c]$. Based on the color map of Figure 16a, u is maximal at the off-shore edge and decelerates while the wave propagates along the dike crest. At the impinging section, the u-value is minimal, correspondingly with the crest surface, and it starts increasing again after the impact. The result of this process is the decreasing/increasing trend of $u_{2\%}$ leading to values of $u_{2\%}(x_c = G_c) \approx u_{2\%}(x_c = 0)$, as shown in Figure 16b.

Figure 18a,b also shows the presence of air pockets entrapped in the water tongue, especially in the area beneath the surging breaker immediately preceding the impinging section. The air entrainment is still thought to be a concurrent cause of the discrepancies between numerical and experimental results and the theoretical approach by Schüttrumpf and Oumeraci [7], as already identified by Guo et al. [17].

The different trends of u associated to the different breaker types affect the value of $u_{2\%}$ at $x_c = G_c$, which in the practice represents one of the most relevant parameters, as it governs the down-wash streaming in the inland area. Figure 19 compares the values of $u_{2\%}(x_c = G_c)$ to the corresponding values of $u_{2\%}(x_c = 0)$ obtained from the experimental (Figure 19a) and numerical tests (Figure 19b) for the different breaking or non-breaking conditions. All the numerical tests (at $R_c/H_s \geq 0$) represent waves reaching the dike crest in breaking ($\cot\alpha_{off} = 4$) or broken ($\cot\alpha_{off} = 6$) conditions, while the experimental tests present both non-breaking ($\cot\alpha_{off} = 2$) and breaking ($\cot\alpha_{off} = 4$) wave conditions.

From Figure 19 it is evident that $u_{2\%}$ effectively decays from $x_c = 0$ to $x_c = G_c$ only in case of c2; i.e., of non-breaking waves (green diamonds). The average decay was of approximately the 30%, ranging between 13% and 45%. The variability of the decay rate was due to the different combinations of crest widths, freeboard conditions, wave steepness and wave heights. However, no explicit or systematic relationship was found between the decay rate and any of the parameters G_c, R_c/H_s, $H_s/L_{m-1,0}$ and H_s. The absence of a direct link between the decay rate and G_c can be explained with the small friction determined by the smooth dike surface and by the limited value of G_c. This result is in agreement with the synthesis by EurOtop [13,28], where a specific indication of negligible decay for flow depths larger than 0.1 m and G_c around 2–3 m is reported.

In cases of c4 (blue circles) and c6 (orange triangles), both the numerical and the experimental results indicate that the decay rates are significantly reduced, being on average of 10% and ranging between −7% and 28%. These results—which are in line with the average trends of Figure 18b—suggest that in case of breaking or broken wave conditions, the decay of $u_{2\%}$ along the dike crest is almost negligible, and in some cases $u_{2\%}(x_c = G_c)$ might result even greater than $u_{2\%}(x_c = 0)$, as already observed by Guo et al. [17]. Generally, the phenomenon of the increasing velocities from $x_c = 0$ to $x_c = G_c$ is more frequently observed with the numerical code (Figure 19b), which also tends to give a slightly lower decay rates with respect to the physical experiments (Figure 19a).

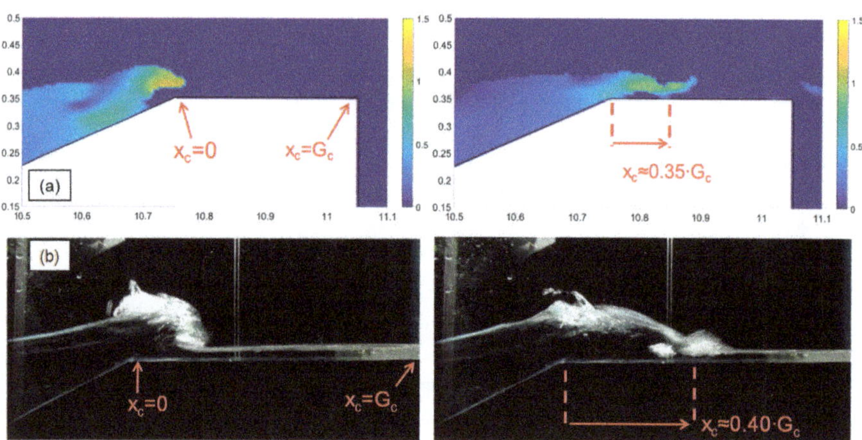

Figure 18. Consecutive frames of an overtopping event propagating over the dike crest during a numerical (**a**) and an experimental (**b**) test (R00H05s2G30c2) relative to non-breaking wave conditions. The color map of panel (**a**) displays the u-velocity field (m/s).

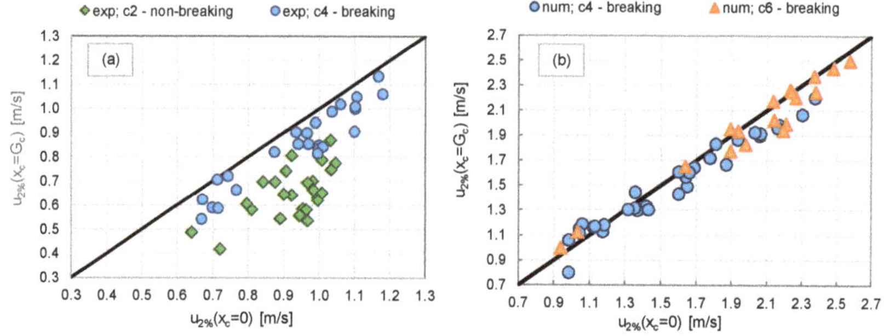

Figure 19. Comparison among values of $u_{2\%}$ measured at the off-shore ($x_c = 0$) and in-shore ($x_c = G_c$) edges of the dike crest in the lab experiments (**a**) and the numerical simulations (**b**). The data are grouped by values of $\cot(\alpha_{off})$. All the tests are at $R_c/H_s \geq 0$.

6.2.2. Effect of the Crest Emergence

In case of $R_c \geq 0$, the decay rate of $u_{2\%}$ from $x_c = 0$ to $x_c = G_c$. was not directly correlated to R_c/H_s (see Section 6.2.1 and Figure 16b). Nevertheless, it was observed that the statistical distribution of the flow velocities was significantly different in cases of $R_c/H_s = 0$ and $R_c/H_s > 0$. To illustrate those outcomes, Figure 20 shows the frequency histograms of the instantaneous depth-averaged u-values measured during an example test case from the laboratory experiment H05s3G30c4 conducted at $R_c/H_s = 0$ (Figure 20a) and $R_c/H_s = 1.0$ (Figure 20b). In both panels of Figure 20a,b, the histograms are provided for the values of u recorded at both D1 (blue shading) and D3 (orange shading); viz., at $x_c = 0$ and $x_c = G_c$, respectively. The probability density function (pdf) for $R_c/H_s = 0$ (Figure 20a) presents a sharp peak corresponding of the lowest values of u (the mode is around ≈ 0.03 m/s for both $x_c = 0$ and $x_c = G_c$); then it decreases almost monotonically with the increasing u-values. The distribution for $R_c/H_s = 1.0$ (Figure 20b) was more flat and symmetrical, showing increasing and decreasing data frequencies for values of u, respectively, lower and greater than the mode, which is ~0.10 m/s for $x_c = 0$ (blue bars) and ~0.10 m/s for $x_c = G_c$ (orange bars). The different mode values in case of $R_c/H_s = 1.0$ indicate that higher values of u are more frequently detected at D3 than at D1 (i.e., the distributions of u at D3 are shifted towards higher values than the distributions at D1); i.e., at $x_c = G_c$ rather than at $x_c = 0$. In other terms, the flow more frequently accelerates than decelerates along the dike crest. This phenomenon is not detectable at $R_c/H_s = 0$ (Figure 20a).

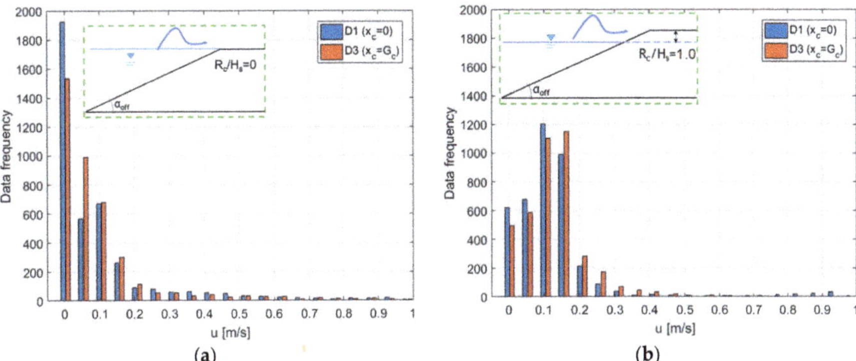

Figure 20. Frequency distribution of the instantaneous depth-averaged values of u at the dike off-shore (blu) and in-shore (orange) edges for the same experimental test (H05s3G30c4) at $R_c/H_s = 0$ (panel (**a**)) and $R_c/H_s > 0$ (panel (**b**)).

From a practical point of view, the analysis of Figure 20 suggests that:

- On average, the flow velocities at both the dike off-shore and in-shore edges are higher at $R_c/H_s > 0$ than at $R_c/H_s = 0$, as already observed for the extreme percentiles $u_{2\%}$ reported in Figure 10;
- In case of $R_c/H_s > 0$, the u-values are more narrowly distributed around the mode, showing a lower variability with respect to the case at $R_c/H_s = 0$;
- At $R_c/H_s > 0$, the flow more frequently accelerates than decelerates from $x_c = 0$ to $x_c = G_c$, in line with the discussion proposed in Section 6.2.1.

The same results here illustrated for the example case of Figure 20 were found for most of the conditions tested.

6.3. Flow Velocities at Negative Freeboard

The phase of decay of $u_{2\%}$ from $x_c = 0$ to $x_c \approx 0.4G_c$ observed for $R_c/H_s \geq 0$ no longer occurred at $R_c < 0$ (see Figure 16b), because the energy dissipation induced by the wave breaking and jumping against the dike crest (see Figure 18 and Section 6.2.1) was reduced or nullified by the presence of the water over the crest. Indeed, at $R_c < 0$, a weir-like constant supercritical flow was established over the dike crest due to the hydraulic gradient between the dike off-shore and in-shore edges. The free outfall condition at the landward slope governed the overflow process, inducing the flow to accelerate and resulting into values of $u(x_c = G_c)$ greater than the values $u(x_c = 0)$. Figure 16b shows, indeed, a monotonic increasing trend of $u_{2\%}$ from $x_c = 0$ to $x_c \approx G_c$ for all the cases at $R_c < 0$.

In many cases, the water layer above the dike crest prevents the wave from breaking at all. For this reason, the effect of the wave breaking presented and discussed for $R_c/H_s \geq 0$ no longer applies. On the contrary, a sharp effect of R_c/H_s was observed, as the increased rate of $u_{2\%}$ increased with the crest submergence. Such an effect was already evident from the average trends of Figure 16b and is displayed by Figure 21, which compares the numerical values of $u_{2\%}(x_c = 0)$ with the corresponding values $u_{2\%}(x_c = G_c)$ for all the available numerical tests at $R_c < 0$ in wet (panel a) and dry (panel b) landward conditions. In that figure, the data are clustered on different levels above the bisector line according to the values of R_c/H_s: the lower the value of R_c/H_s, the farther the data from the bisector; i.e., the lower the R_c/H_s, the larger the ratio $u_{2\%}(x_c = G_c)/u_{2\%}(x_c = 0)$. On average, the values of $u_{2\%}(x_c = G_c)/u_{2\%}(x_c = 0)$ ranged from 5% to 30% for R_c/H_s ranging from -0.2 to -1.5 in dry landward conditions. Similar results were observed for the wet landward conditions, though the increase of $u_{2\%}$ from $x_c = 0$ to $x_c = G_c$ seemed to be more modest (up to 20%–25%) and not sharply related to R_c/H_s. This was probably due to the presence of the water on the landward side, which determines lower hydraulic gradients between the off-shore and in-shore edges, and consequently, more modest flow acceleration.

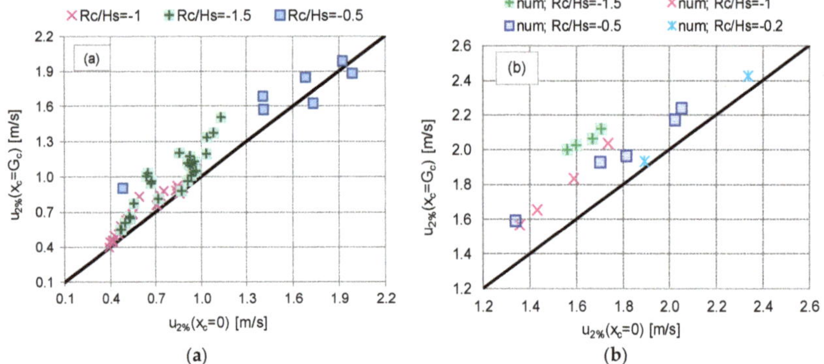

Figure 21. Comparison among values of $u_{2\%}(x_c = 0)$ to corresponding values of $u_{2\%}(x_c = G_c)$ for wet (a) and dry (b) landward conditions. The data belong to the numerical database and are grouped by values of R_c/H_s. For all the data, $R_c/H_s < 0$.

The combined trends of increasing velocities and decreasing thicknesses along the crest fulfill the continuity and the momentum balance equations. The contradiction with the theory [7] holds no more, as the tested conditions by the authors were limited to $R_c > 0$.

6.4. Remarks on the Flow Velocities and Design Recommendations

In agreement with Guo et al. [17], the numerical and experimental results led to the following general remarks about the evolution of the flow velocities along the dike crest.

For smooth surfaces the friction effect is negligible. The decay of u was observed only for very short crest widths for $R_c > 0$; for larger crests and/or negative freeboards, the application of the existing formulae (Equations (8) or (9)) might lead to incautious estimates.

For a conservative approach, it is suggested to neglect any decay of u along the crest in emerged conditions and to assume that $u(x_c = 0) = u(x_c = G_c)$.

In case of $R_c \geq 0$ and non-breaking waves, the minimum or zero velocity is found roughly around the middle section of the dike crest, where the imping jet hits the dike surface, and therefore, the highest loads and stresses are concentrated in this area. Based on this finding, the section subjected to the impinging jet can be considered the most exposed to the risk of scour. That outcome may result in practical interest in the case of permeable structures, and can be combined to the analyses of van der Meer et al. [40], who correlated the start of the scour to the occurrence of a given critical flow velocity.

For $R_c < 0$, the maximum flow velocities are found at the dike in-shore edge. This may result in higher downwash velocities along the dike landward slope and in the landward area. Therefore, it is advised to increase, by approximately 20%, the value of u at the in-shore edge with respect to the value at the off-shore edge (see Figures 16b and 21).

It is important to remark that the analysis of the flow at $R_c < 0$ is available from numerical modelling only and that for $R_c \geq 0$, the experimental data relate to punctual measurements of h and u in regard to the dike crest edges and its middle section.

7. Conclusions

A new database of 60 experimental and 94 numerical tests on wave overtopping against smooth trapezoidal dikes characterized by various slopes, crest levels and landward conditions was created. Overall, the tested configurations consisted of three off-shore slopes ($\cot(\alpha_{off}) = 2$, 4 and 6), relative crest freeboards in the range $R_c/H_s = [-1.5; 1.5]$ and two crest widths G_c (3 and 6 m, in prototype units). The combinations of $H_s/L_{m-1,0}$ and $\cot(\alpha_{off})$ values were such to determine both breaking ($\xi_{m-1,0} \leq 2$) and non-breaking ($\xi_{m-1,0} > 2$) wave conditions within both the numerical and the experimental tests.

The consistency of the new data was checked by comparing the values of the average overtopping discharge (q) and of the wave reflection coefficient (K_r) resulting from the new laboratory and numerical tests, with the corresponding predictions obtained from consolidated methods available from the literature ([28] for q; [34] for K_r; the artificial neural network by [32,33] for both q and K_r). The numerical model was validated by reproducing a subset of 12 experimental tests and by comparing the results in terms of: q, K_r, maximum and mean water depths (h) envelopes and extreme flow depths and velocities ($h_{2\%}$ and $u_{2\%}$) across the dike crest.

The analysis of the values of $h_{2\%}$ and $u_{2\%}$ at the dike off-shore edge ($x_c = 0$) and the comparison with the existing predicting methods [4,5,8] led to the prompting of two new formulations, Equations (5) and (6). These formulae were conceived i) to achieve a better accuracy in the representation of the data; ii) correct some incautious underestimations or excessive overestimations associated to the existing formulae; and iii) extend the field validity of the existing formulae to structures with $\cot(\alpha_{off}) = 2$ and $R_c = 0$. Equations (5) and (6) follow the same formulation of the existing methods (Equations (3) and (4)) but represent a non-linear increase of $h_{2\%}(x_c = 0)$ and $u_{2\%}(x_c = 0)$ with $(R_{u,2\%} - R_c)$ and $(g \cdot (R_{u,2\%} - R_c))^{0.5}$, respectively. Equations (5) and (6) propose also two new coefficients a_h and a_u, which are proportional to $\cot(\alpha_{off})$. The ranges of validity of the method are: $\cot(\alpha_{off}) = 2$–6, $\xi_{m-1,0} = 1$–4 and $R_c \geq 0$ for $h_{2\%}(x_c = 0)$ (Equation (5)); and $R_c > 0$ for $u_{2\%}(x_c = 0)$ and smooth structures, i.e., $\gamma_f = 1$ (Equation (6)).

Equations (5) and (6) were calibrated on numerical and experimental data for modest wave run-up values corresponding to the typical working conditions of the coastal defense structures (($R_{u,2\%} - R_c)/H_s < 2$). Higher run-up heights were also modelled with the numerical code to analyze more severe or catastrophic scenarios (($R_{u,2\%} - R_c)/H_s > 2$–2.5) in a climate change situation. The validity of Equations (5) and (6) were checked against the two sets of experiments, *FlowDike*1 and *FlowDike*2 [39,40], on dikes with $\cot(\alpha_{off}) = 3$ and 6. The new formulae are characterized by at least the same accuracy of the existing methods when applied to *FlowDike*1 and *FlowDike*2, while they provide a remarkably better representation of the new experimental and numerical data.

The trends of $h_{2\%}$ and $u_{2\%}$ along the dike crest, i.e., from the off-shore ($x_c = 0$) to the in-shore ($x_c = G_c$) edge, were analyzed in detail from the numerical model at any crest emergence ($-1.5 \leq R_c/H_s \leq 1.5$) and for both dry, landward conditions. The numerical results are supported by experimental evidence in correspondence of $x_c = 0$, $x_c \approx 0.5$ and $x_c = G_c$ and for values of $R_c/H_s \geq 0$ and dry landward conditions. The analysis resulted in the following criteria for design application.

For each structure slope, crest freeboard and wave attack condition (wave heights and periods), the flow thickness h monotonically decreases from $x_c = 0$ to $x_c = G_c$. The decay of h is strictly dependent on the crest freeboard, varying approximately between the 20% and the 35% for over-washed ($R_c/H_s = -1.5$) and for emerged ($R_c/H_s = 1.5$) conditions respectively. To account for the effect of R_c/H_s, a new coefficient c_3 for the decay formulation of h was proposed in Equation (10). For submerged dikes ($R_c < 0$ and wet landward conditions), the decay of h is almost negligible.

The effect of the friction along the crest on the flow velocity u is negligible in case of smooth dikes, in agreement with the EurOtop manual [13,28]. The trends of u are instead significantly influenced by R_c and by the wave breaking or non-breaking conditions.

For $R_c/H_s \geq 0$ and non-breaking wave conditions, u decreases in the first part, reaches a minimum around the half of the crest, in accordance with the impinging jet hitting the dike surface, and then, increases in the second part, being that the values of u at the landward edge are approximately equal to the seaward edge. The section subjected to the impinging jet is the most exposed to the risk of scour. For a conservative approach, in case of $R_c/H_s \geq 0$ it is suggested to assume $u(x_c = 0) = u(x_c = G_c)$, while the use of the decay trends for u (Equation (8)) is discouraged.

For $R_c/H_s \geq 0$ and breaking or broken waves, u hardly changes along the dike crest. The overtopping flow is governed by the free-outfall boundary condition at the dike off-shore edge, which determines a supercritical accelerated weir-like flow, resulting in values of $u(x_c = G_c)$ which may be even higher than $u(x_c = 0)$.

Based on numerical results only, for $R_c/H_s < 0$ u increases along the crest, and its growth rate increases by increasing the submergence, up to approximately the 30% for $R_c/H_s < -1$, both in the case of wet and dry landward conditions. Overall, it is suggested to assume a value of u at $x_c = G_c$ which is increased of the 20% with respect to the value at $x_c = 0$.

Author Contributions: Conceptualization, S.M.F., G.P. and B.Z.; formal analysis, S.M.F., M.G.G. and M.G.; investigation, G.P. and B.Z.; data curation, S.M.F., M.G.G. and M.G.; writing—original draft preparation, S.M.F.; writing—review and editing, M.G.G. and B.Z.; supervision, B.Z.

Funding: This research was funded by the Horizon 2020 project BRIGAID ("BRIdging the GAp for Innovations in Disaster resilience," www.brigaid.eu), grant number 700699.

Acknowledgments: The first author would like to express her sincere gratitude to Jentsje W. van der Meer for providing his own data and his analyzed results on the FlowDike experiments that we used for the validation of the new fitting. The support of the European Commission through the Horizon 2020 project BRIGAID ("BRIdging the GAp for Innovations in Disaster resilience," www.brigaid.eu) is gratefully acknowledged.

Conflicts of Interest: The authors declare no conflict of interest.

References

1. Van der Meer, J.W.; Hardeman, B.; jan Steendam, G.; Schuttrumpf, H.; Verheij, H. Flow depths and velocities at crest and inner slope of a dike, in theory and with the Wave Overtopping Simulator. In Proceedings of the ICCE 2010 (ASCE), Shanghai, China, 30 June–5 July 2010.

2. Van der Meer, J.W.; Bernardini, P.; Snijders, W.; Regeling, E. The wave overtopping simulator. In Proceedings of the 30th ICCE 2006, San Diego, CA, USA, 3–8 September 2006; Volume 5, pp. 4654–4666.

3. Sumer, B.M.; Fredsøe, J.; Lamberti, A.; Zanuttigh, B.; Dixen, M.; Gislason, K.; Di Penta, A.F. Local scour at roundhead and along the trunk of low crested structures. *Coast. Eng.* **2005**, *52*, 995–1025. [CrossRef]

4. Schüttrumpf, H. Wellenüberlaufströmung bei See-Deichen. Ph.D. Thesis, Technical University Braunschweig, Braunschweig, Germany, 2001.

5. Van Gent, M.R. Wave overtopping events at dikes. In Proceedings of the 28th ICCE 2002, Wales, UK, 7–12 July 2002; Volume 2, pp. 2203–2215.

6. Schüttrumpf, H.; van Gent, M.R. Wave overtopping at seadikes. *Coast. Struct.* **2004**, *2003*, 431–443.

7. Schüttrumpf, H.; Oumeraci, H. Layer thicknesses and velocities of wave overtopping flow at sea dikes. *Coast. Eng.* **2005**, *52*, 473–495. [CrossRef]

8. Bosman, G.; Van der Meer, J.W.; Hoffmans, G.; Schüttrumpf, H.; Verhagen, H.J. Individual overtopping events at dikes. In Proceedings of the 31st ICCE 2008, Hamburg, Germany, 31 August–5 September 2008; pp. 2944–2956.

9. Zanuttigh, B.; Martinelli, L. Transmission of wave energy at permeable low-crested structures. *Coast. Eng.* **2008**, *55*, 1135–1147. [CrossRef]

10. Van Bergeijk, V.M.; Warmink, J.J.; Van Gent, M.R.A.; Hulscher, S.J.M.H. An analytical model of wave overtopping flow velocities on dike crests and landward slopes. *Coast. Eng.* **2019**, *149*, 28–38. [CrossRef]

11. Mares-Nasarre, P.; Argente, G.; Gómez-Martín, M.E.; Medina, J.R. Overtopping layer thickness and overtopping flow velocity on mound breakwaters. *Coast. Eng.* **2019**, *154*, 103561. [CrossRef]

12. Hughes, S.A.; Nadal, N.C. Laboratory study of combined wave overtopping and storm surge overflow of a levee. *Coast. Eng.* **2009**, *56*, 244–259. [CrossRef]

13. Pullen, T.; Allsop, N.W.H.; Bruce, T.; Kortenhaus, A.; Schüttrumpf, H.; van der Meer, J.W. EurOtop. European Manual for the Assessment of Wave Overtopping. August 2007. Available online: www.overtopping-manual.com (accessed on 22 October 2019).

14. Raosa, A.N.; Zanuttigh, B.; Lara, J.L.; Hughes, S. 2DV VOF numerical modelling of wave overtopping over overwashed dikes. *Coast. Eng. Proc.* **2012**, *1*, 62. [CrossRef]

15. Formentin, S.M.; Zanuttigh, B.; van der Meer, J.W.; Lara, J.L. Overtopping flow characteristics at emerged and over-washed dikes. *Coast. Eng. Proc.* **2014**, *1*, 7. [CrossRef]

16. Lara, J.L.; Ruju, A.; Losada, I.J. Reynolds Averaged Navier-Stokes modelling of long waves induced by a transient wave group on a beach. *R. Soc. A* **2011**, *467*, 1215–1242. [CrossRef]

17. Guo, X.; Wang, B.; Liu, H.; Miao, G. Numerical simulation of two-dimensional regular wave overtopping flows over the crest of a trapezoidal smooth impermeable sea dike. *J. Waterw. Port Coast. Ocean* **2014**, *140*, 04014006. [CrossRef]

18. Formentin, S.M.; Zanuttigh, B. A new method to estimate the overtopping and overflow discharge at over-washed and breached dikes. *Coast. Eng.* **2018**, *140*, 240–256. [CrossRef]

19. Zelt, J.A.; Skjelbreia, J.E. Estimating incident and reflected wave field using an arbitrary number of wave gauges. In Proceedings of the 23rd ICCE 1992, Venice, Italy, 4–9 October 1992; Volume I, pp. 777–789.

20. Galvin, C.J. *Wave-Height Prediction for Wave Generators in Shallow Water*; Technical Memorandum No. 4; Army Corps of Engineers: Washington, DC, USA, 1964; pp. 1–20.

21. Wang, S. Plunger-type wavemakers: Theory and experiment. *J. Hydraul. Res.* **1974**, *12*, 357–388. [CrossRef]

22. Takeda, Y. Velocity Profile Measurement by Ultrasound Doppler Shift Method. *Int. J. Heat Fluid Flow* **1986**, *7*, 313–318. [CrossRef]

23. Gaeta, M.G.; Guerrero, M.; Formentin, S.M.; Palma, G.; Zanuttigh, B. Wave-induced flow measurements over dikes by means of ultrasound Doppler velocimetry and videography. under review.

24. Stagonas, D.; Warbrick, D.; Muller, G.; Magagna, D. Surface tension effects on energy dissipation by small scal, experimental breaking waves. *Coast. Eng.* **2011**, *58*, 826–836. [CrossRef]

25. Burcharth, H.F.; Liu, Z.; Troch, P. Scaling of core material in rubble mound breakwater model tests. In Proceedings of the Fifth International Conference on Coastal and Port Engineering in Developing Countries (COPEDECV), Cape Town, South Africa, 16–17 April 1999.

26. Lykke Andersen, T.; Burcharth, H.F.; Gironella, X. Comparison of new large and small scale overtopping tests for rubble mound breakwaters. *Coast. Eng.* **2011**, *58*, 351–373. [CrossRef]

27. Altomare, C.; Gironella, X. An experimental study on scale effects in wave reflection of low-reflective quay walls with internal rubble mound for regular and random waves. *Coast. Eng.* **2014**, *90*, 51–63. [CrossRef]

28. Allsop, N.W.H.; Bruce, T.; DeRouck, J.; Kortenhaus, A.; Pullen, T.; Schüttrumpf, H.; Troch, P.; van der Meer, J.W.; Zanuttigh, B. EurOtop. Manual on Wave Overtopping of Sea Defences and Related Structures. An Overtopping Manual Largely Based on European Research, But for Worldwide Application. Second Edition. 2018. Available online: www.overtopping-manual.com (accessed on 22 October 2019).

29. Franco, L.; Geeraerts, J.; Briganti, R.; Willems, M.; Bellotti, G.; De Rouck, J. Prototype measurements and small-scale model tests of wave overtopping at shallow rubble mound breakwaters: The Ostia-Rome yacht harbour case. *Coast. Eng.* **2009**, *56*, 154–165. [CrossRef]

30. Van Doorslaer, K.; Romano, A.; De Rouck, J.; Kortenhaus, A. Impact on a storm wall cause by non-breaking waves overtopping a smooth dike slope. *Coast. Eng.* **2017**, *120*, 93–111. [CrossRef]

31. Zanuttigh, B.; Formentin, S.M.; van der Meer, J.W. Advances in modelling wave-structure interaction through Artificial Neural Networks. *Coast. Eng. Proc.* **2014**, *1*, 69. [CrossRef]

32. Zanuttigh, B.; Formentin, S.M.; van der Meer, J.W. Prediction of extreme and tolerable wave overtopping discharges through an advanced neural network. *Ocean Eng.* **2016**, *127*, 7–22. [CrossRef]

33. Formentin, S.M.; Zanuttigh, B.; van der Meer, J.W. A neural network for predicting wave reflection, overtopping and transmission. *Coast. Eng. J.* **2017**, *59*, 1750006. [CrossRef]

34. Zanuttigh, B.; van der Meer, J.W. Wave reflection from coastal structures in design conditions. *Coast. Eng.* **2008**, *55*, 771–779. [CrossRef]

35. Muttray, M.; Oumeraci, H.; Ten Oever, E. Wave Reflection and Wave Run-Up at Rubble Mound Breakwaters. *Coast. Eng.* **2007**, *5*, 4314–4324.

36. Formentin, S.M.; Zanuttigh, B. A new fully-automatic procedure for the identification and the coupling of the overtopping waves. *Coast. Eng. Proc.* **2018**, *1*, 36. [CrossRef]

37. Formentin, S.M.; Zanuttigh, B. Semi-automatic detection of the overtopping waves and reconstruction of the overtopping flow characteristics at coastal structures. *Coast. Eng.* **2019**, *152*, 103533. [CrossRef]

38. Lorke, S.; Brüning, A.; Bornschein, A.; Gilli, S.; Krüger, N.; Schüttrumpf, H.; Pohl, R.; Spano, M.; Werk, S. *Influence of Wind and Current on Wave Run-Up and Wave Overtopping*; Hydralab–FlowDike Report; RWTH Aachen, Institute of Hydraulic Engineering: Aachen, Germany, 2009; p. 100. Available online: http://resolver.tudelft.nl/uuid:73cb6cbe-8931-4499-b4d6-aa31548f5dda (accessed on 22 October 2019).

39. Lorke, S.; Brüning, A.; Van der Meer, J.; Schüttrumpf, H.; Bornschein, A.; Gilli, S.; Pohl, R.; Spano, M.; Řiha, J.; Werk, S.; et al. On the effect of current on wave run-up and wave overtopping. *Coast. Eng. Proc.* **2011**, *1*, 13. [CrossRef]

40. Bijlard, R.; Steendam, G.J.; Verhagen, H.J.; van der Meer, J.W. Determining the critical velocity of grass sods for wave overtopping by a grass pulling device. *Coast. Eng. Proc.* **2016**, *1*, 20. [CrossRef]

Article

Laboratory Experimental Investigation on the Hydrodynamic Responses of an Extra-Large Electrical Platform in Wave and Storm Conditions

Dong-Liang Zhang [1,2], Chun-Wei Bi [3,*], Guan-Ye Wu [1,2], Sheng-Xiao Zhao [1,2] and Guo-Hai Dong [3]

[1]   Power China Huadong Engineering Corporation Limited, Hangzhou 310014, China; zhang_dl@ecidi.com (D.-L.Z.); wu_gy@ecidi.com (G.-Y.W.); zhao_sx@ecidi.com (S.-X.Z.)

[2]   Offshore Wind Power R&D Center of Power China Huadong, Hangzhou 310014, China

[3]   State Key Laboratory of Coastal and Offshore Engineering, Dalian University of Technology, Dalian 116024, China; ghdong@dlut.edu.cn

*   Correspondence: bicw@dlut.edu.cn

Received: 8 September 2019; Accepted: 28 September 2019; Published: 29 September 2019

Abstract: The application of electrical platform for converter station in offshore wind farm is highly forward-looking and strategic. The offshore electrical platform is complicated in structure, bulky in volume, and expensive in cost. In addition, the built-in electrical equipment is very sensitive to the acceleration response. Therefore, it is very important to study the hydrodynamic response of the electrical platform exposed in the open sea. Based on the elastic similarity, Froude similarity, as well as the flexural-stiffness similarity of the cross-section, the hydroelastic similarity was derived to guide the model test of a 10,000-ton offshore electrical platform in wind, wave, and current. The hydrodynamic responses including strain and acceleration at key positions of the structure were obtained for different incident angles of external environmental loads. The experimental results showed the increase of water depth can cause more than 10 times increase of strain and acceleration response of the platform. The attack angle of external environmental loads had no definite relationship with strain response of the structure. Therefore, the most dangerous attack angle cannot be determined. The strain of the structure under the combined action of wind, wave, and flow was significantly larger than that under wave load only.

Keywords: electrical platform; hydrodynamic response; strain; acceleration; hydroelastic similarity; laboratory experiment

---

1. Introduction

The wind power is clean renewable energy. The development of wind power, especially offshore wind power, plays a decisive role in accelerating the green energy transformation in coastal areas. In view of the trend of large capacity and long distance of offshore wind farm, the application of electrical platform for offshore converter station is highly forward-looking and strategic. The offshore electrical platform is complicated in structure, bulky in volume, and expensive in cost. In addition, as the load-bearing structure of the electrical equipment in the offshore wind farm, the offshore electrical platform, which is a typical top-heavy structure, is characterized by a large number of electrical equipment arranged on the deck. The dynamic effect of the superstructure will aggravate the vibration of the structure, which will lead to fatigue and even failure of the jacket structure. Therefore, it needs to ensure its own safety exposed in the external environment load in open sea. Wave load is one of the most common environmental loads of offshore electrical platform structure, which is directly related to the selection and design of offshore electrical platform. Especially when the natural vibration frequency of the structure is close to the frequency of dynamic load of wave, it is easy to cause structural damage or collapse. Resonance will aggravate the stress and deformation of the structure and seriously threaten the safety and reliability of offshore electrical platform. Therefore, it is very important to accurately estimate the wave load and the response of the electrical platform structure in waves.

The main supporting structure of the offshore electrical platform is jacket foundation. There have been systematic researches on the dynamic characteristics of jacket foundation. Elshafey et al. [1] studied the dynamic response characteristics of a jacket platform in air and water by combining physical model test with theoretical analysis and focused on the impact of platform weight and wave load on the pile foundation. Dong et al. [2] calculated the dynamic response of the jacket support structure under wind and wave loads using a decoupled procedure and performed long-term fatigue analysis of welded multiplanar tubular joints for a fixed jacket offshore wind turbine. Park et al. [3] used the finite element method and Morison equation to analyze the hydrodynamic response of a jacket platform under wave and current action. The displacement response of the offshore platform under random wave action was analyzed with NEWMARK algorithm. Taylor et al. [4] and Santo et al. [5,6] conducted systematic studies on the shielding effect of the jacket platform and its influence characteristics on the drag force on the platform. Saha et al. [7] obtained short-term extreme values for the dynamic responses of offshore fixed wind turbines by using the aerodynamic software HAWC2 and the hydrodynamic software USFOS. Raheem et al. [8] and Zadeh et al. [9] investigated the effect of geometric nonlinearity and wave nonlinearity on the motion response of the jacket offshore platform, respectively.

Besides the offshore oil platform, the jacket structure is also commonly used as the foundation for the offshore wind turbine. Devaney et al. [10] conducted mathematical modeling of linear and nonlinear waves combined with the Morison equation to check the breaking wave loads and stress analysis of jacket structures supporting offshore wind turbines. Dezvareh et al. [11] suppressed wind/wave-induced vibrations of jacket-type offshore wind turbines by placing a passive vibration absorber. Wang et al. [12] conducted model test and numerical analysis of an offshore wind turbine under seismic, wind, wave, and current loads. Wei et al. [13] studied the effect of structural dynamics on the response of an offshore wind turbine supported by a jacket and subjected to wave loads. Gho et al. [14] presented an eccentric jacket substructure for offshore wind turbines to better withstand intense environmental forces and to replace conventional X-braced jackets in seismically active areas. Liu et al. [15] developed a novel frequency-domain transient response estimation method to obtain reliable estimations of the dynamic responses of offshore wind turbines with obvious nonzero initial conditions.

At present, the research on the dynamic characteristics of jacket structure under wave load tend to be systematic; however, there are few relevant researches on the dynamic characteristics of electrical platform under complex sea conditions. Due to the top-heavy characteristic of the offshore electrical platform, the superstructure may cause significant dynamic amplification effect. Therefore, it is necessary to systematically study the dynamic characteristics of the electrical platform in waves and storm condition.

2. Materials and Methods

2.1. Experimental Model Similarity

Considering the dimensions of the electrical platform, marine hydrological parameters, and laboratory conditions, the geometric scale of this test model was 1:60. The model similarity criterion is the key to this research. Besides, the structural vibration should satisfy the elastic similarity as follows [16]:

$$\lambda_\rho \cdot \lambda_A \cdot \lambda^3 \cdot \lambda_u \cdot \lambda_t^{-2} = \lambda_I \cdot \lambda_E \cdot \lambda_u \cdot \lambda^{-3}, \tag{1}$$

where λ_ρ is density scale, λ is geometric scale, λ_u is deformation scale, λ_E is elastic modulus, λ_t is time scale, λ_A is sectional area scale, and λ_I is sectional moment of inertia scale.

For a structure dominated by bending vibration, the elastic similarity law can be rewritten as follows:

$$\lambda_t{}^2 = \lambda^4 \cdot \lambda_\rho \cdot \lambda_E{}^{-1} \cdot \lambda_r{}^{-2}, \tag{2}$$

where λ_r is radius of inertia scale.

Froude number similarity is expressed as follows [17]:

$$\lambda_v = \lambda^{0.5}, \tag{3}$$

where λ_v is speed scale.

Considering acceleration scale is $\lambda_g = 1$, thus,

$$\lambda_t = \lambda^{0.5}, \tag{4}$$

Combining the Equations (2) and (4), and simultaneously satisfy elastic similarity and Froude number similarity, the hydroelastic similarity can be given as follow:

$$\lambda^3 \cdot \lambda_r{}^{-2} = \lambda_E, \tag{5}$$

This hydroelastic similarity is the main criterion for the design of the experimental model.

2.2. Mechanical Test of the Experimental Material

Based on the hydroelastic similarity and the previous experience on physical model test, plexiglass was selected as the material for the experimental model of the electrical platform. To obtain the basic material parameters of plexiglass, the mechanical test was carried out for the selected plexiglass which was used to make the platform model.

Cuboid specimens with a length of 200 mm, a width of 25 mm, and a thickness of 2 mm, 3 mm, 3.6 mm, and 5 mm were produced. Two pieces of each specimen were duplicated. The density of the specimen was calculated according to the mass and geometric size of the specimen, and the average value of all specimens was taken as the density of the plexiglass. The average density of organic glass materials was 1201.2 kg/m^3. Standard specimens with thickness of 3 mm, 4 mm, 5 mm, and 10 mm were produced with 4 duplicates for each specimen. A universal testing machine was used to determine the static elastic modulus and Poisson ratio of plexiglass by uniaxial tensile test. The average values of static modulus and Poisson ratio of all specimens were taken as the material parameters of plexiglass. The static elastic modulus and Poisson ratio of plexiglass were 2.62 GPa and 0.42, respectively. For the dynamic model test of the electrical platform in this study, the dynamic elasticity modulus of the material should also be considered. The dynamic elasticity modulus can be calculated by measuring the fundamental frequency of a cantilever beam with rectangular section according to the formula as follow.

$$E' = 4\pi^2 \rho A L^4 f^2 / (3.515^2 I), \tag{6}$$

where E' is radius of inertia scale, ρ is the material density, A is the sectional area, L is the length of the cantilever specimen, I is the cross-sectional moment of inertia, and f is the fundamental frequency of the specimen. Rectangular specimen with thickness of 3 mm, 4 mm, 5 mm, 7 mm, and 9 mm were prepared. One end of the specimen was fixed and restrained, and an impact load was applied to the top of the other end of the specimen. The time history of strain response of the specimen was collected, and the fundamental frequency of the specimen was obtained by frequency domain analysis. The dynamic elasticity modulus of plexiglass was 3.91 GPa.

2.3. Experimental Model Design

The prototype structural material is steel with elastic modulus of 206 GPa. The dynamic elastic modulus of plexiglass for the experimental model is 3.91 GPa. Therefore, the elastic modulus scale is

52.69 and the ratio of section radius of inertia is 64.027. The similarity relations of other parameters are summarized in Table 1. Some of the similarities relations related to section are taken platform leg as an example.

Table 1. The similarity relation of basic parameters for the experimental model design.

Parameter	Similarity	Similar Scale
Length	λ	60
Area	λ_A	1120.7
Volume	$\lambda \cdot \lambda_A$	67,241
Density	$\lambda_\rho = 1$	1
Mass	$\lambda_\rho \cdot \lambda \cdot \lambda_A$	67,241
Speed	$\lambda^{0.5}$	7.746
Acceleration	$\lambda_g = 1$	1
Time	$\lambda^{0.5}$	7.746
Frequency	$\lambda^{-0.5}$	0.129
Force	λ^3	60^3
Moment	λ^4	60^4
Moment of area	$\lambda_A \cdot \lambda_r^2$	4.59e6
Moment of mass	$\lambda \cdot \lambda_A \cdot \lambda_r^2$	2.76e8
Stress	$\lambda^4 \cdot \lambda_D \cdot \lambda_A^{-1} \cdot \lambda_r^{-2}$	176.3

The geometric dimensions of each component of the platform model are determined according to the geometric scale and the inertia radius scale. Schematic diagram of the electrical platform is shown in Figure 1. The parameters of the superstructure model are shown in Table 2. The jacket structure mainly consists of cylindrical steel pipes. Both the length and outer diameter of the members and the pile legs are scaled by the geometric scale of 1:60; therefore, the hydrodynamic characteristics of the platform can be simulated accurately. The pipe section of the pile leg was calculated according to the area scale to achieve the accurate simulation of the elastic response of the structure (see Table 3).

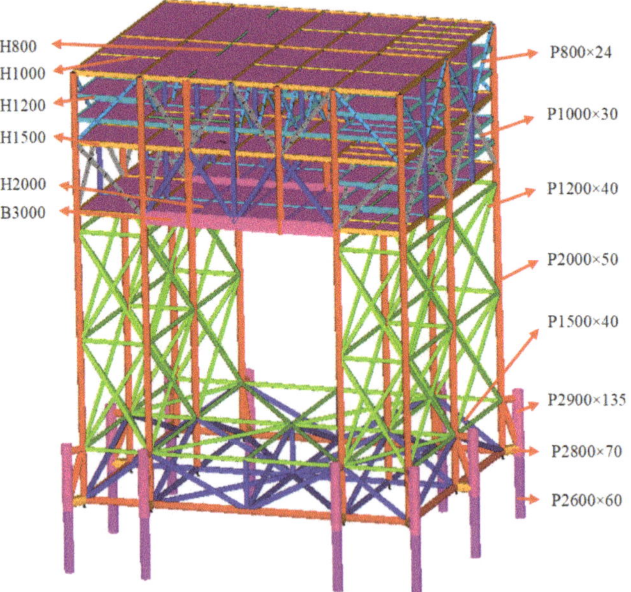

Figure 1. Schematic diagram of the electrical platform.

Table 2. Geometric dimension parameters of the superstructure model.

Member Type	Model Number	Sectional Dimension (mm)	Section for Model
Box beam	B3000	59.1 × 23	
	H2000	45.5 × 8.6	
	H1500	34.2 × 6.8	Rectangular section
I-beam	H1200	27.6 × 6.4	(Height × Width)
	H1000	22.8 × 5.1	
	H800	18.1 × 3.8	
	P2000 × 50	32 × 3	Pipe (Outside diameter ×
Pipe	P1500 × 40	25 × 2	Thickness)
	P1000 × 30	22	Solid bar (Diameter)
	P800 × 24	18	

Table 3. Geometric dimension parameters of the jacket structure model.

Member Type	Model Number	Sectional Dimension (mm)	Section for Model
	P2000 × 50	32 × 3	
	P1500 × 40	25 × 2	
Pipe	P1200 × 40	20 × 2	Pipe (Outside diameter ×
	P2800 × 70	45 × 5	Thickness)
	P2600 × 60	45 × 3	
	P2900 × 135	50 × 5	

Due to the material difference between the prototype and model platform, the weight distribution on the platform is different. Thus, in this study, lead is used to balance each component of the model structure. For the structural mass of the model component, the weight is scaled according to the mass scale. For the nonstructural mass, such as the weight of the equipment, it is scaled as force. In this way, both gravity similarity and hydroelastic similarity are satisfied. Finally, the balance weight was 274.4 kg for the superstructure and 142.83 kg for the jacket model. The total weight of the platform model was 502.5 kg.

In this study, the foundation of the electrical platform consisted of piles. However, in the laboratory test, the soil data is difficult to be determined. Therefore, it is necessary to adopt a simplified approach instead. The equivalent pile method was used to simulate the pile–soil interaction of converter station [18]. Firstly, the finite element model was established to simulate the nonlinear characteristics between pile and soil based on the p-y method. Then, taking the structural basic frequency as the constraint condition, the relationship between the equivalent pile length and pile diameter was determined by adjusting the equivalent pile length. In this study, the equivalent pile length of the electrical platform model is determined to be 5.82 times the pile diameter.

2.4. Experimental Set-Up

The laboratory experiments were conducted in a wave and current flume at the State Key Laboratory of Coastal and Offshore Engineering (SLCOE), Dalian University of Technology, Dalian, China. The flume is 60 m long, 4.0 m wide, and 2.5 m deep. The flume is equipped with wind, current, and wave systems. In the actual sea area, the electrical platform may be subjected to wave, current, and wind from various directions, resulting in the electrical platform being subjected to loads in different directions. In this test, three incident directions of external loads, namely 0°, 45°, and 90°, are adopted to analyze the hydrodynamic response of the electrical platform (see Figure 2). The electrical platform model and its layout in wave flume are shown in Figure 3.

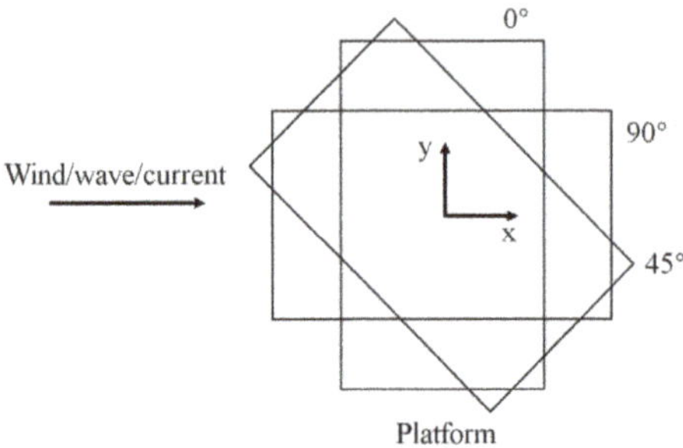

Figure 2. The experimental layout of the electrical platform model.

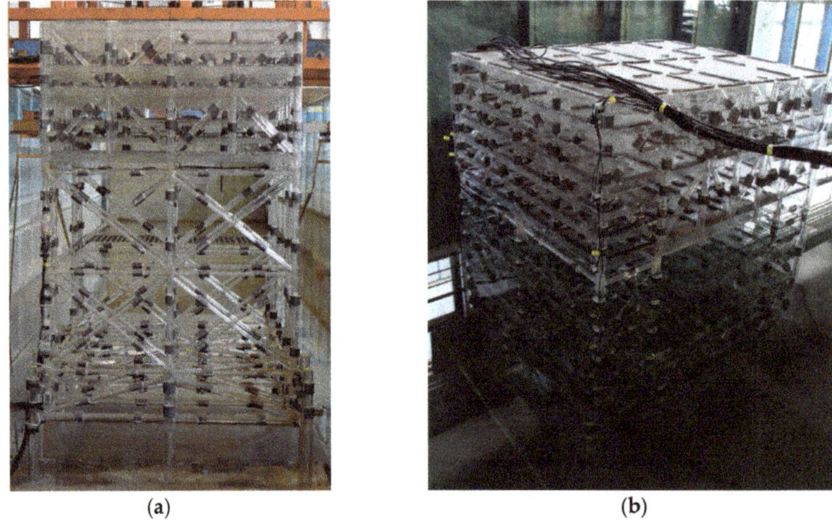

Figure 3. The electrical platform model in the wave flume. (**a**) In air; (**b**) In water.

In the physical model test, wave parameters were measured by capacitive wave gauge with a measuring range of 35 cm and a relative error of 0.5%. Wave height and period can be collected and calculated automatically with the wave height acquisition system. The ADV current meter of Nortek company was used for velocity collection. The sampling rate of this instrument was 200 Hz, the measuring range was 1 m/s, and the measured accuracy was 0.5%. The laboratory adopts blowers to generate wind, and the measurement of wind speed is carried out by a thermal anemometer. The measuring range is 0.3~30 m/s, and the measuring precision can be controlled within 1%. A NI (National instruments) data acquisition instrument was used to collect strain data at key positions of the structure in combination with strain gauge. The waterproof acceleration sensor is adopted to measure the acceleration response of key positions at the platform structure (see Figure 4). The measuring range of the acceleration sensor is 1.0 g and the accuracy is approximately 0.3%. The measuring frequency is up to 1000 Hz, which has high stability and anti-interference ability. The arrangement of measuring points in physical model test is very important for obtaining effective data. The data of measuring points can reflect the load on the weak position of the electrical platform. If the measuring points are not arranged reasonably, the validity and reference value of the data will be reduced. This study determined key nodes and positions of the electrical platform using the finite element analysis. The strain gauge and the acceleration sensor were arranged as shown in Figure 5.

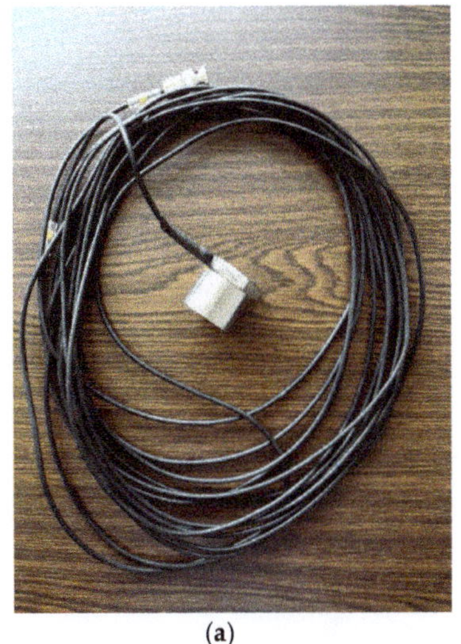 

(a)   (b)

Figure 4. The experimental layout of acceleration sensors. (**a**) Acceleration sensor; (**b**) Acceleration sensors on the platform.

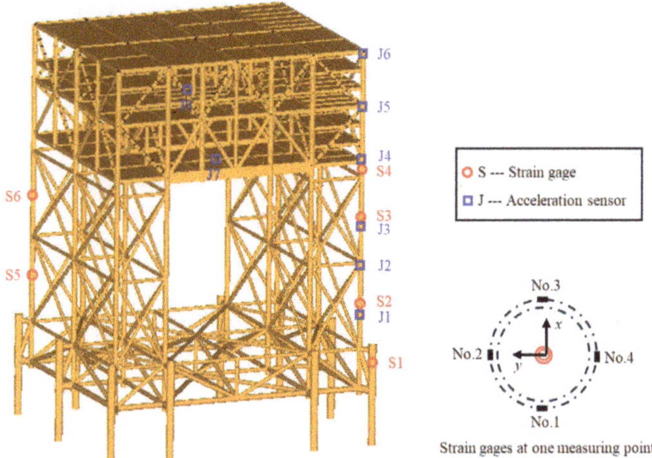

Figure 5. Measurement positions of dynamic response at key nodes of the electrical platform.

2.5. Experimental Conditions

The normal water depth of the electrical platform is 52 m, corresponding to the water depth during the model experiment of 0.87 m. Considering the length of the equivalent pile and the thickness of the fixed bottom plate, the actual water depth is set as 1.17 m in the experiment. In order to obtain the hydrodynamic responses of the electrical platform under storm condition, a test water depth of 1.45 m, corresponding to the extreme depth of 68.8 m, was added to carry out the storm condition test.

To analyze the hydrodynamic characteristics of the electrical platform, a series of regular waves were designed in the test, as shown in Table 4. The waves include five different wave heights and five different wave periods. The storm condition was modeled using combined action of wind, current, and irregular wave (see Table 5) with deeper water depth. In this way, the hydrodynamic characteristics of the electrical platform under extreme load can be considered. This experiment was conducted with the three kinds of environmental loads acting in the same direction, which was considered as most dangerous condition for the platform structure. The wind speed, current speed, significant wave height, and wave period for an irregular wave are shown in Table 6. During the test, the current was generated first using the circulation flow system. After the current speed reached a stable level, the wind was generated. Finally, wave was generated.

Table 4. Regular waves for the physical model experiment.

Test Number	Wave Height (m)	Wave Period (s)
A1	0.05 (3) [1]	1.6 (12.4)
A2	0.10 (6)	1.6 (12.4)
A3	0.15 (9)	1.0, 1.2, 1.4, 1.6, 1.8 (7.7, 9.3, 10.8, 12.4, 13.9)
A4	0.20 (12)	1.6 (12.4)
A5	0.25 (15)	1.6 (12.4)

[1] The values in parentheses represent the prototype values.

Table 5. Regular waves for the physical model experiment.

Test Number	Significant Wave Height (m)	Significant Wave Period (s)
B1	0.15 (9)	1.6 (12.4)
B2	0.20 (12)	1.8 (13.9)

Table 6. Storm condition for the physical model experiment.

Test Number	Wind Speed (m/s)	Current Speed (m/s)	Irregular Wave
C1	4.6 (36)	0.226 (1.75)	$H_s = 0.15$ m $T_p = 1.6$ s
C2	6.6 (51.5)	0.258 (2)	$H_s = 0.20$ m $T_p = 1.8$ s

3. Results

3.1. Strain Response of Electrical Platform

3.1.1. The Electrical Platform in Regular Waves

Under the water depth of operating condition, the dynamic responses of the electrical platform in regular waves are studied. Strain gauges were used to collect the strain response of structural piles. Taking the measurement results of four strain gauges at measuring point S5 at 0° angle of attack as an example, the time-history curves of structural strain response measured by different strain gauges along the circumference of a pile leg are shown in Figure 6. It can be seen from the experimental results that the strain-response value of strain gauge No. 1 at the measuring point S5 is greater than that of the other three strain gauges at the same measurement position. Similarly, the strain response of strain gauge No. 1 at position S6 is greater than that of other strain gauges. The strain response of strain gauge No. 3 at positions S1, S2, S3, and S4 is larger than that of other strain gauges. This is mainly due to the relative position of the measuring points. Measuring points S1, S2, S3, and S4 are on the wave side of the platform, while S5 and S6 are on the leeward side of the structure.

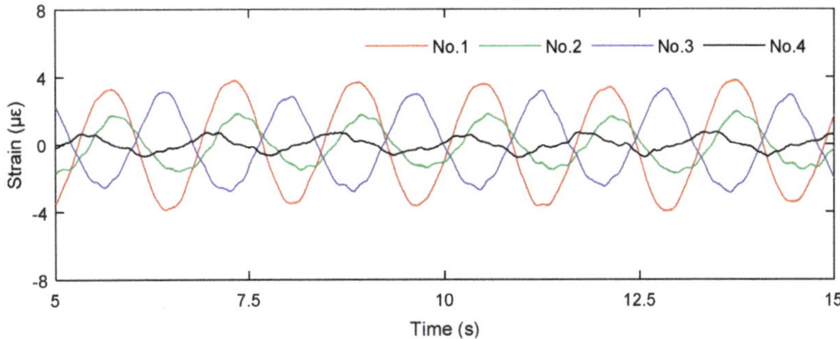

Figure 6. Time-series of the strain response of strain gauges at measurement position S5 in waves. Wave condition: $H = 0.10$ m and $T = 1.6$ s.

The time-history responses of strain at different positions on the pile leg of the electrical platform are shown in Figure 7. The maximum value of strain at measuring points S2, S3, and S4 decreases along with the height of the structure. The maximum strain at measuring point S1 is less than that at S2. Take the maximum value of strain peaks of four strain gauges at the same measuring point as the peak value of strain response at a measurement position. The experimental results show that the strain-response values at the measuring points 2, 3, and 4 on the same pile leg gradually decrease with the increase of the height (see Table 7).

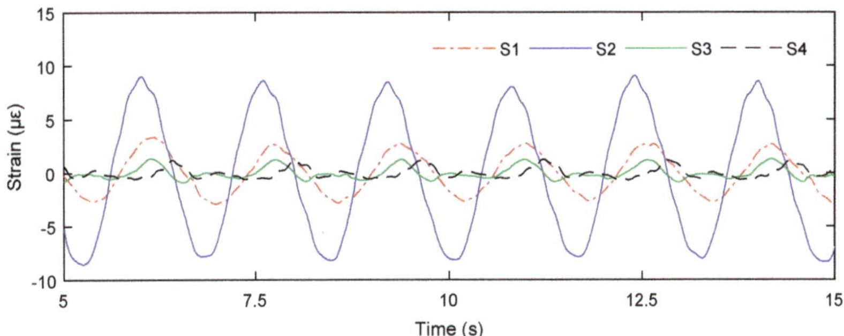

Figure 7. Time-series of the strain response at various measurement positions in waves. Wave condition: $H = 0.15$ m and $T = 1.4$ s.

Table 7. Strain-response values at various measurement positions in waves (unit: $\mu\varepsilon$).

Wave Condition	Measuring Point					
	1	**2**	**3**	**4**	**5**	**6**
H0.05_T1.6	1.7	2.3	0.5	0.13	1.9	0.53
H0.10_T1.6	3.4	5.2	0.65	0.35	3.6	1.2
H0.15_T1.0	1.8	5.1	1.2	0.8	3.7	2.8
H0.15_T1.2	2.1	6.5	0.9	0.5	3.7	3.3
H0.15_T1.4	3.9	7.7	1.1	0.6	4.9	3
H0.15_T1.6	5.3	8.2	1.2	0.81	6	2.4
H0.15_T1.8	6.4	9.2	1.7	1.3	6.7	2.2
H0.20_T1.6	8.2	12.1	1.9	1.5	8.8	3.8
H0.25_T1.6	10.6	16.2	3.2	2.5	11.8	5.3

Taking the test conditions with a period of 1.6 s as an example, the maximum strain at measuring points 1, 2, 3, and 4 changes with wave height, as shown in Figure 8a. Within the range of wave height of 0.25 m, the structural strain response increases with increasing wave height. For wave height of 0.15 m, the results of maximum strain at measuring points 1, 2, 3, and 4 are shown in Figure 8b. Overall, the strain value of the structure increases with the increase of wave height. The experimental results show that the strain response at the measuring points 1 and 2 increases with the increase of the wave period from 1.0 to 1.8 s. However, the strain response at measuring points 3 and 4 is not sensitive to the wave period.

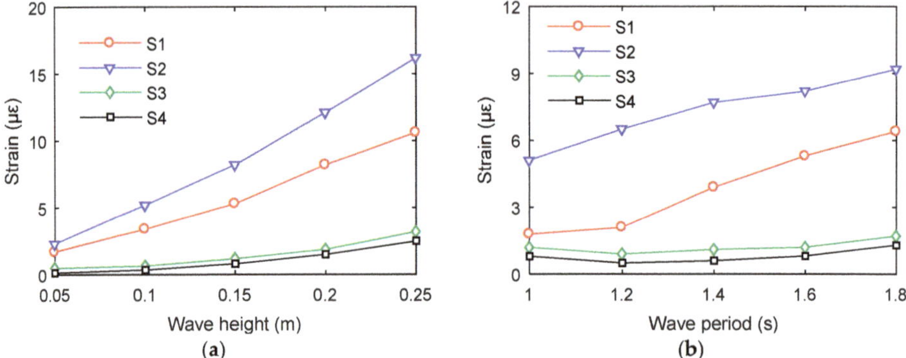

Figure 8. The structural strains at various measurement positions in waves. (**a**) Variation with wave height when wave period $T = 1.6$ s; (**b**) Variation with wave period when wave height $H = 0.15$ m.

3.1.2. The Electrical Platform in Storm Condition

The dynamic response test of the electrical platform is carried out under the combined actions of wind, wave, and current and considering a storm surge elevation. The structural strain response of the electrical platform under three different attack angles, 0°, 45°, and 90°, was simulated under storm condition. Taking the platform with an attack angle of 90° as an example, the time-history curves of structural strain response at the four strain gauges around the measurement point S1 on the pile leg under storm condition C2 are shown in Figure 9. By analyzing the measured results of four strain gauges around the circumference of the pile leg at different positions, the results show that the measured results of the strain gauge at the outside of the platform structure are larger than those of the other three strain gauges.

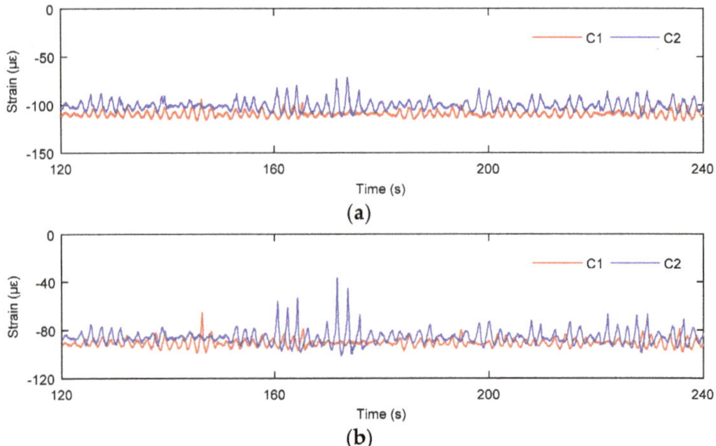

Figure 9. *Cont.*

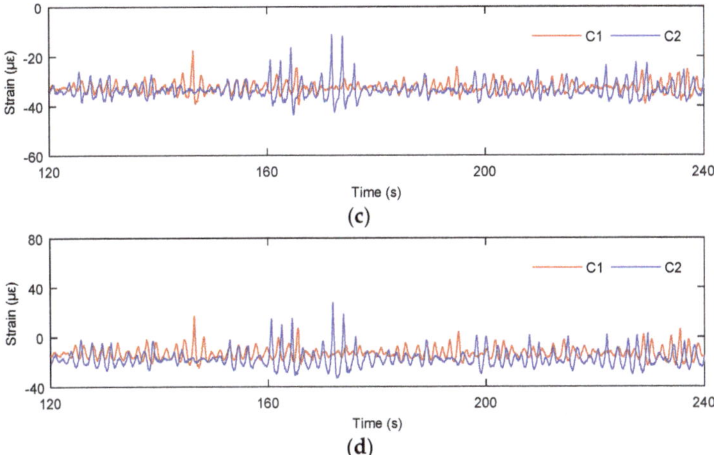

Figure 9. Time-series of the strain response at various measurement positions in storm conditions. (a) S1; (b) S2; (c) S3; (d) S4.

The strain response of the pile leg was analyzed under storm conditions of C1 and C2. The comparison results of the time history of strain response at different measuring points are shown in Figure 9. According to the analysis results, the strain at different measuring points in storm condition C2 is significantly larger than that in C1. Under the same environmental load, the strain values of strain gauges No. 2 and No. 4 were very large, followed by strain gauge No. 1, and strain gauge No. 3 had a minimum strain.

3.2. Acceleration Response of Electrical Platform

3.2.1. The Electrical Platform in Regular Waves

In this study, the peak value of the acceleration response of the structure is taken as the key index to reflect the motion response of the electrical platform. Under the action of different regular waves, the acceleration response of the structure at different positions is shown in Table 8. Figure 10 shows the variation of the peak value of acceleration response with wave height and wave period at various measurement points of the structure. It is indicated that when the wave period is constant, the acceleration response of the measurement point of the structure increases with the increase of wave height. While the wave height is constant, the acceleration response of the measuring point decreases with the increase of the period. Overall, the peak value of acceleration response at measuring point J3 is the largest under different wave conditions. However, the acceleration response at measuring points J5, J6, J7, and J8 is not sensitive to the changes of wave height and period.

Through frequency domain analysis of acceleration time-history curve, the corresponding frequency-domain curves of various measuring points are obtained. The peak value of frequency domain curve appears at 5.6 Hz which is approximately the frequency of the first-order mode in the normal direction of the platform (see Figure 11).

According to the experimental results, the acceleration response at different positions of the electrical platform was analyzed. Comparison between time series of the acceleration response at superstructure, jacket structure, and sensitive positions for internal electrical equipment is shown in Figure 12. To observe the change of the acceleration response more clearly, the time series at various measurement positions are presented with a range of 3.2 s, respectively. Among the eight acceleration sensors arranged on the structure, acceleration sensor J3 had the largest response. This is because J3 is located on the lower jacket structure and near the water surface. This part of the structure is directly affected by wave load; thus the peak value is higher than that of other positions. When wave acts on the structure, the instantaneous response of acceleration at each measuring point is quite different. The acceleration responses at all measuring points decay synchronously after wave excitation, and the attenuation trend is basically the same. It can be concluded that the wave excitation propagates upward and downward from the middle part of the platform structure at the same time. When the whole structure is excited, the response at each measuring point attenuates synchronously.

Table 8. Peak values of acceleration response at various measurement positions in waves (unit: m/s^2).

Wave Condition	Measuring Point							
	J1	J2	J3	J4	J5	J6	J7	J8
H0.05_T1.6	0.0197	0.024	0.0251	0.0121	0.0044	0.0047	0.0035	0.0075
H0.10_T1.6	0.0336	0.0644	0.0691	0.0368	0.0082	0.0083	0.0065	0.0055
H0.15_T1.0	0.1078	0.2110	0.2087	0.1063	0.0226	0.0219	0.0208	0.0174
H0.15_T1.2	0.0722	0.1510	0.167	0.0619	0.0173	0.0146	0.0119	0.0155
H0.15_T1.4	0.0630	0.1420	0.1561	0.056	0.0132	0.0126	0.0095	0.0118
H0.15_T1.6	0.0577	0.1296	0.1313	0.0601	0.0134	0.0141	0.0124	0.0111
H0.15_T1.8	0.0510	0.1081	0.1118	0.0552	0.0112	0.0146	0.0106	0.0104
H0.20_T1.6	0.0715	0.1369	0.1624	0.077	0.0163	0.0146	0.0126	0.0131
H0.25_T1.6	0.078	0.1505	0.1901	0.075	0.0173	0.0149	0.0152	0.0178

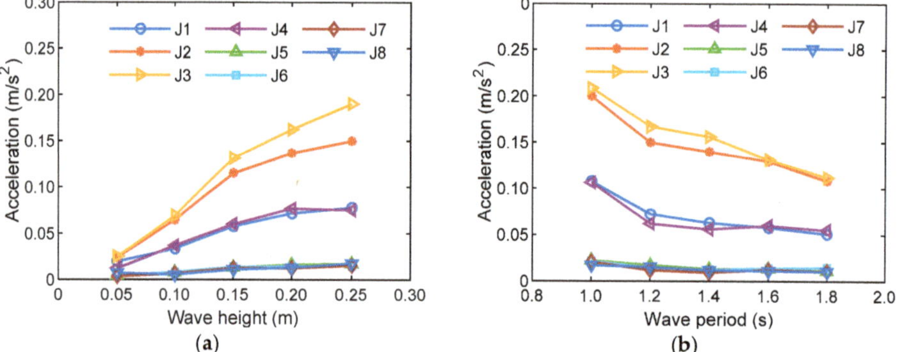

Figure 10. The peak accelerations at various measurement positions in waves. (**a**) Variation with wave height when wave period T = 1.6 s; (**b**) Variation with wave period when wave height H = 0.15 m.

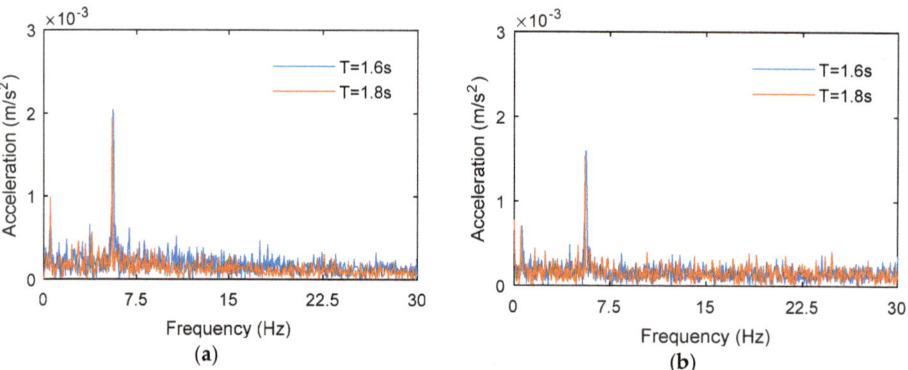

Figure 11. Frequency-domain comparison of acceleration responses at measurement positions J7 and J8 in waves. (**a**) J7; (**b**) J8.

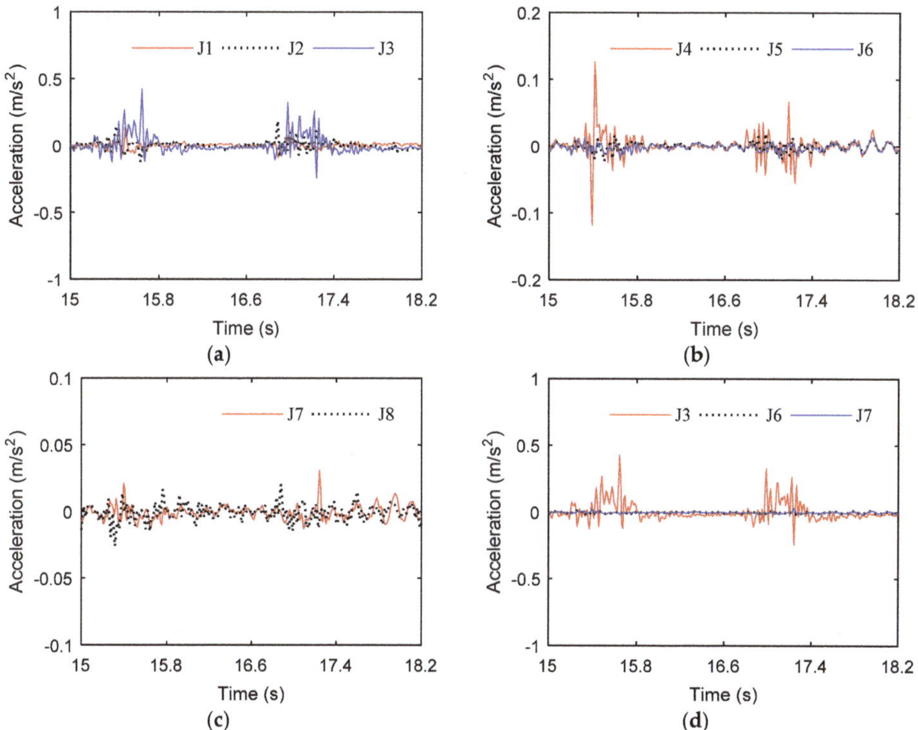

Figure 12. Time-series of the acceleration response at various measurement positions in wave condition A5. (**a**) J1, J2, J3; (**b**) J4, J5, J6; (**c**) J7, J8; (**d**) J3, J6, J7.

3.2.2. The Electrical Platform in Storm Condition

The acceleration response of some key nodes of the electrical platform under storm condition was measured. Under the combined action of wind, wave, and current, the acceleration response of the structure increases with increasing wind speed, current speed, and wave height. According to the acceleration time series at different positions, the acceleration response of the structure is analyzed in frequency domain (see Figure 13). The results show that the spectral peaks of the structure at different positions appear at approximately 6 Hz, which is close to the first-order mode frequency of the structure in the normal direction. In addition, it can be found from the spectrum curve that the wave slamming on the lower deck of the platform excites the high-order mode of the structure.

Based on the experimental results, the acceleration responses of the jacket structure, the superstructure, and sensitive positions of the electrical equipment inside the platform were compared and analyzed. To observe the variation of the acceleration more clearly, each response time series was presented with a range of 8.0 s (see Figure 14). The results show that the trend of acceleration response at three measuring points on the jacket is consistent. However, there are significant differences in both the time history and the peak value of acceleration response at the measurement points in the superstructure and electrical equipment. This is mainly due to that the phase difference occurs in the transmission process after the external environmental load stimulates the hydrodynamic response in the structure. At the same time, this phenomenon is related to the high-order mode of the structure due to wave excitation.

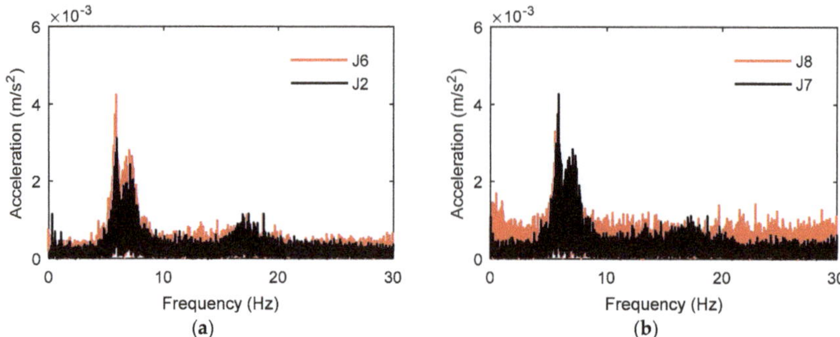

Figure 13. Frequency domain comparison of acceleration responses at various measurement positions in storm condition C2. (**a**) J2 and J6; (**b**) J7 and J8.

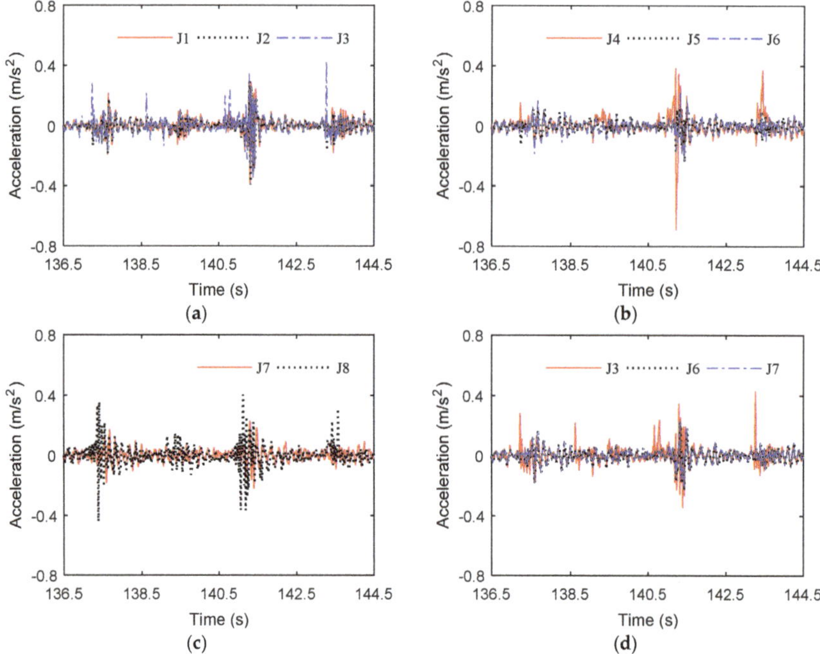

Figure 14. Time-series of the acceleration response at various measurement positions in storm condition C2. (**a**) J1, J2, J3; (**b**) J4, J5, J6; (**c**) J7, J8; (**d**) J3, J6, J7.

4. Discussion

4.1. Effect of Water Depth on Strain Response

The electrical platform structure presents different characteristics of the hydrodynamic response under normal and storm conditions. When the water depth changes, the hydrodynamic response of the structure will be different, even if the wave state remains the same. The structural strain peaks at different positions under the action of regular waves with an attack angle of 0° are shown in Figure 15. When the wave height increases, the strain of the structure increases in both extreme and normal water depths. Moreover, the strain peak of the structure in extreme depth is larger than that in normal depth.

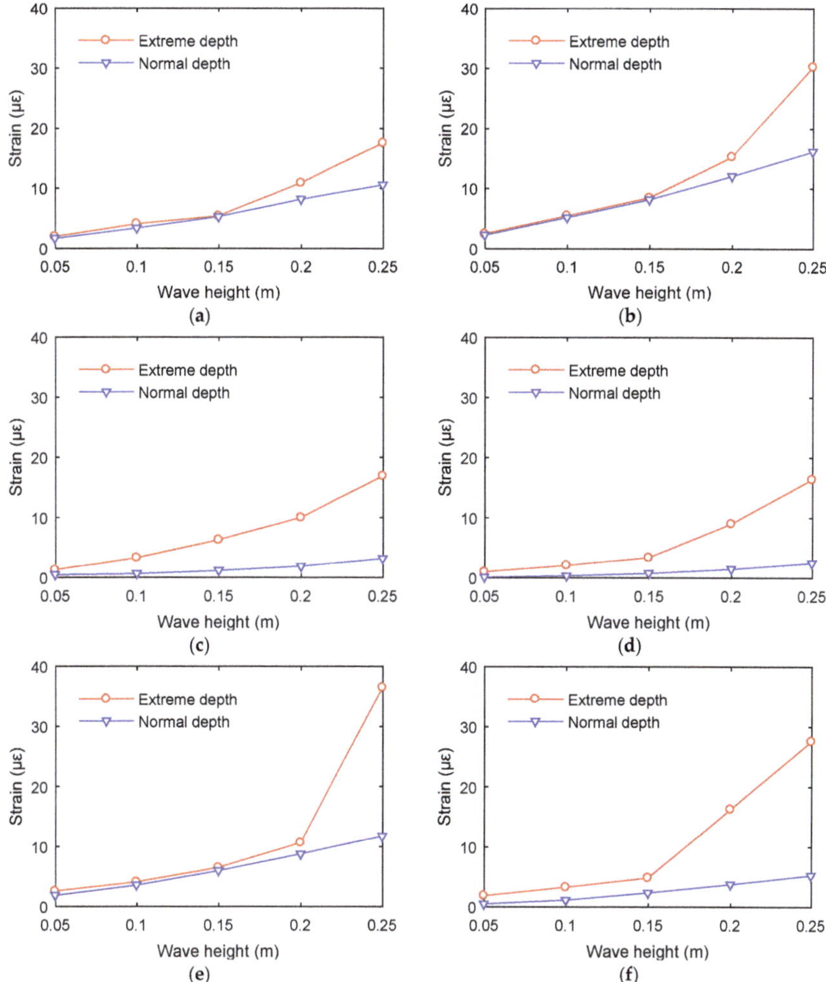

Figure 15. Structural strain responses at various measurement positions of the electrical platform with different water depths in waves for wave period of 1.6 s. (**a**) S1; (**b**) S2; (**c**) S3; (**d**) S4; (**e**) S5; (**f**) S6.

In the experiment, when the wave height is greater than 0.20 m, the structural strain in extreme depth is significantly greater than that in normal depth. This is mainly due to the following two reasons: (i) When the water depth increases, the area of interaction between water and the jacket structure of the electrical platform increases. Therefore, the loads acting on the structure increases, and the local stress and strain increase correspondingly. However, jacket members are slender and the increase of effective area is limited. Therefore, the strain increment is small under waves with lower wave height (less than 0.15 m). (ii) When the water level rises, larger waves (wave height larger than 0.20 m) will interact with the superstructure. On the one hand, waves will impact on the windward side of the superstructure of the platform. On the other hand, waves interact with the lower deck of the platform, which creates a complex wave-slamming effect. This will change the mechanism of dynamic response of the structure significantly, and the stress and acceleration responses will show strong nonlinearity. Moreover, the environmental load and response of the structure increase significantly.

4.2. Effect of Attack Angle on Strain Response

In this study, three attack angles between the environmental loads and the platform structure were considered. Under the same environmental loads, the attack angle has a significant influence on the structural strength and motion response. However, from a perspective of practical engineering, structural strain is more closely related to structural safety. Therefore, the strain response under different angles of attack was analyzed in this section. With the increase of wave height, the strain response of the structure under the action of waves at all angles tends to increase. However, the variation trends of structural strain are different at various positions under three angles of attack. The electric platform structure is a statically indeterminate frame structure. No matter which direction the wave load impacts on the platform, the strain at different positions has a good consistency; in addition, the strain values are comparable. There was no significant difference in strain value between the wave side and the leeward side (see Figure 16).

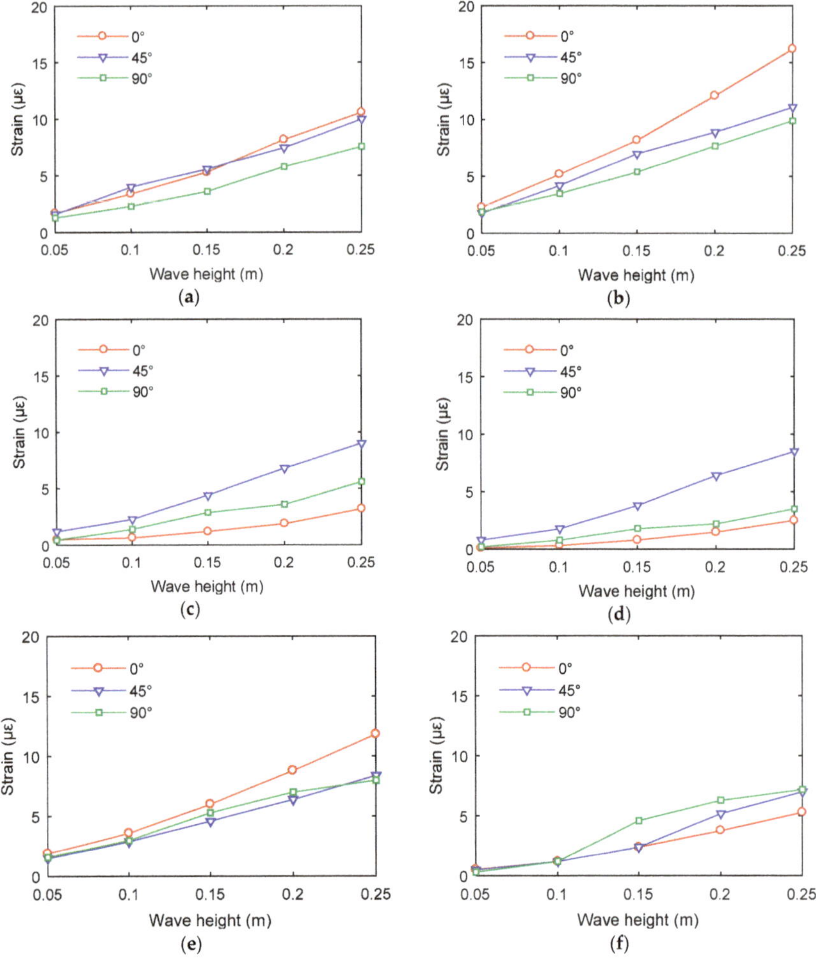

Figure 16. Structural strain responses at various measurement positions of the electrical platform with various attack angles in waves for wave period of 1.6 s. (**a**) S1; (**b**) S2; (**c**) S3; (**d**) S4; (**e**) S5; (**f**) S6.

4.3. Effect of Environmental Loads on Strain Response

Various combined modes of environmental loads have different effects on strain response of the platform. Taking the time series of strain at the measuring point S2 at 0° attack angle as an example, the strain response of the platform structure under irregular wave only and the combined action of wind-wave-current is shown in Figure 17. It can be seen from the experimental results that the strain response of the structure under waves with the wave height of 0.20 m is larger than that when the wave height is 0.15 m, regardless of the combined action of wind, wave, and current or wave only. Overall, the strain response of the structure under combined loads is larger than that under wave load only. Under wave action only, the structural strain changes periodically around the equilibrium position, with a maximum value of approximately 10 με. However, due to the actions of wind and current, the equilibrium position of the time series of strain increases to approximately 30 με. Under the combined action of wind, wave, and current, the maximum value of strain can exceed 40 με. Due to the action of wind and water, the structure experiences an initial constant value of strain. Therefore, the strain of the structure under the combined action of wind, wave, and current is significantly larger than that under the action of irregular waves.

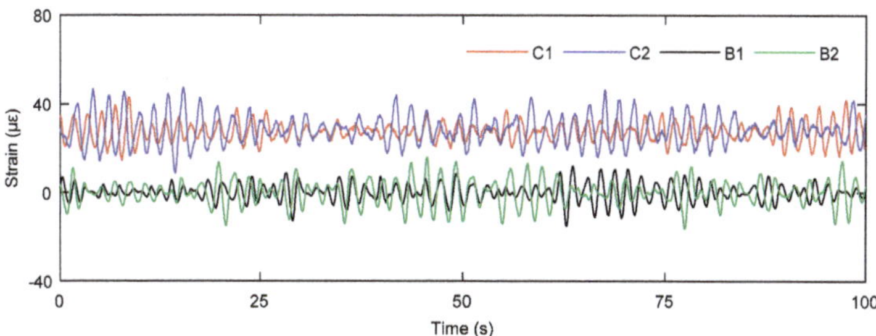

Figure 17. Time-series of the strain response at measurement position S2 in both wave and storm conditions.

4.4. Effect of Water Depth on Acceleration Response

Similar to the strain response, the influence of water depth on the acceleration response of the structure is mainly related to the following two reasons. On the one hand, the area of interaction between water and the jacket structure and electrical platform increases with the increase of water depth. Therefore, the structural load increases and the acceleration response of the structure increases correspondingly. On the other hand, as the water rises, waves with larger wave height slam directly on the superstructure. It can greatly increase the wave load and acceleration response of the structure. According to the experimental results, when the wave height increases, the peak acceleration of the structure in extreme and normal water depths presents a nonlinear increasing trend (see Figure 18). When the wave height increases, the peak acceleration of the structure increases in both extreme and normal water depths. When the wave height is small (less than 0.10 m), the peak acceleration of each measuring position of the structure in extreme and normal water depths has limited difference. However, when the wave height is greater than 0.15 m, the acceleration response of the structure in extreme depth is significantly greater than that in normal depth. This phenomenon is consistent with the variation of wave load. The results of acceleration response in this study can provide reference for the layout of electrical equipment on the platform.

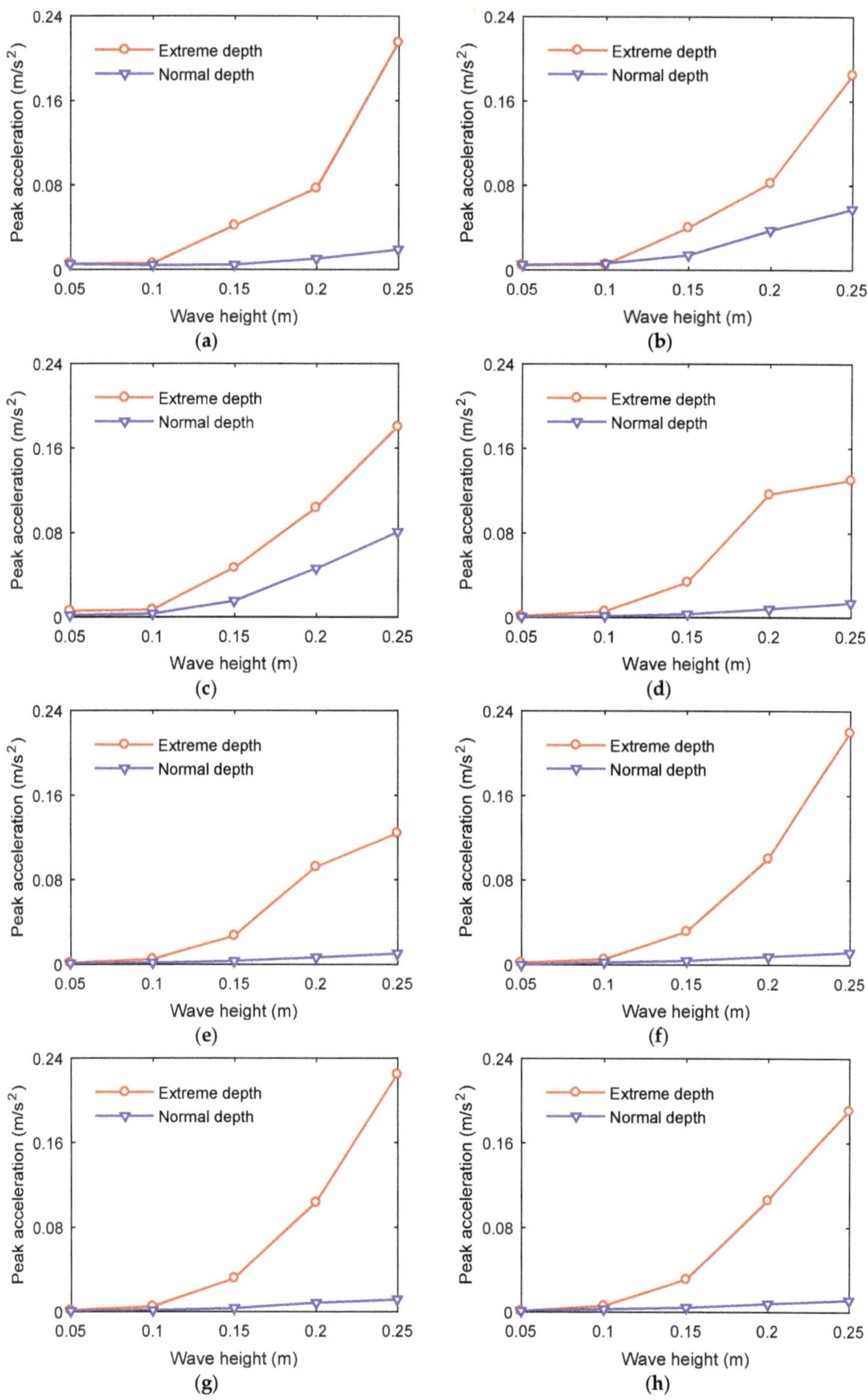

Figure 18. Structural peak accelerations at various measurement positions of the electrical platform with different water depths in waves for wave period of 1.6 s at attack angle of 90°. (**a**) J1; (**b**) J2; (**c**) J3; (**d**) J4; (**e**) J5; (**f**) J6; (**g**) J7; (**h**) J8.

The present study is primarily based on physical model test. Wind, wave, and current acting on the platform in the same direction is considered as the most dangerous sea state. In addition, both normal wave and storm conditions are considered in this study. The present results could reflect the dynamic responses of the offshore electrical platform. Due to high efficiency and accuracy, numerical simulation has been widely used in the fields of coastal and offshore engineering [19–22]. In the future work, the results of physical model test will be taken as the benchmark for the development of finite element model. On the basis of validation, the numerical simulation of the electrical platform will be carried out to further enrich the structural responses of strain, stress, and motion. The study of wind, wave, and current loads acting on the structure with different incident directions will be conducted and provide a supplement for the results of the physical model test. Besides, the effect of extreme loads, such as earthquake, on the dynamic response of an electrical platform can be investigated by numerical simulation [23].

5. Conclusions

In this study, a physical model experiment was carried out to study the dynamic characteristics of a 10,000-ton offshore electrical platform under the combined actions of wind, wave, and current. Two water depth conditions, extreme and normal depths, were simulated in the experiments. Through the physical model test, the acceleration and stress responses of the offshore electrical platform in three typical attack angles were obtained. The main conclusions are as follows:

(i) The increase of water depth caused a rapid increase of strain and acceleration response of the electrical platform. Especially in waves with high wave height, the peak value of strain and acceleration responses could increase by more than 10 times, if waves directly impacted on the superstructure of the electrical platform. Therefore, in practical engineering, the wave height and storm surge should be fully considered to avoid wave-slamming effect on the superstructure.

(ii) With the increase of wave height, the strain response of the structure under waves at various attack angles tended to increase. However, there was no definite relationship between attack angle and peak strain at different locations. It was suggested that the peak values of strain and acceleration response caused by different external environmental loads should be fully considered in structural design.

(iii) The experimental results showed that the maximum strain value is about 10 $\mu\varepsilon$ under waves. When wind, wave, and current acted on the platform together, the maximum strain value can exceed 40 $\mu\varepsilon$. The strain of the structure under the combined action of wind, wave, and flow was significantly larger than that under wave load only.

Author Contributions: Conceptualization, D.-L.Z.; formal analysis, C.-W.B.; funding acquisition, G.-Y.W. and S.-X.Z.; investigation, D.-L.Z.; methodology, C.-W.B. and G.-H.D.; resources, G.-Y.W.; supervision, G.-Y.W. and G.-H.D.; writing—original draft, C.-W.B.; writing—review and editing, D.-L.Z. and S.-X.Z.

Funding: This research was funded by National Key R&D Program of China, Grant No. 2017YFC1404200; National Natural Science Foundation of China (NSFC), Grant No. 31972843; Research Fund of Power China Huadong Engineering Corporation, Project No. KY2018-ZD-03; and Fundamental Research Funds for the Central Universities, Project No. DUT18RC(3)076.

Acknowledgments: The authors acknowledge the technical support from Power China Huadong Engineering Corporation Limited.

Conflicts of Interest: The authors declare no conflict of interest.

References

1. Elshafey, A.A.; Haddara, M.R.; Marzouk, H. Dynamic response of offshore jacket structures under random loads. *Mar. Struct.* **2009**, *22*, 504–521. [CrossRef]
2. Dong, W.; Moan, T.; Gao, Z. Long-term fatigue analysis of multi-planar tubular joints for jacket-type offshore wind turbine in time domain. *Eng. Struct.* **2011**, *33*, 2002–2014. [CrossRef]

3.   Park, M.S.; Koo, W.; Kawano, K. Numerical analysis of the dynamic response of an offshore platform with a pile-soil foundation system subjected to random waves and currents. *J. Waterw. Port Coast. Ocean Eng.* **2012**, *138*, 275–285. [CrossRef]

4.   Taylor, P.H.; Santo, H.; Choo, Y.S. Current blockage: Reduced Morison forces on space frame structures with high hydrodynamic area, and in regular waves and current. *Ocean Eng.* **2013**, *57*, 11–24. [CrossRef]

5.   Santo, H.; Taylor, P.H.; Day, A.H.; Nixon, E.; Choo, Y.S. Current blockage and extreme forces on a jacket model in focused wave groups with current. *J. Fluid. Struct.* **2018**, *78*, 24–35. [CrossRef]

6.   Santo, H.; Taylor, P.H.; Day, A.H.; Nixon, E.; Choo, Y.S. Blockage and relative velocity Morison forces on a dynamically-responding jacket in large waves and current. *J. Fluids Struct.* **2018**, *81*, 161–178. [CrossRef]

7.   Saha, N.; Gao, Z.; Moan, T.; Naess, A. Short-term extreme response analysis of a jacket supporting an offshore wind turbine. *Wind Energy* **2014**, *17*, 87–104. [CrossRef]

8.   Raheem, A.; Shehata, E. Study on nonlinear response of steel fixed offshore platform under environmental loads. *Arabian J. Sci. Eng.* **2014**, *39*, 6017–6030. [CrossRef]

9.   Zadeh, S.M.G.; Baghdar, R.S.; Olia, S.V.K. Finite element numerical method for nonlinear interaction response analysis of offshore jacket affected by environment marine forces. *Open J. Mar. Sci.* **2015**, *5*, 422–442. [CrossRef]

10.   Devaney, L.C. *Breaking Wave Loads and Stress Analysis of Jacket Structures Supporting Offshore Wind Turbines*; University of Manchester: Manchester, UK, 2012.

11.   Dezvareh, R.; Bargi, K.; Mousavi, S.A. Control of wind/wave-induced vibrations of jacket-type offshore wind turbines through tuned liquid column gas dampers. *Struct. Infrast. Eng.* **2016**, *12*, 312–326. [CrossRef]

12.   Wang, W.H.; Gao, Z.; Moan, T.; Li, X. Model Test and Numerical Analysis of a Multi-Pile Offshore Wind Turbine under Seismic, Wind, Wave and Current Loads. *J. Offshore Mech. Arct. Eng.* **2016**, *139*, 031901. [CrossRef]

13.   Wei, K.; Myers, A.T.; Arwade, S.R. Dynamic effects in the response of offshore wind turbines supported by jackets under wave loading. *Eng. Struct.* **2017**, *142*, 36–45. [CrossRef]

14.   Gho, W.M.; Yang, Y. Ultimate strength of completely overlapped joint for fixed offshore wind turbine jacket substructures. *J. Mar. Sci. Appl.* **2019**, *18*, 99–113. [CrossRef]

15.   Liu, F.; Li, X.; Tian, Z.; Zhang, J.; Wang, B. Transient response estimation of an offshore wind turbine support system. *Energies* **2019**, *12*, 891. [CrossRef]

16.   Lin, G.; Zhu, T.; Lin, B. Dynamic model test similarity criterion. *J. Dalian Univ. Technol.* **2000**, *40*, 1–8. (In Chinese)

17.   Chakrabarti, S. *Handbook of Offshore Engineering*; Elsevier Press: Amsterdam, The Netherlands, 2005.

18.   API. *Recommended Practice for Planning, Designing, and Constructing Fixed Offshore Platforms-Working Stress Design*, 22nd, ed.; American Petroleum Institute Recommended Practice 2A-WSD: Washington, DC, USA, 2007.

19.   Gaeta, M.G.; Bonaldo, D.; Samaras, A.G.; Carniel, S.; Archetti, R. Coupled wave-2D hydrodynamics modeling at the Reno river mouth (Italy) under climate change scenarios. *Water* **2018**, *10*, 1380. [CrossRef]

20.   Hu, X.; Yang, F.; Song, L.; Wang, H. An unstructured-grid based morphodynamic model for sandbar simulation in the Modaomen Estuary, China. *Water* **2018**, *10*, 611. [CrossRef]

21.   Petti, M.; Bosa, S.; Pascolo, S. Lagoon sediment dynamics: A coupled model to study a medium-term silting of tidal channels. *Water* **2018**, *10*, 569. [CrossRef]

22.   He, Z.; Hu, P.; Zhao, L.; Wu, G.; Pähtz, T. Modeling of breaching due to overtopping flow and waves based on coupled flow and sediment transport. *Water* **2015**, *7*, 4283–4304. [CrossRef]

23.   Sun, Z.Z.; Bi, C.W.; Zhao, S.X.; Dong, G.H.; Yu, H.F. Experimental analysis on dynamic responses of an electrical platform for an offshore wind farm under earthquake load. *J. Mar. Sci. Eng.* **2019**, *7*, 279. [CrossRef]

Article

Application of an Analytic Methodology to Estimate the Movements of Moored Vessels Based on Forecast Data

José Sande [1], Andrés Figuero [1,*], Javier Tarrío-Saavedra [2], Enrique Peña [1], Alberto Alvarellos [2] and Juan Ramón Rabuñal [3]

[1] Water and Environmental Engineering Group (GEAMA), Universidade da Coruña, Campus Elviña S/n, 15071 A Coruña, Spain
[2] Centre for Research of Information and Communication Technologies (CITIC), Universidade da Coruña, Campus Elviña S/n, 15071 A Coruña, Spain
[3] Department of Computer Science, Universidade da Coruña, Campus Elviña S/n, 15071 A Coruña, Spain
* Correspondence: andres.figuero@udc.es

Received: 21 June 2019; Accepted: 1 September 2019; Published: 4 September 2019

Abstract: A port's operating capacity and the economic performance of its concessions are intimately related to the quality of its operational conditions. This paper presents an analytical methodology for estimating the movements of a moored vessel based on field measurements and forecast data, specifically including ship dimensions and meteorological and maritime conditions. The methodology was tested and validated in the Outer Port of Punta Langosteira, A Coruña, Spain. It was determined that the significant wave height outside the port, and the ratio of the vessel's length divided by its beam (L/B), are the variables that most influence movements. Furthermore, heave and surge are the movements with a better value of the coefficient of determination (R^2 values of 0.71 and 0.67, respectively), the sway ($R^2 = 0.30$) and roll ($R^2 = 0.27$) being the worst when using the available forecast variables of the Outer Port of Punta Langosteira. Despite their low R^2 values, sway and roll models are able to estimate the main trends of these movements. The obtained estimators provide good predictions with assumable error values (root mean square error—RMSE and mean absolute error—MAE), showing their potential application as a predictive tool. Finally, as a consequence, the A Coruña Port Authority has included the results of the methodology in its port management system allowing them to predict moored vessel behavior in the port.

Keywords: ship motions; in-situ observations; port operation; transfer functions; meteorological and ocean conditions; vessel dimensions

1. Introduction and Objectives

In one respect, the quality of port operations can be defined by the maxim "the better a vessel's stay in port, the greater the economic returns". An important aspect that affects this process is the movements of the moored vessels. These movements are divided into three rotations (roll, pitch, and yaw) and three linear displacements (heave, surge, and sway). Each of these degrees of freedom is dependent on many variables, including climatic conditions, the vessel load cargo configuration, the vessel type, its location in the dock, the available defenses (fenders, bollards, etc.), and the mooring system employed [1].

On the other hand, decisions relating to the number of mooring lines, the ropes material (steel wire, synthetic fiber ropes, etc.), and the mooring arrangement depend on harbor pilot considerations, the mooring service providers, the mooring equipment on the berths, and the vessel captain. Finally, the vessel's cargo configuration during operations modifies its center of gravity. This variation is difficult to ascertain with precision and would require a continuous record [2].

At present, there are a number of general recommendations regarding threshold values for movements during vessel loading and unloading operations [3,4]. Although these regulations establish movement criteria for safe working conditions, they do not clearly specify what type of statistical value of the movement they refer to (maximum, average or significant motion amplitudes). Moreover, because they are general recommendations, their specific application to each individual port requires a separate study [5].

Studies relating to operational capacity are traditionally conducted using three methodologies: numerical models, physical models, and field campaigns. Small-scale physical models [6–8] allow the simplified reproduction of port characteristics, vessel dimensions, mooring configuration, and different climatic conditions, but do not permit the accurate analysis of the variation in cargo configuration which occurs during operations. In addition, for a physical model to be reliable, it is important to assure that the model is accurate and realistic, which is achieved by costly construction and intense calibration [9,10]. On the other hand, although the advancement of numerical models facilitates the analysis of the behavior of a moored vessel and the influence on it of the mooring configurations or the effect of passing ships with lower computational and economic costs [11–13], these tools also have similar limitations as the physical models, such as the disadvantage of not reproducing the variations experienced by the position of the vessel's center of gravity during the cargo operation. Therefore, using these two methodologies it is possible to analyze a specific loading condition (ship fully loaded, ballasted, etc.) but not the continuous variation of the same. Finally, studies conducted through field campaigns allow a comprehensive analysis of this process and its influence on the dynamic behavior of moored vessels. However, the current measurement techniques and data processing technology have limitations in terms of accuracy, the resolution of the instrumentation, temporary data logging, information storage, and computational cost. Nevertheless, at present there are studies in which some of the degrees of freedom are analyzed, together with the equations that define them and the loads that moored vessels are subjected to in specific situations, such as the swell generated by a vessel navigating in the port [14,15].

The objective of the present work was the development of an analytical methodology to predict the movements of moored vessels based on the data available by the Port Authority forecast and the vessel movements measured in a field campaign. This methodology has been applied and validated at the facilities of the Outer Port of Punta Langosteira, in A Coruña, Spain (Figure 1a). Each of the degrees of freedom was correlated to climatic variables and vessel dimensions, by means of multivariate linear approximation (transfer functions). These results allowed the A Coruña Port Authority to develop a management system to determine the port's operating capacity, based on forecast data. With this system, it will be possible to evaluate the quality of the port's operational facilities, determine the ideal working windows, and optimize the use of the port's spaces. Furthermore, this methodology could be exportable to other ports if an analysis of the influential and available forecast variables is made, as well as a record of the movements of the moored vessels. Despite the influence of mooring lines on the behavior of vessels at berth, the mooring system information (material, initial pretension, and mooring arrangement) was not introduced as a variable to obtain the transfer functions, since no forecast data on these parameters would be available to subsequently feed the obtained models. In addition, as a results of the characteristics and layout of the port mooring equipment, vessels use two mooring arrangements (Figure 1b): 4-2-2-4 for large bulk carriers (4 bow lines–2 bow spring–2 stern springs–4 stern lines) and 3-2-2-3 for general cargo ships (3 bow lines–2 bow spring–2 stern springs–3 stern lines). Therefore, there is no variability in the number of moorings lines within the same vessel type.

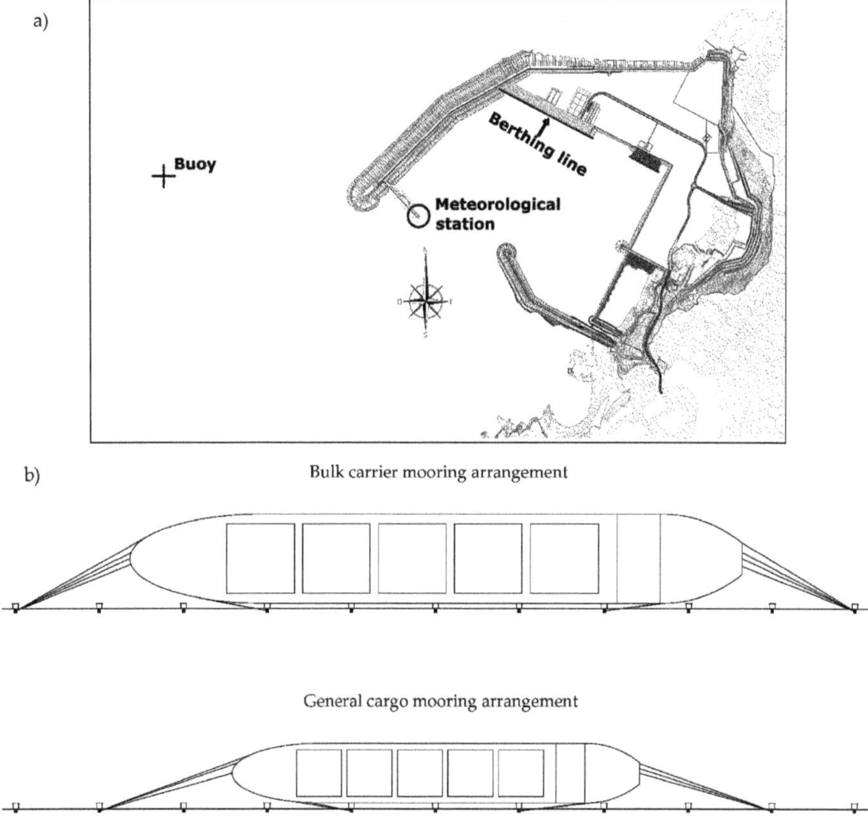

Figure 1. The Outer Port of Punta Langosteira, A Coruña, Spain. The berthing line and the location of the wave buoy and the weather station used in this study are highlighted (**a**). Mooring arrangement of each vessel type (**b**).

2. Field Campaign and Forecast Data

Three field campaigns were conducted at the facilities of the Outer Port of Punta Langosteira in A Coruña for the measurement of the variables involved, lasting a total of 18 months (October–March 2015–2016, 2016–2017, and 2017–2018). The port is located 10 km southeast of the city of A Coruña in Spain and is protected by a 3360 m-long main breakwater and a spur breakwater 1320 m long. The current berthing line is 914 m long with an average depth of 22 m. The orientation of the dock is N62.7W and reaches a crest height 6.5 m above the zero datum of the port. A set of bollards spaced 31 m apart with a load capacity of 200 t is situated 0.75 m from the dock. In addition, there is a double-fender system protected by a shield to streamline vessel operations.

In order to compute the transfer functions, the six degrees of freedom of 27 vessels (15 Bulk carriers and 12 General cargo vessels) were recorded under different climatic conditions (Table 1). These vessels were located along the entire berthing line and represent a typical harbor fleet in this port.

Table 1. Characteristics of the vessels measured during the field campaign.

Vessel	Type	Deadweight Tonnage DWT (t)	Length (m)	Beam (m)
Fu Da	Bulk carrier	71,330	224.9	32.2
Avax	Bulk carrier	87,030	225.0	32.2
Yannis	Bulk carrier	50,792	189.9	32.2
Western Boheme	Bulk carrier	37,000	186.9	28.6
Pina Cafiero	Bulk carrier	75,668	225.0	32.2
Jing Jin Hai	Bulk carrier	77,872	225.0	32.2
Lowlands Saguenay	Bulk carrier	37,152	179.9	30.0
Aloe	Bulk carrier	30,618	178.7	28.0
CSK Unity	Bulk carrier	77,105	225.0	32.2
Topaz	Bulk carrier	75,499	225.0	32.2
Walsall	Bulk carrier	58,018	189.9	32.3
Kyzicos	Bulk carrier	92,598	229.5	36.9
Nautical Lucia	Bulk carrier	63,548	199.9	32.2
Nord Saturn	Bulk carrier	77,288	225.0	32.2
Orange Harmony	Bulk carrier	81,837	229.0	32.2
Marc	General cargo	4135	89.8	13.6
Celine	General cargo	8600	129.4	15.8
Dominica	General cargo	13,022	127.3	21.2
Kelly C	General cargo	6250	106.0	15.5
Notos	General cargo	8049	125.1	16.4
Don Juan	General cargo	21,057	158.0	23.0
Eems River	General cargo	4066	89.9	12.5
Linau	General cargo	3699	88.0	12.8
Fortune	General cargo	12,692	138.9	21.3
Moraime	General cargo	7300	118.0	16.5
Onego Capri	General cargo	10,273	138.9	15.9
Oppland	General cargo	9273	107.0	18.4

The methodology used for the measurement of the movements was validated in other studies by the authors of this paper [16]. The measurement equipment consists of three systems that continuously record each of the vessel's degrees of freedom with a frequency of 1 Hz. The first of these systems is an inertial measurement unit (IMU) [17] consisting of three accelerometers and three gyroscopes that record the roll and the pitch (Figure 2, Left). The second system comprises two electronic distance-measuring lasers for the sway and yaw measurements (Figure 2, Right). Finally, photogrammetric techniques were employed to measure the heave and surge movements [18] (Figure 2, Center).

Figure 2. Left: IMU (Inertial measurement unit), **Center**: Photogrammetric techniques, **Right**: Laser distance sensor.

The climatic variables were measured using the available instrumentation in the Spanish Port Authority network and the A Coruña Port Authority. The location of the instruments is shown in Figure 1. This decision was made since the port's own meteorological forecasting system collects data at these points. In first place, the hydrodynamic variables were measured at the outer buoy of the Port of Punta Langosteira, located at 43°20′58.34″ N–8°33′41.32″ W at a depth of 60 m. During the first 20 min of each hour it recorded the following variables: significant wave height (H_s (m)), maximum wave height (H_{max} (m)), peak wave period (T_p (s)), average wave period (T_m (s)), and wave direction (Dir_W (°)). Second, the weather station located near the roundhead of the main breakwater was used to record wind speed and direction. The instrumentation continuously records the average wind speed (V_w (km/h)), wind gust speed (V_g (km/h)), average direction, and wind gust direction (Dir_{Vw} (°) and Dir_{Vg} (°)). However, the port weather forecast system only provides 72 h in advance data of the variables H_s (m), T_p (s), and Dir_W (°) at the buoy location, and, V_w (km/h) and Dir_{Vw} (°) at the weather station, so these variables were finally used in this study. This forecasting system was developed by the Spanish government agency Puertos del Estado in collaboration with the State Meteorological Agency (AEMET). This system is driven by wind fields supplied by AEMET from the High Resolution Limited Area Model (HIRLAM). The system starts a new execution twice a day providing data outputs with a time resolution of 1 h. To ensure good initial conditions, before the forecast starts, the model is forced using wind analyzed fields of the 12 h prior to forecast initialization [19].

The seasonal wind and wave roses (winter and summer) at the buoy position for the period 2015–2018 are shown in Figure 3, in order to clarify the values of the main forcings acting in the Outer Port of Punta Langosteira.

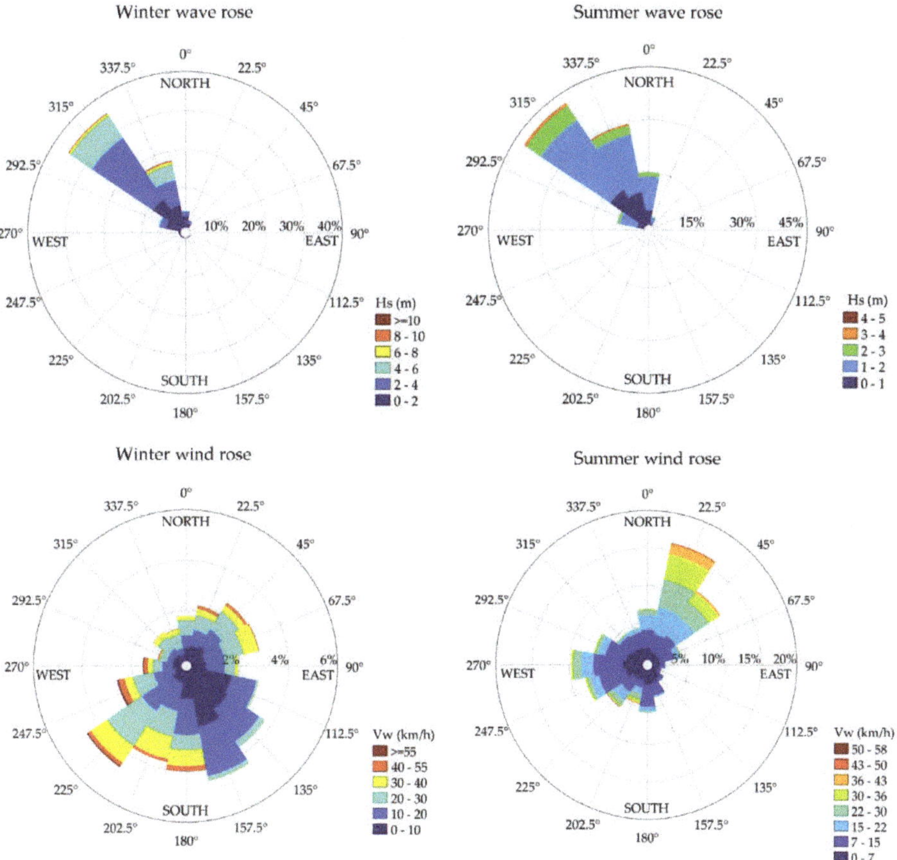

Figure 3. Seasonal wind and wave roses (winter and summer) at the buoy position for the period 2015–2018.

Although the models for estimating the movements of moored vessels were obtained with observational meteorological and ocean data (recorded by the wave buoy and the weather station), they are run with forecast data. Therefore, it is important to know the differences between both sources of information (observational data vs. forecast data). To this end, an analysis of the estimation errors of each variable (H_s (m), T_p (s), Dir_w (°), V_w (km/h), and Dir_{Vw} (°)) was conducted. Table 2 shows the obtained results.

Table 2. Coefficient of determination (R^2), root mean square error (RMSE) and mean absolute error (MAE) of forecast and observed meteorological and ocean variables.

Forecast Variable vs. Observed Variable	R^2	RMSE	MAE
H_s (m) forecast vs. H_s (m) observed	0.86	0.39 m	0.29 m
T_p (s) forecast vs. T_p (s) observed	0.63	1.9 s	1.2 s
Dir_w (°) forecast vs. Dir_w (°) observed	0.77	18.2°	11.3°
V_w (km/h) forecast vs. V_w (km/h) observed	0.64	6.8 km/h	5.2 km/h
Dir_{Vw} (°) forecast vs. Dir_{Vw} (°) observed	0.51	55.1°	36.5°

On the one hand, H_s (m), T_p (s), Dir_w (°), and V_w (km/h) variables present a better approximation to the observed value, showing acceptable prediction errors (MAE values of 0.29 m, 1.2 s, 11.3°, and 5.2 km/h, respectively). On the other hand, Dir_{Vw} (°) shows the largest deviation between the observed and the forecast value (MAE value of 36.5°). Since the models are fed with forecast data, having an accurate weather forecasting system will provide similar results in terms of accuracy to those obtained by these models in their development stage.

As previously mentioned, the wave buoy employed in this study characterizes the main ocean variables of each sea state (1-h duration) using the records obtained during the first 20 min of each hour. For this reason, the joint analysis of the data was carried out using only the concomitant data of both wind, waves, and vessel movements. Regarding moored vessels behavior, the analysis of motion time series was based on a zero crossing technique. A peak-to-peak criterion was applied to each movement to obtain their amplitudes, except in the case of sway motion, for which the zero-peak criterion was used. Complete time series of each motion were split into blocks of 1-h duration, obtaining the maximum motion amplitude, average motion amplitude, and significant motion amplitude calculated as the average of the highest third for each block [16]. Figure 4 shows a sample of roll motion time series recorded on the bulk carrier vessel *WESTERN BOHEME*. The analysis of the maximum, average, and significant amplitudes of each motion and their concomitant climatic forcings recorded on the same vessel are shown in Figure 5.

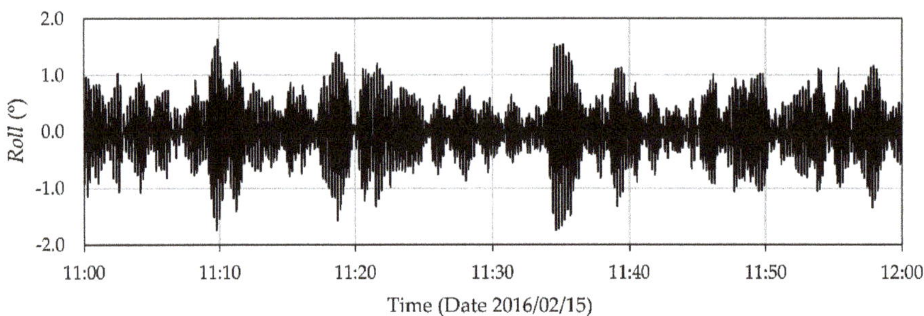

Figure 4. A 1 h sample of roll time series recorded on the bulk carrier vessel *WESTERN BOHEME*.

As can be seen in Figure 5, a moored ship that under specific meteorological and ocean conditions moves with certain amplitudes may experience a maximum punctual movement much larger than its significant or average movement. This value that stands out from the main trend of the movement could be occasionally caused by the action of other external agents such as the waves generated by passing vessels or the punctual modification of the mooring lines tension to adapt them to the tidal variations.

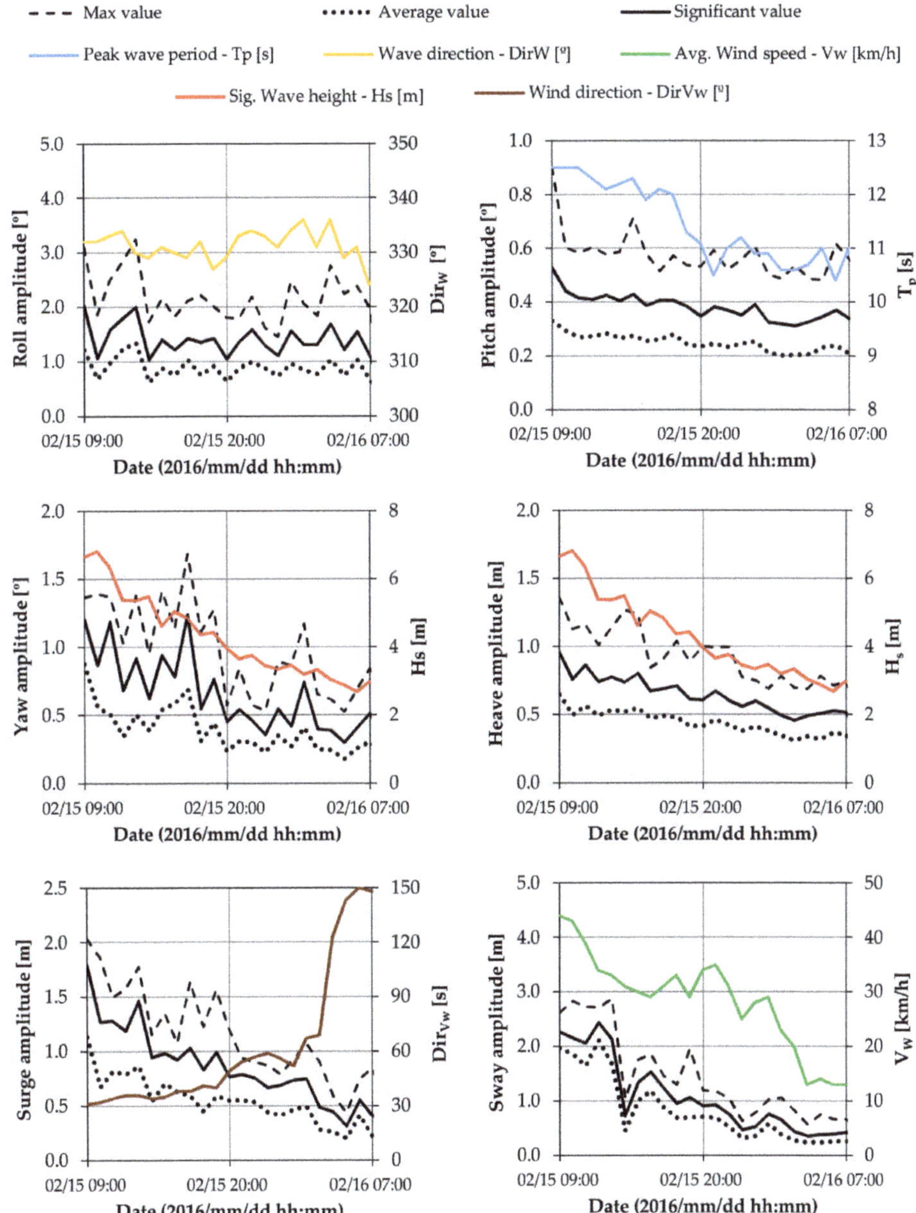

Figure 5. Maximum, average and significant amplitudes of each motion recorded during the *WESTERN BOHEME* bulk carrier vessel's stay in port and its concomitant climatic forcings.

3. Analytic Methodology

This section presents the analytical methodology developed in this study for the calculation of each of the transfer functions. The analysis was performed for significant movements. This parameter has commonly been used in similar studies to evaluate the dynamic behavior of a moored vessel during its stay in port [16,20,21]. Also, the use of the significant value assures that motions and wave parameters (significant wave height and peak wave period) are obtained following the same statistical analysis. This can contribute to achieving better relations with the main forcings of moored vessels' behavior. In addition, the range and independence of each of the variables used were also calculated.

3.1. Dataset Creation

First, a dataset was created to handle the calculation of the transfer functions. The range of validity is shown in Table 3 and defines the values of each of the variables recommended for the equations calculated in this study.

Table 3. Range of the variables considered for the calculation of the transfer functions.

Range	Length (m)	Beam (m)	H_s (m)	Dir_W (°)	T_p (s)	V_W (km/h)	Dir_{Vw} (°)
Min.	88.0	12.5	1.04	1.0	5.5	0.1	4.0
Max.	229.5	36.9	8.95	359	18.2	80.0	351.0

Abbreviations: H_s (m)—Significant wave height; Dir_W (°)—Wave direction; T_p (s)—Peak wave period; V_W (km/h)—Average wind speed; Dir_{Vw} (°)—Wind direction.

The homogeneity in the distribution of variables of the dataset was analyzed. To this end, each data of a given variable was dimensionalized with the highest value of the same (Variable value (i)/Maximum variable value). In this way, the spectrum of values was contained between 0 and 1. Figure 6 shows the cumulative frequency of each parameter (Y-axis) within its range (X-axis). Therefore, in a situation of homogeneous distribution of the data, the plot would correspond with the bisector of the graph.

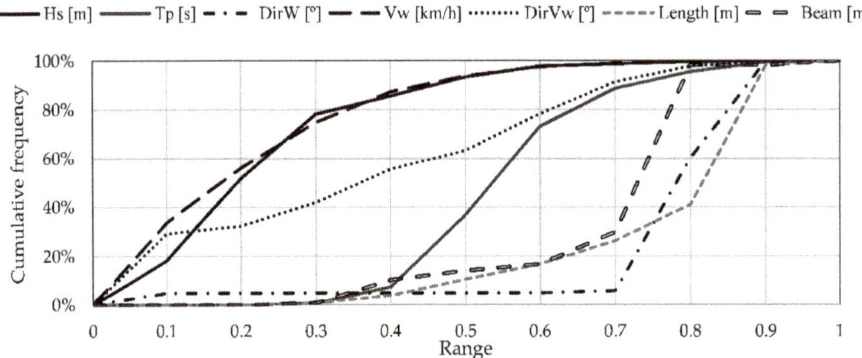

Figure 6. Cumulative frequency distribution of each of the recorded variables (Dimensionless variable range (Variable value (i)/Maximum variable value) vs. Accumulated frequency for each of the variables).

The results show that the significant wave height (Hs (m)) and wind velocity (V_W (km/h)) variables are concentrated between 0.1 and 0.4 of their range. The wind direction (Dir_{Vw} (°)) presents a homogeneous distribution in its range. However, the variables representing the peak wave period (T_p (s)), wave direction (Dir_W (°)), length (m), and beam (B [m]), are all concentrated between 0.6 and 1 for the majority of the data. With the aim of more precisely quantifying the variables without dimensionalizing them, Table 4 shows the frequency ranges and values for each of the variables.

An amount of 83% of the significant wave height data (Hs (m)) is concentrated in the range $1.0 \leq H_s$ (m) ≤ 4. For the peak period, 93% of the data lies between $8 \leq T_p$ (s) ≤ 16. Moreover, 38% of the data is concentrated within a 2-s range (10 s–12 s). Most of the data pertaining to the wave direction come from the NW direction (81%). Regarding the ship's dimension, 59% of the length values and 70% of the beam values are for the largest vessels (200–250 m length and > 30 m beam). In view of the results it can be seen that some of the possible combinations between the variables are not defined by a very high number of data points.

Table 4. Frequency of the data recorded during the field campaign for each of the variables.

H_s (m)		T_p (s)		Dir_W (°)		V_w (km/h)		Dir_{Vw} (°)		Length (m)		Beam (m)	
Range	%	Range	%	Range	%	Range	%	Range	%	Range	%	Range	%
<1.0	0%	<4	0%	N [337.5–22.5]	16%	0–10	14%	N [337.5–22.5]	2%	<100	1%	<10	0%
1–2	27%	4–6	0%	NE [22.5–67.5]	0%	10–20	35%	NE [22.5–67.5]	27%	100–150	13%	10–15	1%
2–3	37%	6–8	2%	E [67.5–112.5]	0%	20–30	24%	E [67.5–112.5]	4%	150–200	27%	15–20	9%
3–4	19%	8–10	19%	SE [112.5–157.5]	0%	30–40	16%	SE [112.5–157.5]	19%	200–250	59%	20–25	6%
4–5	8%	10–12	38%	S [157.5–202.5]	0%	40–50	7%	S [157.5–202.5]	8%	>250	0%	25–30	13%
5–6	7%	12–14	21%	SW [202.5–247.5]	0%	50–60	3%	SW [202.5–247.5]	19%			>30	70%
6–7	1%	14–16	15%	W [247.5–292.5]	2%	60–70	1%	W [247.5–292.5]	14%				
7–8	1%	16–18	4%	NW [292.5–337.5]	81%	>70	0%	NW [292.5–337.5]	6%				
8–9	1%	>18	1%										
>9.0	0%												

3.2. Statistical Response, Variables and Predictors

Next, the analytical methodology employed, the variables used, and their influence on each of the degrees of freedom is presented.

The variables were selected taking their a priori possible influence on vessel movements into account. They were divided into three groups, depending on whether they were climatic variables, vessel dimensions, or dimensionless vessel size features. The latter were obtained by scaling the vessel size measurements with the following wave characteristics: significant wave height (H_s (m)), and wave length in deep water (L_{op} (m)). Table 5 shows the description of all the predictor and response (vessel movements) variables obtained, studied, and modeled in this work.

Table 5. Response and predictor variables, with corresponding tags.

Movement (y_i)	Name	Variables (X_m)	Name	Typology
Roll	y_1	Wave height (H_s (m))	X_1	
Pitch	y_2	Wave period (T_p (m))	X_2	
Heave	y_3	Wave length (L_{op} (m))	X_3	Meteorological and
Surge	y_4	Wave steepness (s)	X_4	ocean variables
Sway	y_5	Wave direction (Dir_W (°))	X_5	
Yaw	y_6	Wind velocity (V_w (km/h))	X_6	
		Wind direction (Dir_{Vw} (°))	X_7	
		Length (L (m))	X_8	Vessel dimensions
		Beam (B (m))	X_9	
		Length/Beam (L/B)	X_{10}	
		Length/H_s	X_{11}	
		Length/L_{op}	X_{12}	Dimensionless
		Beam/H_s	X_{13}	
		Beam/L_{op}	X_{14}	

The transfer functions were calculated and analyzed using statistical correlation studies and multivariate linear regression techniques [22]. This methodology has recently been applied to various different engineering domains, including naval and oceanic engineering [23–25]. In the case of ocean engineering, following a similar methodology, Carral-Couce et al. [23] developed nonlinear and multivariate linear regression models to estimate the traction of towing and anchor-handling winches. Additionally, the transit time to cross the Panama Canal's new locks was estimated using multivariate linear regression [24], and the effect of vessel dimensions, type, and fishing ground were also modeled to estimate net drum and winch traction for trawler design tasks [25]. These techniques were also used to forecast wave height [26] and vessel traffic flow [27], among various other applications. For the present case, the proposed multivariate regression model can be expressed as Equation (1):

$$y_i = \hat{\beta}_0 + \sum_{m=1}^{M} \hat{\beta}_m x_m + \hat{\varepsilon}_i, \text{ with } i = 1, 2, \ldots, 6 \text{ and } m = 1, 2, \ldots, 14 \tag{1}$$

where y_i represents the sample values of the response variable (vessel movement) corresponding to the multivariate linear model, x_m represents the m predictor variables (there were up to $M = 14$ variables analyzed), and $\hat{\varepsilon}_i$ represents the model's residuals or the discrepancy between the real y_i and the model estimates, $\hat{y}_i = \hat{\beta}_0 + \sum_{m=1}^{M} \hat{\beta}_m x_m$. The i index accounts for the vessel's degrees of freedom (roll, pitch, heave, surge, sway, and yaw). $\hat{\beta}_0$ represents the constant term of each model, and $\hat{\beta}_m$ represents the model's parameter estimates corresponding to each of the independent variables. They account for the linear effect of each predictor on the response.

4. Results and Discussion

This section includes the descriptive analysis, including the predictor correlation study, the multivariate linear model's estimation, and the model's predictions of vessel movements obtained from the previously described dataset.

4.1. Correlation Analysis

The predictors of a multivariate linear model should be uncorrelated in order to obtain reliable model parameter estimations, and, hence, accurate and precise predictions [23–25,28]. Indeed, the existence of multicollinearity leads to estimates of model parameters being highly dependent on sample data, preventing an analysis of the effect of each predictor or covariate on the response, and limiting the model's ability to generate accurate predictions. Accordingly, a pairwise dependence relationship analysis should be performed prior to including the predictors in the final model [28]. The most widely used measurement for goodness of fit is the Pearson coefficient (r). At this point, it is important to note that the inclusion of new additional predictors to the model always increases the Pearson coefficient. Nevertheless, those predictors must be uncorrelated to prevent spurious dependence relationships and inaccurate models. Accordingly, the dependence structure of the predictors was analyzed by calculating the Pearson coefficient, r (Table 6).

Table 6. Pairwise Pearson linear correlation coefficients for the predictors (in gray when r ≥ 0.6 or r < −0.6).

		H_s (m)	T_p (s)	L_{op} (m)	s	Dir_W (°)	V_w (km/h)	Dir_{Vw} (°)	L (m)	B (m)	L/B	L/H_s	L/L_{op}	B/H_s	B/L_{op}
	H_s (m)	1.0	0.3	0.3	0.7	0.1	0.3	0.1	−0.1	−0.1	0.1	−0.8	−0.3	−0.8	−0.3
	T_p (s)	0.3	1.0	1.0	−0.4	0.1	−0.1	−0.1	0.1	0.1	0.1	−0.2	−0.7	−0.2	−0.8
	L_{op} (m)	0.3	1.0	1.0	−0.4	0.1	−0.1	−0.1	0.1	0.1	0.1	−0.2	−0.7	−0.2	−0.7
	s	0.7	−0.4	−0.4	1.0	0.1	0.3	0.1	−0.1	−0.1	0.1	−0.6	0.4	−0.6	0.4
	Dir_W (°)	0.1	0.1	0.1	0.1	1.0	−0.1	0.1	0.1	0.1	0.0	0.0	0.0	0.0	0.0
	V_w (km/h)	0.3	−0.1	−0.1	0.3	−0.1	1.0	0.1	−0.1	0.0	−0.2	−0.3	0.0	−0.3	0.0
r	Dir_{Vw} (°)	0.1	−0.1	−0.1	0.1	0.1	0.1	1.0	−0.1	−0.1	−0.1	−0.1	0.0	−0.1	0.0
	L (m)	−0.1	0.1	0.1	−0.1	0.1	−0.1	−0.1	1.0	0.9	0.2	0.5	0.4	0.4	0.4
	B (m)	−0.1	0.1	0.1	−0.1	0.1	0.0	−0.1	0.9	1.0	0.0	0.5	0.4	0.5	0.4
	L/B	0.1	0.1	0.1	0.1	0.0	−0.2	−0.1	0.2	0.0	1.0	0.1	0.1	0.0	0.0
	L/H_s	−0.8	−0.2	−0.2	−0.6	0.0	−0.3	−0.1	0.5	0.5	0.1	1.0	0.4	1.0	0.4
	L/L_{op}	−0.3	−0.7	−0.7	0.4	0.0	0.0	0.0	0.4	0.4	0.1	0.4	1.0	0.4	1.0
	B/H_s	−0.8	−0.2	−0.2	−0.6	0.0	−0.3	−0.1	0.4	0.5	0.0	1.0	0.4	1.0	0.4
	B/L_{op}	−0.3	−0.8	−0.7	0.4	0.0	0.0	0.0	0.4	0.4	0.0	0.4	1.0	0.4	1.0

Table 6 shows that the wave period (T_p (s)) and wave length (L_{op} (m)) present a direct linear relationship ($r = 1$) due to their definition. In addition, the wave height (H_s (m)) and steepness (s) are also correlated ($r > 0.6$). Additionally, the vessel size predictors are also significantly correlated.

This is the case for vessel length (L (m)) and beam (B (m)), which are very strongly correlated ($r \geq 0.9$). A similar dependence structure is obtained when the size dimensionless predictor variables are studied. Taking into account the fact that the dimensionless variables were derived from the vessel size and meteorological and ocean variables, Table 6 shows that they are strongly correlated with both size and meteorological and ocean predictors. On the other hand, it can be observed that the dimensionless variable Length/Beam is independent, and this allows the influence of the size of the vessel to be introduced into the analysis.

On the basis of the results depicted in Table 6, linear regression models were developed using variables that were independent of each other. Thus, these models were constructed using five hydrodynamic predictors (H_s (m), T_p (s), $\mathrm{Dir_W}$ (°), V_w (km/h), $\mathrm{Dir_{Vw}}$ (°)) and the dimensionless variable Length/Beam (Table 7).

Table 7. Predictors involved in fitting regression models.

Selected Predictors
Wave height (H_s (m))
Wave period (T_p (s))
Wave direction ($\mathrm{Dir_W}$ (°))
Wind velocity (V_w (km/h))
Wind direction ($\mathrm{Dir_{Vw}}$ (°))
Length/Beam (L/B)

The variables H_s (m) and T_p (s) were selected instead of s (wave steepness) and L_{op} (m) since they are the main parameters that define the characteristics of a sea state (together with $\mathrm{Dir_W}$ (°)). In addition, their values are directly provided by both the wave buoy and the weather forecasting system of the Port, facilitating the data acquisition and the implementation of the models. Regarding vessel dimensions, neither L (m) nor B (m) was selected to participate as an independent variable since their information was already included in the dimensionless variable L/B.

4.2. Regression Modelling of Vessel Movements

Once the variables that could potentially participate in the generation of the models were selected, the next step consisted in identifying those that had an important influence on the prediction provided by each model. To this end, an Akaike criterion (AIC) was used [29]. First, the multivariate linear regression models were calculated including all selected predictors. The parameters corresponding to each predictor, $\hat{\beta}_m$ were estimated from the data base by means of the least squares method. Then, a statistical significance analysis of each variable was carried out, selecting those with a level of significance $\alpha \leq 0.01$ (Table 8).

Table 8. Summary of the selected predictors for each response variable. Variables that have an effect on the response significantly different from zero are indicated by a cross (significance level $\alpha \leq 0.01$).

	Roll (y_1)	Pitch (y_2)	Heave (y_3)	Surge (y_4)	Sway (y_5)	Yaw (y_6)
H_s	x	x	x	x	x	x
T_p	x	x			x	
$\mathrm{Dir_W}$	x		x	x		x
V_w	x		x		x	
$\mathrm{Dir_{Vw}}$				x	x	
L/B	x	x	x	x	x	x

Finally, models were re-calculated using only the most influential predictors in each vessel movement, obtained from the significance analysis (Table 8). Adopting this methodology ensured that the models would provide predictive results. The following expressions show the structure and selected variables for each transfer function:

$$y_1 \, (\text{Roll}) = \beta_{0_{\text{Roll}}} + \beta_{1_{H_s}} \cdot H_s + \beta_{1_{T_p}} \cdot T_p + \beta_{1_{\text{Dir}_W}} \cdot \text{Dir}_W + \beta_{1_{V_W}} \cdot V_W + \beta_{1_{\frac{L}{B}}} \cdot \frac{L}{B} \tag{2}$$

$$y_2 \, (\text{Pitch}) = \beta_{0_{\text{Pitch}}} + \beta_{2_{H_s}} \cdot H_s + \beta_{2_{T_p}} \cdot T_p + \beta_{2_{\frac{L}{B}}} \cdot \frac{L}{B} \tag{3}$$

$$y_3 \, (\text{Heave}) = \beta_{0_{\text{Heave}}} + \beta_{3_{H_s}} \cdot H_s + \beta_{3_{\text{Dir}_W}} \cdot \text{Dir}_W + \beta_{3_{V_W}} \cdot V_W + \beta_{3_{\frac{L}{B}}} \cdot \frac{L}{B} \tag{4}$$

$$y_4 \, (\text{Surge}) = \beta_{0_{\text{Surge}}} + \beta_{4_{H_s}} \cdot H_s + \beta_{4_{\text{Dir}_W}} \cdot \text{Dir}_W + \beta_{4_{\text{Dir}_{V_W}}} \cdot \text{Dir}_{V_W} + \beta_{4_{\frac{L}{B}}} \cdot \frac{L}{B} \tag{5}$$

$$y_5 \, (\text{Sway}) = \beta_{0_{\text{Sway}}} + \beta_{5_{H_s}} \cdot H_s + \beta_{5_{T_p}} \cdot T_p + \beta_{5_{V_W}} \cdot V_W + \beta_{5_{\text{Dir}_{V_W}}} \cdot \text{Dir}_{V_W} + \beta_{5_{\frac{L}{B}}} \cdot \frac{L}{B} \tag{6}$$

$$y_6 \, (\text{Yaw}) = \beta_{0_{\text{Yaw}}} + \beta_{6_{H_s}} \cdot H_s + \beta_{6_{\text{Dir}_W}} \cdot \text{Dir}_W + \beta_{6_{\frac{L}{B}}} \cdot \frac{L}{B} \tag{7}$$

Each multivariate linear regression model was adjusted with 80% of the composed data sample. The rest of the data was reserved for external validation of the transfer functions calculated by the models.

In order to quantify the importance of each variable for vessel movements, a relative frequency descriptive analysis was performed (Figure 7). From this analysis, the wave height (H_s (m)) and the dimensionless variable Length/Beam (L/B) effect on the response was found to be significant in all (100%) of the regression models performed (transfer functions), while the wave direction (Dir_W (°)). effect was non-zero in 66.67% of the transfer functions performed. In addition, the wave period (T_p (s)) and wind velocity (V_W (km/h)) were significant in 50% of the movements. Finally, wind direction (Dir_{V_W} (°)) effect was only significant for surge and sway movements.

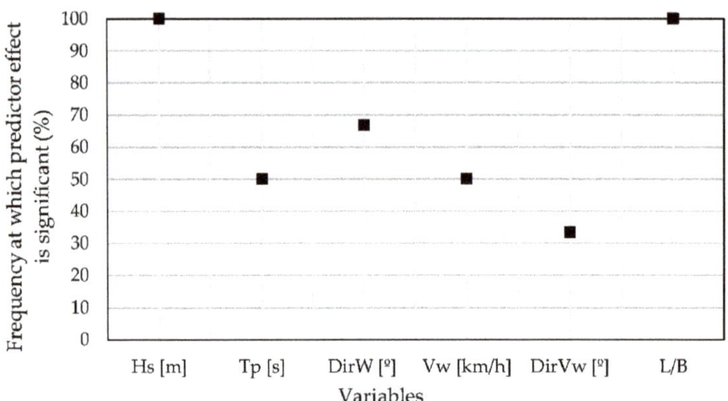

Figure 7. The relative frequency corresponding to the models where the effect of each predictor on the response is significantly different from zero.

Figure 8 shows the results obtained with each of the models constructed. This data visualization provides information about the goodness of fit, the ability to predict vessel movements using a variable-dependent model, and the model's accuracy and precision.

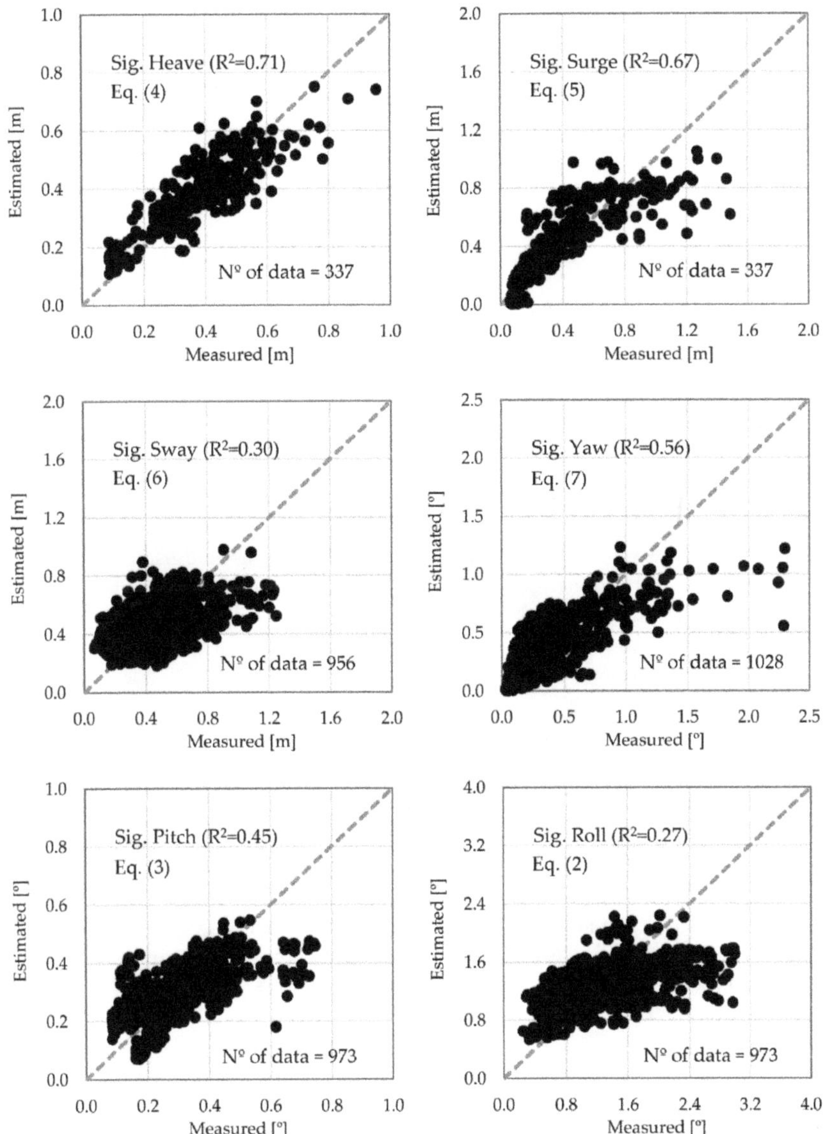

Figure 8. Scatter plots of measured values versus predicted values for the multivariate linear models fitted to significant vessel movements.

As can be observed, the models with the highest accuracy and precision were those that estimate the heave and the surge. This trend was also observed for the yaw and pitch movements. The lowest accuracy was obtained for the sway and the roll. These two movements had greater variability over time, as well as an inertial component from the vessel and the cargo, so their accuracies were lower. Accordingly, the fittings for these latter vessel movements were less precise (a greater dispersion of points around the diagonal). However, although these models did not allow the real motion amplitude to be predicted accurately, they were able to estimate the main trends of these movements.

In addition, the R^2 coefficients and the root mean square error (RMSE) provided a quantitative measure of each model's goodness of fit (Table 9). The best goodness of fit was produced for the heave movement, with an R^2 value of 0.71 and an RMSE of 0.08 m.

The surge movements fitted with $R^2 = 0.67$, while the yaw and pitch movements had R^2 values of 0.56 and 0.45, respectively. In addition, the RMSE is 0.18 m for the first, 0.21° for yaw, and 0.09° for the pitch. Finally, the movements with the lowest goodness of fit values were the sway ($R^2 = 0.30$) and the roll ($R^2 = 0.27$). In these two cases it was verified that the RMSE of the sway was about 0.18 m, while for the roll it was 0.46°.

Table 9. Values of the R^2 coefficient and the root mean square error (RMSE) of the calculated transfer functions.

Movement	R^2	RMSE
Heave	0.71	0.08 m
Surge	0.67	0.18 m
Sway	0.30	0.18 m
Yaw	0.56	0.21°
Pitch	0.45	0.09°
Roll	0.27	0.46°

Additionally, the error for each function was quantified. This was done using the mean absolute error (MAE) parameter (Table 10). The objective was to estimate the deviation of the functions, because all the variables involved in the process were not taken into account. The joint analysis of these three parameters allows for a determination to be made as to whether the error obtained was acceptable for use in a port operational management system.

Table 10. Mean absolute error (MAE) for each of the six degrees of freedom analyzed using transfer functions.

	Heave (m)	Surge (m)	Sway (m)	Yaw (°)	Pitch (°)	Roll (°)
Mean Absolute Error	0.06	0.14	0.14	0.15	0.07	0.36

The results show that despite not having all the variables referenced in the model, it is possible to estimate with a mean precision of at least 0.36° the rotations, and 14 cm the displacements. From Table 10 it can be seen that, coinciding with the values of R^2, the largest errors were produced in the case of the roll, and the smallest for the heave.

4.3. Model Validation

An external validation procedure was implemented in order to evaluate the predictive ability of the transfer functions compared in the previous section. For this purpose, 20% of the data obtained in the field campaigns was applied to the transfer functions and the results were compared (Figure 9).

As can be observed in Figure 9, heave, surge, yaw, and pitch estimated and measured movements conform to the bisector of the first quadrant. Sway and roll movements present a similar fit, but in a less precise way. This fact demonstrates that the proposed models achieve their objective. However, as before, the existing differences were produced by climatic characteristics, the mooring lines and the cargo configuration. Figure 10 shows the comparison between the measured heave and roll motions, and those estimated by the transfer functions from the validation data.

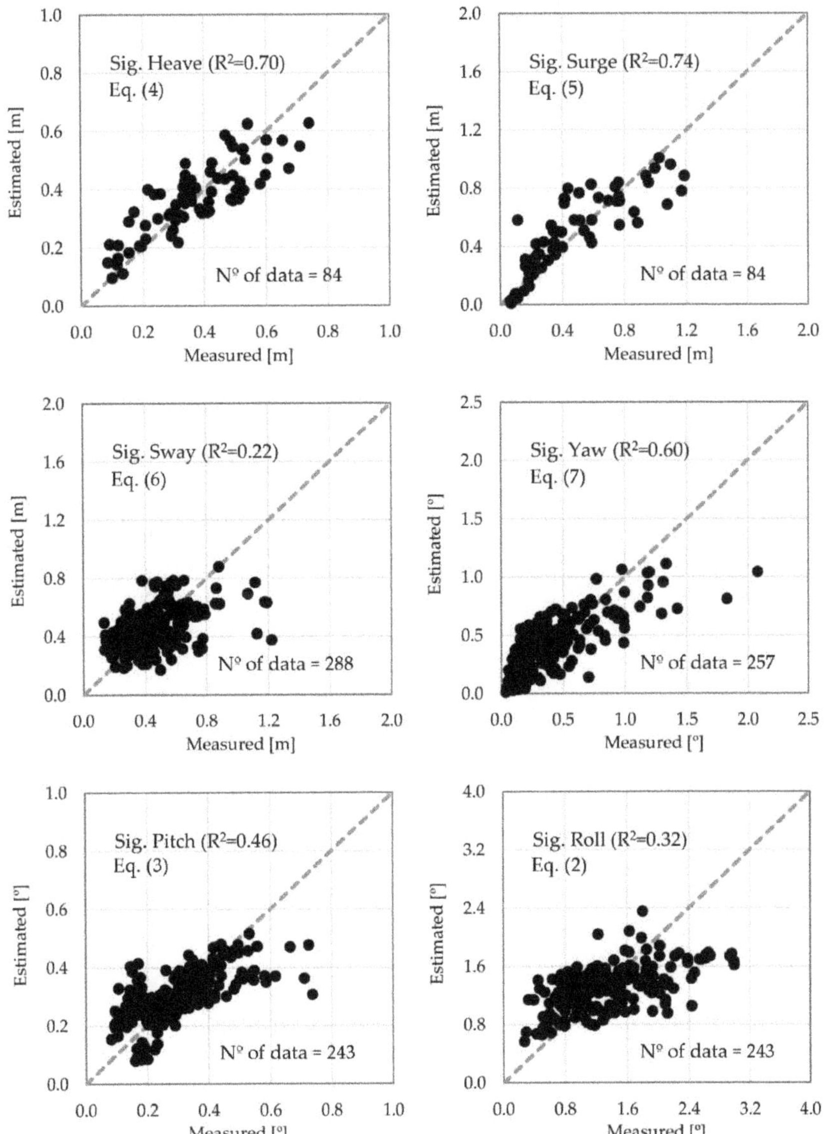

Figure 9. Validation of the multivariate linear models.

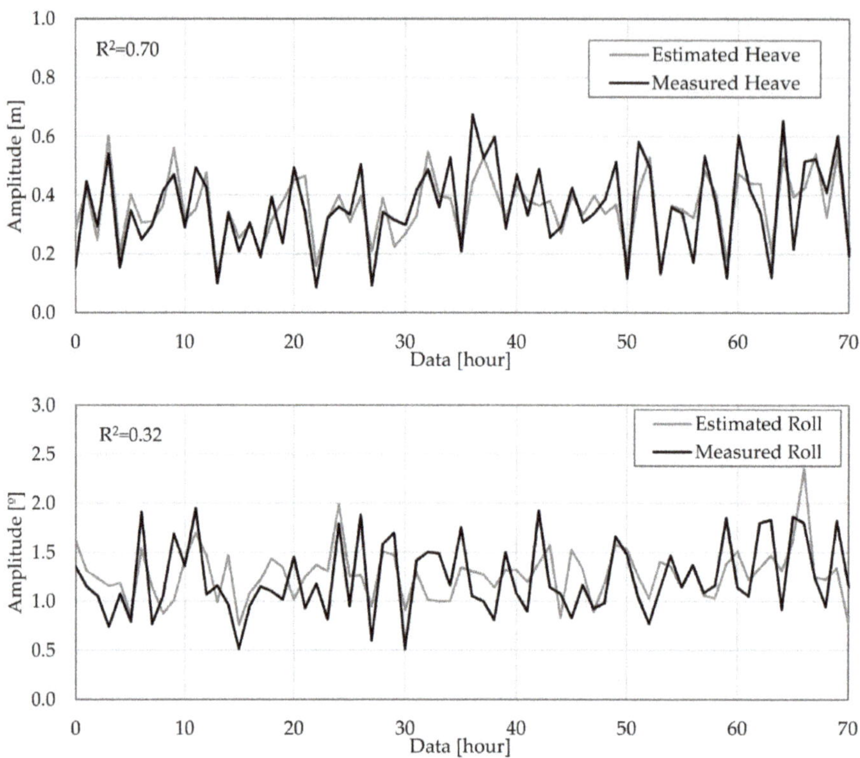

Figure 10. Comparison between measured and estimated values of heave and roll motions.

To quantify the accuracy of the estimations, the determination coefficient (R^2) and the root mean square error (RMSE) of each movement were analyzed (Table 11).

Table 11. Obtained values of the R^2 coefficient and the root mean square error (RMSE) in transfer functions validation.

Movement	R^2	RMSE
Heave	0.70	0.08 m
Surge	0.74	0.16 m
Sway	0.22	0.18 m
Yaw	0.60	0.20°
Pitch	0.46	0.09°
Roll	0.32	0.44°

Comparing Tables 9 and 11, the obtained validations reflect the same pattern as the calculated transfer functions. Similarly, both the determination coefficient (R^2) and the root mean square error (RMSE) obtained were shown to be of the same order of magnitude. Therefore, it can be concluded that the accuracy of the validation is similar to that of the calculated functions. Moreover, it was also verified that the mean absolute error (MAE) had a similar value to that calculated by the models: 0.35° for the rotations, and 14 cm for the displacements (Table 12). However, as mentioned in Section 2, these tools will be fed with weather forecast data, so their accuracy will also be conditioned by the port's own forecasting system (Table 2).

Table 12. Mean absolute error (MAE) for each of the six degrees of freedom studied in the validation of the transfer functions.

	Mean Absolute Error					
	Heave (m)	Surge (m)	Sway (m)	Yaw (°)	Pitch (°)	Roll (°)
Validation	0.07	0.12	0.14	0.14	0.07	0.35

Finally, the application of this methodology and the implementation of the obtained models in a port management system would provide reasonable predictions of the expected movements of moored vessels from weather forecast data. Comparing this information with the movement thresholds specified by the different standards would detect possible operational downtimes and risk situations in the berthing area. Therefore, this tool would help to identify operational windows for ships, facilitating decision making on port berth occupancy planning.

5. Conclusions

This paper presents an analytical methodology to relate the movements of moored vessels using the variables available in forecast data including specifically, ship dimensions and climatic conditions. This work was applied and validated for 27 moored vessels (15 Bulk carrier and 12 General cargo) at the facilities of the Outer Port of Punta Langosteira, A Coruña, Spain. The results obtained are currently incorporated in its port management system.

The results show that this methodology can be used to predict the six degrees of freedom of moored vessels. These models were obtained assuring that the variables used were independent of each other. The values of the determination coefficient (R^2) and of the root mean square error (RMSE) indicate that the equations calculated allow a reasonable prediction of the movements. Even models with lower R^2 values (sway and roll movements) are able to estimate the main trend of the expected movements. In addition, the mean absolute error reveals that the errors are less than 14 cm for the displacements, and less than 0.36° for the rotations.

As a conclusion, it can be verified that the methodology proposed facilitates an advance towards a better understanding of the factors that influence port operations in the Outer Port of Punta Langosteira. This is the first step in order to generate warnings that assist port management and help to optimize the use of the port's resources and facilities. Also, this methodology could be exportable to other ports providing an analysis of the influential and available forecast variables is made, as well as a record of the movements of the moored vessels.

Author Contributions: Investigation and data acquisition: J.S., A.F. and A.A.; conceptualization and methodology: J.S., A.F., and J.T.-S.; writing—original draft: J.S. and A.F.; writing—review and editing: A.F.; supervision and critical revision: E.P. and J.R.R.

Funding: This research was funded by the Spanish Ministry of Economy, Industry and Competitiveness, R & D National Plan, within the project BIA2017-86738-R.

Acknowledgments: The authors gratefully acknowledge the financial support of the Spanish Ministry of Economy, Industry and Competitiveness, R & D National Plan, within the project BIA2017-86738-R. In addition, special thanks to the Port Authority of A Coruña, Spain.

Conflicts of Interest: The authors declare no conflict of interest.

References

1.	Kumar, P.; Batra, G.; Kim, K.I. A Moored Ship Motion Analysis in Realistic Pohang New Harbor and Modified PNH. In Proceedings of the 1st International Conference on Modern Mathematical Methods and High Performance Computing in Science and Technology, M3HPCST, Ghaziabad, India, 27–29 December 2015; Springer International Publishing: Ghaziabad, India, 2016; pp. 207–214.

2. Peña, E.; Figuero, A.; Sande, J.; Guerra, A.; Perez, J.D.; Maciñeira, E. Integrated system to evaluate moored ship behavior. In Proceedings of the 36th International Conference on Offshore Mechanics and Arctic Engineering-OMAE, Trondheim, Norway, 25–30 June 2017; Volume 7A-2017.

3. PIANC. *Criteria for Movements of Moored Ships in Harbours: A Practical Guide*; PIANC General Secretariat: Brussels, Belgium, 1995.

4. Puertos del Estado. *ROM 2.0-11. Recomendaciones para el Proyecto y Ejecución en Obras de Atraque y Amarre*; Ministerio de Fomento: Madrid, Spain, 2011; ISBN 9788488975409.

5. Figuero, A.; Peña, E.; Sande, J.; Costa, F. Analysis of the movements and operational limits of moored vessels in outer and inner ports of A Coruña (Spain). In Proceedings of the 3rd International Conference on Maritime Technology and Engineering, MARTECH 2016, Lisbon, Portugal, 4–6 July 2016; Volume 1.

6. Natarajan, R.; Ganapathy, C. Model experiments on moored ships. *Ocean Eng.* **1997**, *24*, 665–676. [CrossRef]

7. Rosa-Santos, P.; Taveira-Pinto, F.; Veloso-Gomes, F. Experimental evaluation of the tension mooring effect on the response of moored ships. *Coast. Eng.* **2014**, *85*, 60–71. [CrossRef]

8. Baker, S.; Frank, G.; Cornett, A.; Williamson, D.; Kingery, D. Physical Modelling and Design Optimizations for a New Container Terminal at the Port of Moin, Costa Rica. In *Ports 2016: Port Engineering Proceedings of the 14th Triennial International Conference, New Orleans, LA, USA,12–15 June 2016*; Oates, D., Burkhart, E., Grob, J., Eds.; American Society of Civil Engineers (ASCE): Reston, VA, USA, 2016; pp. 560–569.

9. Taveira Pinto, F.; Veloso Gomes, F.; Rosa Santos, P.; Guedes Soares, C.; Fonseca, N.; Santos, J.A.; Moreira, A.P.; Costa, P.; Brógueira Dias, E. Analysis of the Behavior of Moored Tankers. In Proceedings of the 27th International Conference on Offshore Mechanics and Arctic Engineering (OMAE2008), Estoril, Portugal, 15–20 June 2008; American Society of Mechanical Engineers (ASME): New York, NY, USA, 2008; Volume 4, pp. 755–764.

10. Cornett, A. Physical modelling of moored ships for optimized design of ports and marine. In Proceedings of the 5th International Conference on the Application of Physical Modelling to Port and Coastal Protection, Varna, Bulgaria, 29 September–2 October 2014.

11. Pawar, R.; Bhar, A.; Dhavalikar, S.S. Numerical prediction of hydrodynamic forces on a moored ship due to a passing ship. *Proc. Inst. Mech. Eng. Part. M J. Eng. Marit. Environ.* **2019**, *233*, 575–585. [CrossRef]

12. Kwak, M. Numerical Simulation of Moored Ship Motion Considering Harbor Resonance. In *Handbook of Coastal and Ocean Engineering: Expanded Edition*; Kim, Y.C., Ed.; World Scientific Publishing Company: Singapore, 2017; Volume 2, pp. 1081–1110.

13. Pinheiro, L.; Fortes, C.J.; Santos, J.A.; Rosa-Santos, P. Numerical simulation of the motions and forces of a moored ship in Leixões harbour. In Proceedings of the 3rd International Conference on Maritime Technology and Engineering, MARTECH 2016, Lisbon, Portugal, 4–6 July 2016; Volume 1, pp. 217–226.

14. Varyani, K.S.; Vantorre, M. New generic equation for interaction effects on a moored containership due to a passing tanker. *J. Ship Res.* **2006**, *50*, 278–287.

15. Aksnes, V. A simplified interaction model for moored ships in level ice. *Cold Reg. Sci. Technol.* **2010**, *63*, 29–39. [CrossRef]

16. Figuero, A.; Sande, J.; Peña, E.; Alvarellos, A.; Rabuñal, J.R.; Maciñeira, E. Operational thresholds of moored ships at the oil terminal of inner port of A Coruña (Spain). *Ocean Eng.* **2019**, *172*, 599–613. [CrossRef]

17. Figuero, A.; Rodríguez, A.; Sande, J.; Peña, E.; Rabuñal, J.R. Field measurements of angular motions of a vessel at berth: Inertial device application. *Control Eng. Appl. Inform.* **2018**, *20*, 79–88.

18. Figuero, A.; Rodriguez, A.; Sande, J.; Peña, E.; Rabuñal, J.R. Dynamical study of a moored vessel using computer vision. *J. Mar. Sci. Technol.* **2018**, *26*, 240–250.

19. Gómez Lahoz, M.; Carretero Albiach, J.C. Wave forecasting at the Spanish coasts. *J. Atmos. Ocean Sci.* **2005**, *10*, 389–405. [CrossRef]

20. López, M.; Iglesias, G. Long wave effects on a vessel at berth. *Appl. Ocean Res.* **2014**, *47*, 63–72. [CrossRef]

21. Van der Molen, W.; Scott, D.; Taylor, D.; Elliott, T. Improvement of Mooring Configurations in Geraldton Harbour. *J. Mar. Sci. Eng.* **2016**, *4*, 3. [CrossRef]

22. Zelterman, D. *Applied Multivariate Statistics with R*; Springer International Publishing: Cham, Switzerland, 2015; ISBN 978-3-319-14093-3.

23. Carral-Couce, L.; Naya, S.; Álvarez Feal, C.; Lamas Pardo, M.; Tarrío-Saavedra, J. Estimating the traction factor and designing the deck gear for the anchor handling tug. *Proc. Inst. Mech. Eng. Part. M J. Eng. Marit. Environ.* **2016**, *231*, 600–615. [CrossRef]

24. Carral, L.; Tarrío-Saavedra, J.; Naya, S.; Bogle, J.; Sabonge, R. Effect of Inaugurating the Third Set of Locks in the Panama Canal on Vessel Size, Manoeuvring and Lockage Time. *J. Navig.* **2017**, *70*, 1205–1223. [CrossRef]

25. Carral Couce, L.; Couce, J.C.; Tarrío-Saavedra, J.; Formoso, J.A.F. Net winch design in trawlers, influence of vessel size and fishing ground. *Proc. Inst. Mech. Eng. Part. M J. Eng. Marit. Environ.* **2017**, *233*, 108–123. [CrossRef]

26. Jain, P.; Deo, M.C. Artificial Intelligence Tools to Forecast Ocean Waves in Real Time. *Open Ocean Eng. J.* **2008**, *1*, 13–20. [CrossRef]

27. Wang, D.; Xiong, X.L. Multivariate Linear Regression Model for Ship Traffic Flow Prediction Based on Influence Factors Analysis. *Ship Ocean Eng.* **2010**, *3*, 178–180.

28. Kutner, M.H.; Nachtsheim, C.; Neter, J.; Li, W. *Applied Linear Statistical Models*, 5th ed.; McGraw-Hill/Irwin: New York, NY, USA, 2005; ISBN 0072386886.

29. Burnham, K.P.; Andreson, D.R. *Model. Selection and Multimodel Inference. A Practical Information-Theoretic Approach*, 2nd ed.; Springer: New York, NY, USA, 2002; ISBN 978-0-387-22456-5.

Article

Dynamic Calculation of Breakwater Crown Walls under Wave Action: Influence of Soil Mechanics and Shape of the Loading State

Enrique Maciñeira *, Enrique Peña, José Sande and Andrés Figuero

Water and Environment Engineering Group (GEAMA), Universidade da Coruña, 15071 A Coruña, Spain; enrique.penag@udc.es (E.P.); jose.sande@udc.es (J.S.); andres.figuero@udc.es (A.F.)
* Correspondence: enrique.macineira@udc.es; Tel.: +34-636-987-800

Received: 6 May 2019; Accepted: 24 May 2019; Published: 31 May 2019

Abstract: As a consequence of the action of waves on rubble mound breakwaters, there are loads—both on the vertical and horizontal sides of the crown walls—which modify the conditions of their stability. These loads provoke dynamic impulses that generate movements that are not possible to be analyzed by static calculation. This study presents the results obtained using a simplified method of dynamic calculation of the crown walls, presented in Appendix A, based on the variation of the forces acting against the structure in the time domain and the soil characteristics. It provides results of the expected movements of the structure and the deformations produced in the foundation. With this, traditional static calculation is improved and knowledge about the phenomenon is enhanced, highlighting the uncertainties in the system.

Keywords: breakwater; crown wall failure; dynamic response; sliding; overturning; bearing capacity

---

1. Introduction

Within the modes of failure of breakwaters, those related to the crown wall are some of the most important. The main and specific ones, which usually are the object of practical calculations, are sliding, rigid and plastic overturning, and bearing capacity of the foundation (Figure 1). Other breakwater modes of failure, such as global geotechnical failure, may affect or be affected by the crown wall [1].

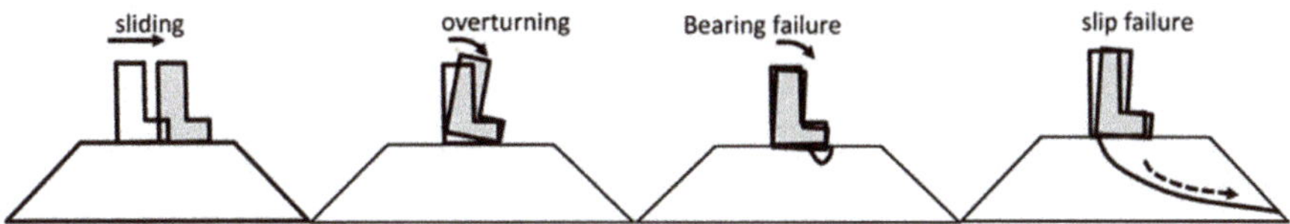

Figure 1. Breakwater modes of failure (adapted from Pedersen [2]).

The calculation of these modes of failure is usually done by integrating the instantaneous forces acting against the crown wall, in the consideration of maximum static forces. However, the forces are usually of very short duration, especially in the case of crown walls not fully protected by the slope's main armor units. The broken wave hits directly against them, generating impulsive pressures depending on the wave's characteristics and its impact against the structure.

To consider this effect, it is common to truncate the signal obtained in laboratory tests (by pressure cells or by dynamometers) in different ways: by eliminating 1‰ or 2.5‰ of the waves that generate greater loads, by eliminating 1‰ or 0.1‰ of the data records, or other laboratory practices [1–4]. This is done in order to eliminate some instantaneous data records that, if we considered them, would generate excessively large structures.

For this reason, is convenient to improve the knowledge of the performance of the crown wall and its movements in the time domain by proposing the dynamic calculation of the structure, particularly in cases where impulsive pressures are present [5] and considering the variation of the conditions of the core. Thus, the entire data can be used, even those data that produce instantaneous forces over the static equilibrium conditions, which could generate permissible movements of the crown wall by port operations and that do not produce the failure of the structure.

In the present study, the movements expected by the crown wall of the main breakwater of the outer port of La Coruña in Punta Langosteira are analyzed by using a new proposed simplified model. The structure is subjected to four theoretical load signals (permanent loads, sinusoidal, and two impulsive signals) varying the mechanical characteristics of the foundation. A total of 752 simulations were carried out varying the parameters involved (Young's modulus and soil model).

2. Proposed Simplified Model

The movement of the crown wall can be estimated by solving the general equation of the dynamics of the rigid solid (see Equation (A1) in Appendix A), with six degrees of freedom (three movements and three turns. However, the analytical resolution of the system becomes very complex [6] and requires, in addition to the knowledge of the instantaneous acting forces, a deep knowledge of the constituent materials of the breakwater; in particular, the core and its characteristics: elasticity, stiffness, damping, permeability, heterogeneity, etc. There are some approximations by numerical model [7,8] that allow point-to-point calculations of the interaction between the structure, the soil, and the water flow inside the breakwater in the time domain. However, the dynamic problems of interacting with an elasto-plastic soil are still not well solved, so in practice, simplifications are proposed for their resolution [7].

The time-domain resolution of the system of equations can be simplified notably by canceling the matrix of damping and introducing two simplifications: on one hand, introducing the damping of the rotation as a response to the deformation existing at the immediately previous instant in the iterative process of calculation; on the other hand, replacing the reaction of the soil to the displacement by the friction in the contact between structure and foundation. The detail of these approximations of the proposed model can be found in Appendix A of this article (see Equation (A15) and Figure A2 in Appendix A). The system of equations can be solved iteratively, so that movements of the crown wall and the deformations of the foundation can be obtained in the time domain. So, the instability or situations incompatible with port operations are determined with enough accuracy.

3. Case Study: Breakwater of Punta Langosteira, Spain

3.1. Breakwater and Crown Wall Description

The breakwater of the Outer Port of Punta Langosteira (A Coruña, Spain) is 3.3 km in length, built in a maximum draft of 40 m with respect to lowest astronomical tide (LAT), protected by an armor of two layers of 150 T concrete cubic units of 2.4 T/m^3 density, and the slope of the front part is 1 V (vertical):2 H (horizontal) and 1 V:1.5 H in the inner (Figure 2).

The crown wall rests on the core at +10 m with respect to LAT and is crowned at +25 m over LAT. It is protected in all its height by the main armor of the breakwater and its filters. The crown wall is formed by three solid bodies of different dimensions. The first is 10 m wide at the base and 6 m high; the second is above the first and is 8.5 m wide and 4.5 m high; and the last, above the second, is 5.5 m wide and 4.5 m high. The material of the crown wall is concrete, with a density of 2.37 T/m^3, so the mass per linear meter is 275.51 T.

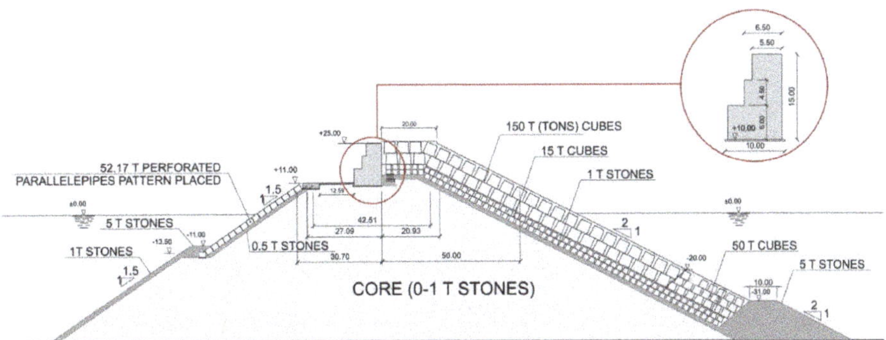

Figure 2. Cross section of the main breakwater of the outer port of Punta Langosteira (A Coruña, Spain).

3.2. Soil Mechanics Characteristics

For this study, two models of soil were analyzed to find out the influence of its plasticity:

1. Elastic model, in which the response of the soil is represented assuming homogeneous, elastic, and isotropic constituent materials, with constant characteristics over time.

    - Type of material: gravel and sand;
    - Fine fraction granulometry: D_{n50} (cm) < 0.10;
    - Five Young's modulus values varying from E (MPa) = 10–400 to analyze the influence of this parameter. The case of an absolutely rigid foundation with an E (MPa) = 27,000 (concrete) is also analyzed;
    - Poisson coefficient ν = 0.30 for permanent loads. In the case of pulsating loads, it is considered to be ν = 0.50 [9];
    - soil–concrete friction coefficient μ_t = 0.6 (static); μ_d = 0.48 (dynamic);
    - soil friction angle θ = 38°.

2. Elasto-plastic model: in this case, Young's modulus and shear modules vary depending on the state of loads and the deformation according to a hyperbolic elasto-plastic model of soil response, including the hysteresis of the materials [10,11]. The details of application of the model can be found in Appendix B of this article.

3.3. Loading States

To analyze the performance of the proposed model, the crown wall was tested under three different loading states. One was used to validate the results obtained with the simplified model, and the other two to analyze the different modes of failure of the structure (sliding, overturning, and bearing failure):

Load state A (contrast case): Application of the worse loads determined in the physical model tests (2D large wave flume of the Ports and Coasts Laboratory of the Spanish Public Works Ministry, Madrid, with an active wave absortion system, scale: 1:25) was carried out for the official project [12], obtained for a significant wave height, H_s (m) = 15.1 and peak period, T_p (s) = 20 (Figure 3). A horizontal water load and an overload of the water mass are introduced at the top. The permanent loads against the crown wall due to the pressure of the blocks and materials of the different layers and filters are not considered. The results of the proposed simplified model are compared with the study carried out [13] using the FLAC2D 7.0 code. The worse soil data and parameters obtained by the Ports and Coasts Laboratory of the Spanish Public Works Ministry [14] have been used, taking into account the great variability existing in the functioning of the analysis technique used to determine them.

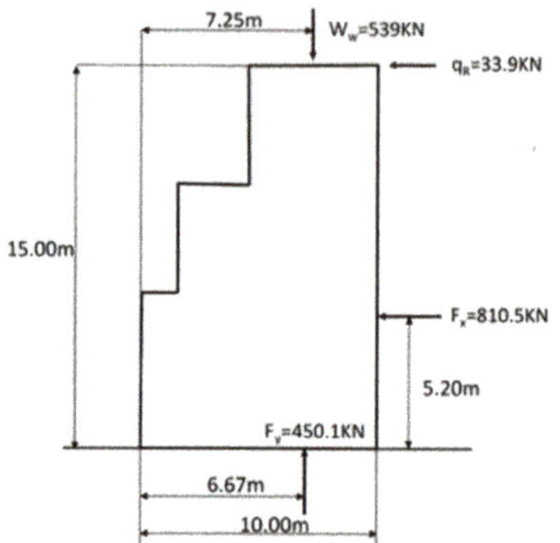

Figure 3. Load state A, used for crown wall design (Wave height Hs (m) = 15.1; Peak period Tp (s) = 20), [11].

This load state, although it is demanding, generates high safety coefficients, so limited movements of the crown wall are expected (Table 1).

Table 1. Safety coefficients obtained for load state A.

Mode of Failure	Safety Coefficient	Verification
Sliding safety coefficient (SSC)	1.93	
Rigid overturning safety coefficient (OSC)	4.61	
Bearing capacity safety coefficient (BCSC)	6.24	Brinch-Jansen [15]

Load state B: This was used to analyze the performance of the structure before the sliding failure. The horizontal load is increased from F_x (KN) = 1000 to 2200 with a determined arm (5.2 m) so that the instability due to overturning does not occur. The existence of vertical sub-pressure loads will not be considered in order to not introduce a component that distorts the slip analysis (Figure 4).

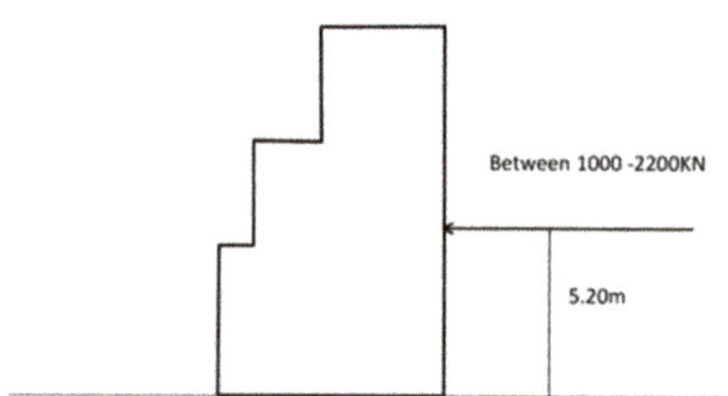

Figure 4. Load state B.

Sliding safety coefficients less than unity are obtained by the application of this load state (Table 2). Thus, by introducing this information in the proposed model, it will be possible to know the influence of the shape of the signal in the time domain and the influence of the characteristics of the core material in this mode of failure.

Table 2. Safety coefficients obtained for load state B.

F_x (KN) (Horizontal Force)	1000	1200	1400	1600	1800	2000	2200
Sliding Safety Coefficient (SSC)	1.62	1.35	1.16	1.01	0.90	0.81	0.74
Overturning Safety Coefficient (OSC)	3.27	2.55	2.18	1.91	1.70	1.53	1.39
Bearing Capacity Safety Coefficient (BCSC) [15]	4.34	2.86	1.83	1.14	0.69	0.40	0.22

Load state C: This state corresponds to a state of static instability due to overturning and bearing capacity failure of the foundation. Maintaining the horizontal load F_x (KN) = 1200, the application arm is increased from 5 m to 16 m, so that the instability due to overturning occurs. The existence of vertical sub-pressure loads is not considered in order to not introduce a component that distorts the problem of overturning and failure of the foundation (Figure 5).

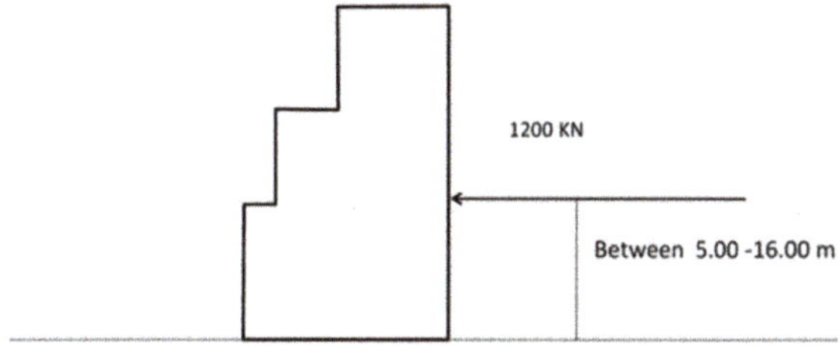

Figure 5. Load state C.

The bearing capacity safety coefficients obtained by the application of this load state are substantially less than unity (Table 3). Introducing this load state in the proposed simplified model, the influence of the shape of the signal in the time domain and the influence of the characteristics of the core material can be analyzed.

Table 3. Coefficients obtained for load state C.

M (KNm) (Overturning Moment)	4800	7200	9600	10,800	12,000	14,400	15,600	16,800	19,200
Sliding Safety Coefficient (SSC)	1.35	1.35	1.35	1.47	1.35	1.35	1.35	1.35	1.35
Overturning Safety Coefficient (OSC)	3.31	2.21	1.66	1.35	1.33	1.10	1.02	0.95	0.83
Bearing Capacity Safety Coefficient (BCSC) [15]	3.44	2.49	1.66	1.29	0.96	0.35	0.07	0.01	0

3.4. Shape of the Loading State Signal in the Time Domain

A fundamental aspect for the practical applicability of the model and its sensitivity is the typology of the acting force. Therefore, the performance of the model is analyzed with four different theoretical signals in each of the loading states, with an interval of T(s) = 20, where T—wave period (Figures 6 and 7), and repetition of 10 cycles of loads.

1. Permanent: The signal (F_x, F_y) is constant and equal to the maximum force.
2. Sinusoidal: The signal (F_x, F_y) follows a sinusoidal law in a semi-period of the wave with a maximum amplitude equal to the maximum force.

3. Impulsive 1: The signal for the horizontal force (F_x) presents a maximum equal to the maximum force at a rise time of 0.05 T (1 s), subsequently reducing to 75% and disappearing in a half-period of the wave. With regards to the vertical force F_y, in the case of load state A, a sinusoidal law follows in a semi-period of the wave, with a maximum amplitude equal to the maximum F_y. In the case of loading states B and C, F_y is not considered.

4. Impulsive 2: Same as Impulsive 1, with a value of 50% for F_x.

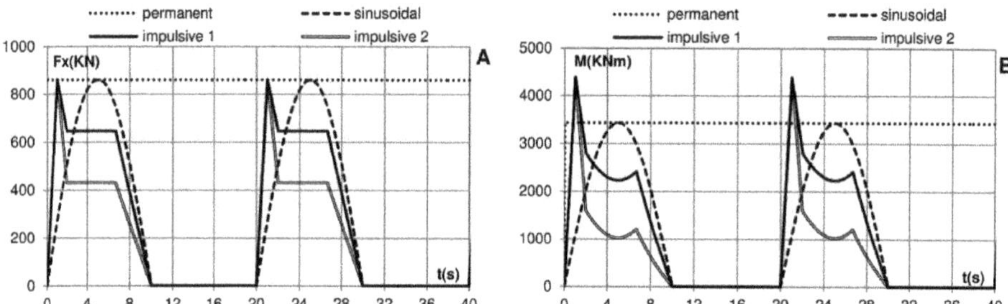

Figure 6. Shape of the load state A in the time domain: (**A**) horizontal force (F_x); (**B**) overturning moment (M).

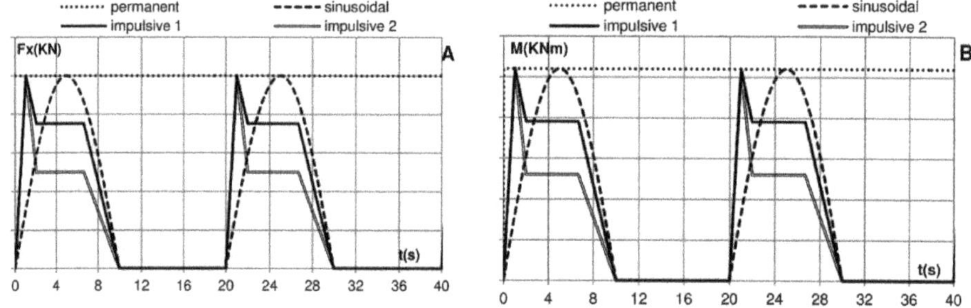

Figure 7. Shape of the load states B and C in the time domain: (**A**) horizontal force (Fx); (**B**) overturning moment (M).

The calculation made with the permanent actions is comparable to the static calculation of the crown wall, introducing the mechanic characteristics of the soil as an element that determines the turning movement of the crown wall in adapting to the overturning stress that it experiences. The sinusoidal signal has the shape of a theoretical sine wave. The impulsive actions respond to church-roof signals [1] with impulsive pressures of short duration against the crown wall. These types of signals occur in many cases if there is a direct impact of the wave against the crown wall, in the case that it is not fully protected by the armor. In the first impulsive action defined (tip factor of 1.33), the influence of the impulsive pressure is much less than in the second one (tip factor of 2.0).

4. Results Obtained with the Simplified Model Proposed

4.1. Load State A

A total of 56 simulations were carried out, varying the shape of the loading state signals (4 shapes), the elasticity of the soil (7 elasticity modules), and the model of the soil (2 models).

4.1.1. Elastic Soil

The movements produced on the top have a direct relationship with the modulus of elasticity of the foundation material. At a higher stiffness, a smaller movement corresponds, and vice versa. In Figure 8, the influence of the shape of the signal can be observed. Impulsive actions produce less movement since the maximum forces are of very short duration. This fact is more accentuated in the case of the Impulsive 2 signal with a tip factor of 2.0.

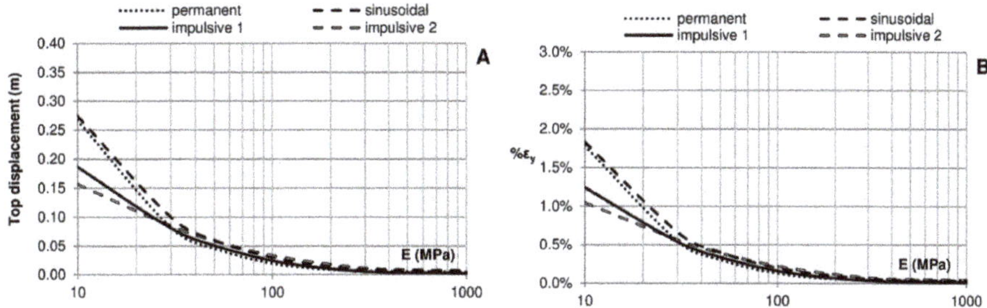

Figure 8. (**A**) Movement on the top of the crown wall; (**B**) maximum vertical deformation of the foundation.

The movements of the top, for the case of E (MPa) = 10, are presented in Figure 9. It can be observed that the maximum movement occurs in the case of sinusoidal acting forces due to the larger Poisson's coefficient considered in a pulsating loading shape compared with permanent loads. Also, it can be observed that, as Pedersen [2] pointed out, when the tip factor is equal to or greater than 2.0, the constant part of the force has very little relevance in the movement.

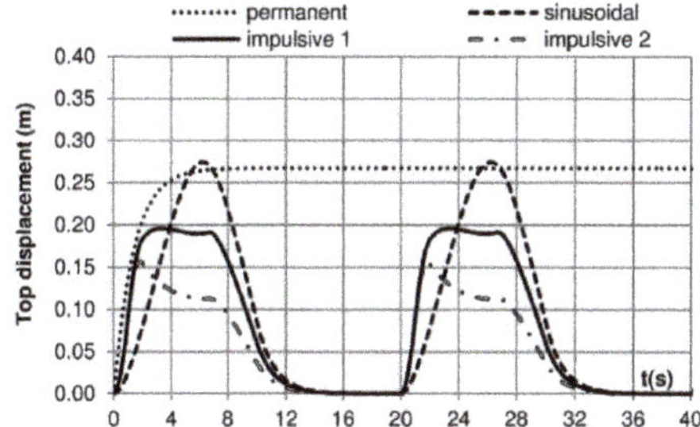

Figure 9. Expected movement on the top of the foundation: E (MPa) = 10; 2 load cycles.

The estimated deformations of the foundation are far from those considered inadmissible (See Table A2 in Appendix B). For the values of E (MPa) = 10 and E (MPa) = 100, the maximum deformation does not exceed 2.0% and 0.25%, respectively.

Figure 10 shows the movement on the top and the angular velocity in the time domain obtained with a sinusoidal action, related to the overturning moment. It can be observed that the largest movement occurs with a certain delay in relation to the maximum loading moment. In the case of the sinusoidal signal, in which the maximum loading moment occurs at the instants t (s) = 5 and t (s) = 25, the maximum movement occurs at the values of t (s) = 6.4 and t (s) = 26.4. This delay is produced by the inertia of the movement. In the analysis of the speed, the passing of "0" or change of direction of rotation indicates the point of the maximum amplitude of the movement.

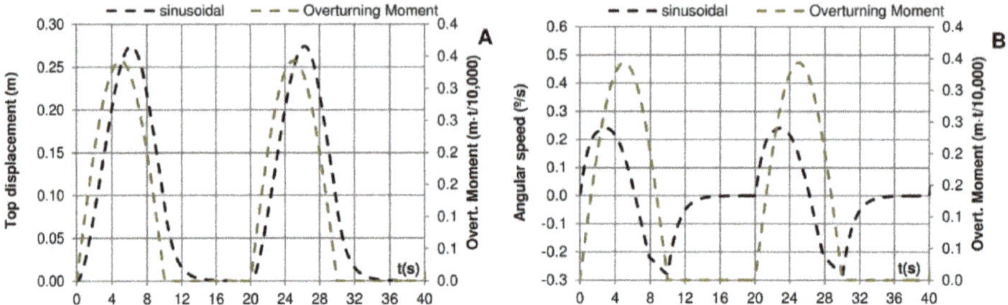

Figure 10. Sinusoidal force. (**A**) Movement on the top of the crown wall; (**B**) angular velocity, related to the overturning moment. E (MPa) = 10; 2 load cycles.

Figure 11 shows the movement on the top and angular velocity of the crown wall for the case of Impulsive 1 (tip factor = 1.33). It is observed that the maximum moment coincides in time with the maximum positive angular velocity. However, throughout the load reduction phase of the impulsive pressure, the crown wall continues to rotate in a positive direction, although the velocity is slowed. The passing of "0", which coincides with the maximum displacement of the top, occurs in this case once the impulse pressure has finished. This fact causes a reduction of the total movement of the crown wall with respect to the expected one in the case of permanent actions applied, which stabilizes approximately at t (s) = 14 (Figure 9).

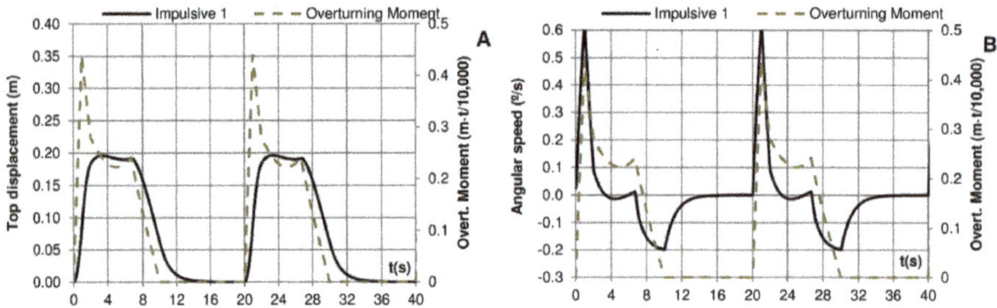

Figure 11. Impulsive 1. (**A**) Movement on the top of the crown wall; (**B**) angular velocity, related to overturning moment. E (MPa) = 10; 2 load cycles.

4.1.2. Elasto-Plastic Soil

In this section, the influence of model plasticity is analyzed. As in the elastic case, the influence of the load shape on the movement can be observed (Figure 12). However, it can be verified by analyzing Figures 8 and 12 that the movements and deformations in this second case (elasto-plastic) are larger than those in the first (elastic), demonstrating the good performance of the model.

The expected movement in the case of the sinusoidal action is larger than that expected in the case of a permanent load. This fact is derived from the different Poisson coefficients considered in both cases ($v = 0.3$ for permanent load and $v = 0.5$ for sinusoidal load [9]. On the other hand, the inertia of the movement, produced in the case of cyclic signals, produces a change in the conditions of deformation.

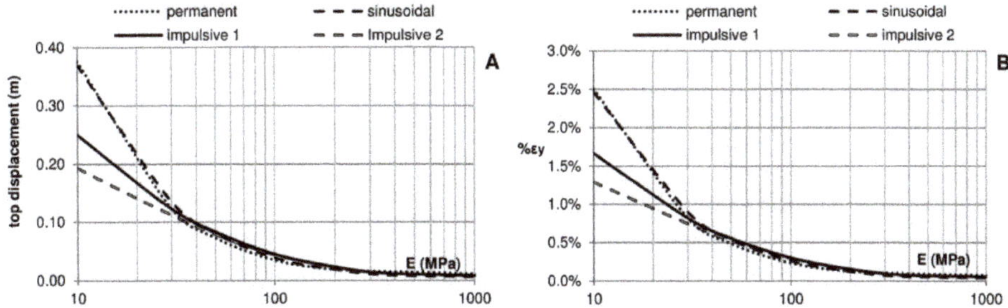

Figure 12. (**A**) Movement on the top of the crown wall; (**B**) maximum vertical deformation of the foundation.

The estimated deformations in the foundation are again far from those considered inadmissible (See Table A2 in Appendix B). In the case of this model of soil, a residual deformation of the foundation between 0.24% and 0.64% occurs as a function of the load signal, as a consequence of the nonlinearity of the stress and deformation (Figure 13).

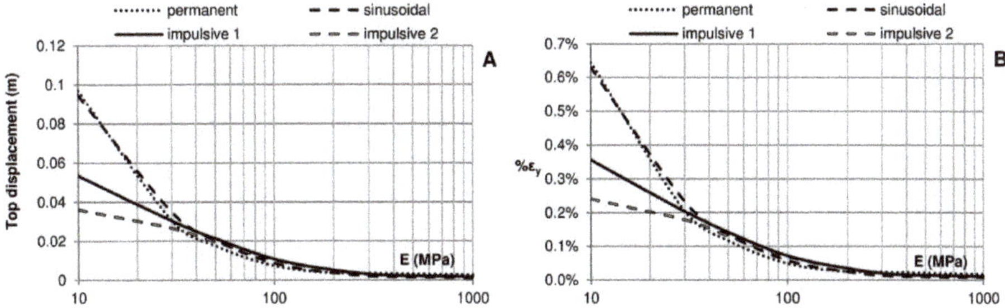

Figure 13. (**A**) Residual movement on the top of the crown wall; (**B**) residual vertical deformation.

Analyzing the movement that occurs in the first 2 cycles with E (MPa) = 10, for each of the excitation signals, it is observed that there is a residual deformation generated in the first load cycle, which remains until the occurrence of a larger value than that which produced it, according to Atkinson's model [16]. This fact is more visible in the stress–strain graph of the sinusoidal load, with a vertical residual deformation of 0.64% (Figure 14).

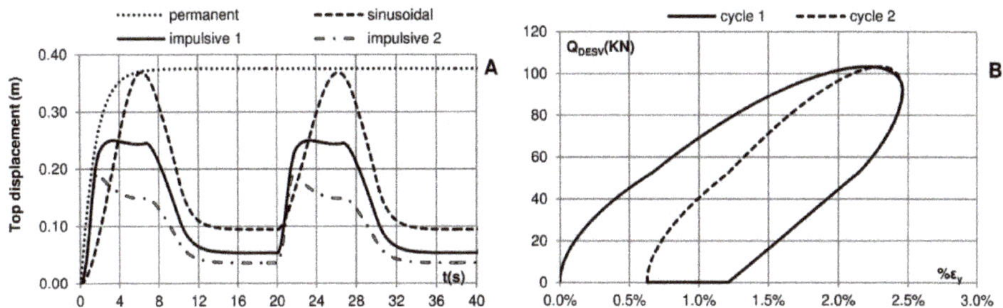

Figure 14. (**A**) Movements on the top of the crown wall; (**B**) stress–strain graph showing the relationship between the load deviator and vertical deformation produced in the first and second cycles of loading.

The hardening process of the soil can also be seen in Figure 14. The expected movement is less in the second cycle than in the first one. Due to the consideration of Atkinson's model [16], the behavior of the soil will be elastic if the load deviator does not increase over the maximum that had occurred (more information about the load deviator (Q_{DESV}) is provided in Appendix B).

4.1.3. Comparison with FLAC2D 7.0 Code

The results obtained with the simplified model and those obtained with the FLAC2D 7.0 code are presented below (Table 4). Using the model hypothesis made with respect to the damping, no horizontal movements are recorded, since the sliding threshold is not reached (Coulomb hypothesis [17] has been considered: the resistance to the movement over a plane is proportional to the normal force exerted). However, it is verified that net movements of the same order of magnitude are obtained, providing a case for validation of the proposed model.

Table 4. Results obtained with the simplified model and with FLAC2D 7.0 code, comparison without considering transverse deformation of the soil.

E (MPa) Considering the Worst Values for the Core	Poisson Coefficient	Shape of the Load's Signal	Simplified Model (Elastic)		FLAC2D 7.0 (Net Movements)	
			Top Displacement (m)	$\varepsilon_y\%$	Top Displacement (m)	$\varepsilon_y\%$
10	0.30	1	0.27	1.8	0.30	2.0
30	0.30	1	0.08	0.5	0.07	0.5

The comparison with the model FLAC2D 7.0 code considering the maximum possible transverse deformation of the soil (see Equation (A16) in Appendix A) is presented in Table 5:

Table 5. Results obtained with the simplified model and with FLAC2D 7.0 code, comparison considering transverse deformation of the soil.

E (MPa) Considering the Worst Values for the Core	Poisson Coefficient	Shape of the Load's Signal	Simplified Model (Elastic)			FLAC2D 7.0 (Total Movements)		
			Top Displacement (m)	$\varepsilon_y\%$	$\varepsilon_x\%$	Top Displacement (m)	$\varepsilon_y\%$	$\varepsilon_x\%$
10	0.30	1	0.49	1.8	2.2	0.46	2.0	1.6
30	0.30	1	0.16	0.6	0.7	0.13	0.5	0.6

Where $\varepsilon_x\%$— soil deformation in X-axis; $\varepsilon_y\%$—soil deformation in Y-axis.

Again, similar results are obtained, with a small variation in the horizontal deformation for E (MPa) = 10 due to the hypothesis considered.

Even though there are very few data, there has been analyzed the correlation between the displacements of the top of the crown wall obtained using the proposed simplified model and those obtained using FLAC2D code. The correlation obtained is quite good. The root-mean-square error (RMSE) obtained is 0.03 m. The Pearson's coefficient of correlation is $R^2 = 0.97$.

4.1.4. Influence of Tip Factor and Rise Time in the Church-Roof Signals

The tip factor can be larger than 2.5 [3]. There have been used in the calculations two smaller ones because, as the maximum forces and moments were maintained in all shapes of loading state, the use of a larger tip factor should produce a faster decrease in the total movement of the crown walls.

On the other hand, the rise time has also an influence on the dynamic response. Typically, the values obtained in laboratory are in the range of 0.01–0.2 s [2], and in the prototype, in the range of 0.1–1 s [1] (depending on the scale: approximately between 0.01 T_p and 0.1 T_p). In our case, the chosen rise time was equal to 0.05 T in the church-roof signals (Impulsive 1 and Impulsive 2).

To analyze the influence of these two parameters in the performance of the crown wall, its dynamic response was calculated under three additional church-roof theoretical signals:

1. Impulsive 3: Same as Impulsive 1, with a value of 33% for F_x;
2. Impulsive 4: Same as Impulsive 2, with a rise time of 0.025 T (0.5 s);

3. Impulsive 5: Same as impulsive 2, with a rise time of 0.075 T (1.5 s).

The influence on the dynamic response of tip factors can be seen in Figure 15. It can be observed that the reduction in the flat part of the signal has a large influence on the movement of the structure. As stated by Pedersen [2], if the tip factor is larger than 2.0, the constant part of the wave loading following the peak has little influence on the dynamic response.

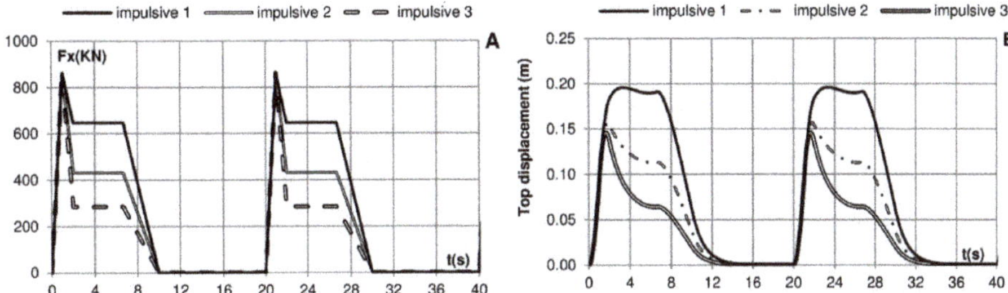

Figure 15. Tip factor influence: (**A**) shape of the signals; (**B**) movements on the top of the foundation. 2 cycles.

The influence of rise time on dynamic response can be seen in Figure 16. It can be observed that, if the rise time is reduced from 0.075 T (Impulsive 5) to 0.025 T (Impulsive 4), the maximum movement of the crown wall is also reduced. This effect is produced due to the inertia of the structure. The movement is very fast at the beginning in the case of Impulsive 4 (0.025 T). Then, when the impulse stops, the movement continues, but it does not reach the maximum of Impulsive 2 or 5.

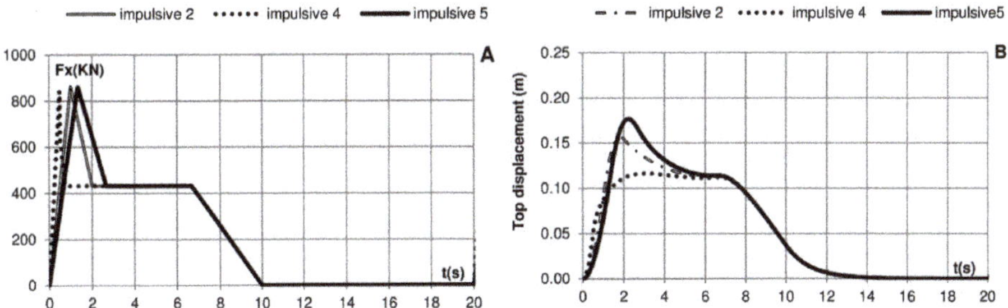

Figure 16. Rise time influence: (**A**) shape of the signals; (**B**) movements on the top of the foundation.

In these theoretical signals, there is no reproduction of other kinds of impulses that could occur in the prototype or in the laboratory. However, the peak force generated by the impact of the wave crest could be followed by other force oscillations due to the air pockets entrapped within [1]. These oscillations could be in the range of the natural period of oscillation of the structure and could amplify its movement.

4.2. Load State B

The influence of soil mechanics and the shape of the loading state in sliding were studied in the analysis of load state B.

A total of 336 simulations were carried out, varying the forces acting against the crown wall (7 horizontal forces), the shape of the loading state signal (4 shapes), the elasticity of the soil (6 elasticity modules), and the model of the soil (2 models).

Figure 17 shows the sliding of the crown wall, considering an elastic soil foundation in each of the four excitation forces as a function of the sliding safety coefficient for different Young's modulus values of the soil.

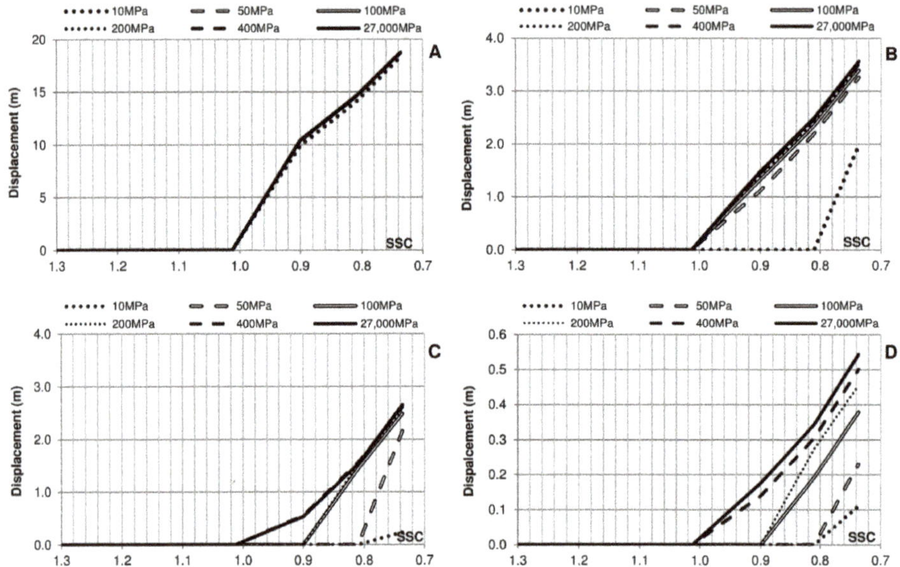

Figure 17. Crown wall displacement as a function of Sliding Safety Coefficient (SSC) and Young's modulus of the soil, assuming elastic soil, for the 4 different theoretical signals: (**A**) Permanent; (**B**) Sinusoidal; (**C**) Impulsive 1; (**D**) Impulsive 2.

In the case of permanent loads, the situation can be assimilated to the static calculation where the crown wall begins to slide and the safety coefficient is less than 1.0. On the other hand, the displacement produced is the same, being independent of the soil characteristics. Therefore, these factors do not influence sliding in the static calculation.

In the case of a very rigid foundation, such as the crown wall resting directly on concrete (E (MPa) = 27,000), or even with values of E (MPa) > 400, the displacement occurs whenever the static equilibrium situation is exceeded, whatever the shape of the action. This result is especially important and is a central element of the present study, since it is verified that in these cases, the calculation must be made using the maximum impulsive pressures recorded without truncating the laboratory or other data records (Figure 17). The reason for this performance is that, when the safety coefficient factor falls below 1.0, there is not any absorption of energy in the foundation soil, and all the wave energy is used to "move" the crown wall. Further, when the movement begins, the friction coefficient is reduced dramatically.

This type of breakage is very common in crown walls that can be stable as a whole, constructed over the core of the breakwater that behaves as a dumping absorber of the movement. However, the upper parts, if they are not joined to the rest of the body of the crown wall, could become unstable. Figure 18 shows a breakage of this type in the breakwater of A Garda (Spain) in February 2017. Other cases have occurred in recent years, particularly in locations in the north and northwest of Spain.

Figure 18. Displacement of the upper part of the crown wall in the breakwater of A Garda, Spain (2017).

For the rest of the signals analyzed the less rigid the soil, the smaller the expected sliding of the crown wall. Moreover, the applied horizontal load could exceed the static equilibrium conditions without starting the sliding. This is because although the soil is deformable, the horizontal impulse is initially transformed into rotation and, later, into displacement. If the duration of the load that exceeds the static equilibrium condition is very small, the crown wall does not mobilize and it recovers the original position in the load reduction phase.

The influence of the load type of signal on crown wall sliding has been compared for 3 different Young's modulus values (Figure 19). It can be verified that in the case of very rigid soil (E (MPa) = 27,000), in all cases, the sliding starts if the sliding safety coefficient (SSC) is less than 1.0. On the other hand, it has been proved that regardless of the type of soil, in the case of permanent loads, the sliding occurs if the load exceeds the static equilibrium conditions (SSC < 1), reflecting the good performance of the proposed simplified model. Likewise, it can be observed that if the soil has a certain flexibility, in the case of impulsive loads, sliding may not occur even with SSC less than 1.0.

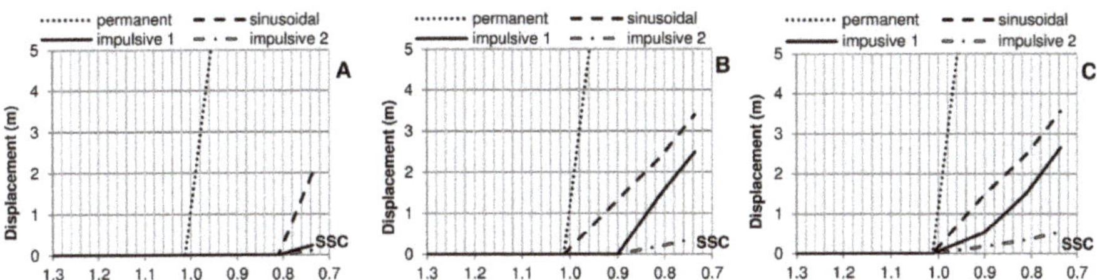

Figure 19. Wall sliding as a function of SSC and the signal shape, assuming elastic soil. (**A**) E (MPa) = 10; (**B**); E (MPa) = 100; (**C**) E (MPa) = 27,000.

Comparing the results obtained for load state B with the bearing capacity safety coefficient (BCSC) obtained (Figure 20), it can be deduced that, independently of the amplitude of the deformations, especially in the case of the soil with E (MPa) = 10, the soil does not collapse (See Table A2 in Appendix B). In the case of this loading state, the failure of the structure occurs by sliding.

Table 6 presents, for the load state B, the sliding occurring after 10 cycles of load for each of the simulations and each of the signal shapes. It can be seen that in the case of a Young's modulus above E (MPa) = 400, if the sliding condition is exceeded (SSC < 1.0), displacement of the crown wall occurs. However, in the case of more flexible soils, depending on the load shape, the sliding condition can be overcome without displacement of the crown wall (e.g., for E (MPa) = 100 with SSC = 0.90, sliding does not occur in the case of impulsive loads). In the case of very rigid foundations (E (MPa) = 27,000), the elasto-plastic soil model was not considered.

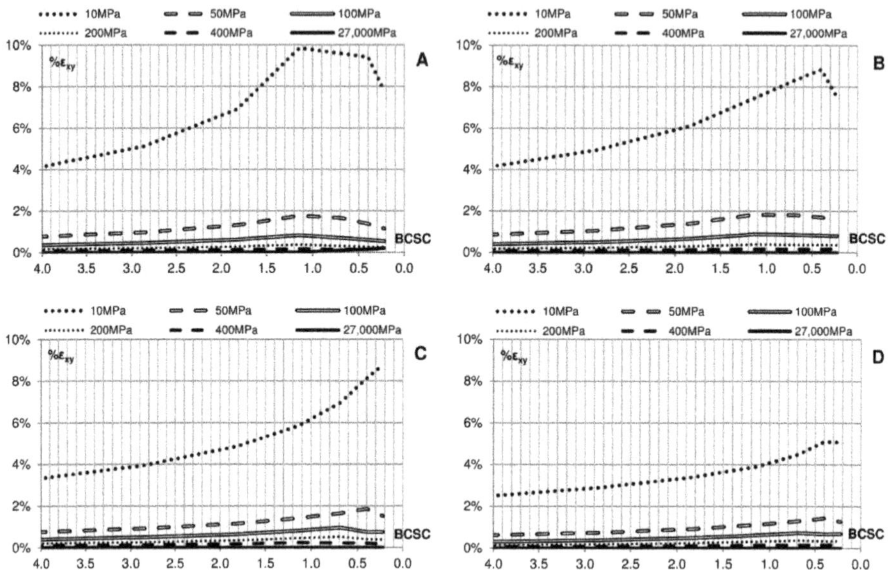

Figure 20. Maximum deformation as a function of the bearing capacity safety coefficient (BCSC) and Young's modulus of the soil, assuming elastic soil. (**A**) Permanent; (**B**) Sinusoidal; (**C**) Impulsive 1; (**D**) Impulsive 2.

Table 6. Crown wall sliding in simulations under load state B.

		Sliding Failure (m of Displacement, 10 Cycles of Load)											
Sliding Safety Coefficient		**1.35**	**1.16**	**1.01**	**0.90**	**0.81**	**0.74**	**1.35**	**1.16**	**1.01**	**0.90**	**0.81**	**0.74**
Young's Modulus	**Signal**	**Elastic Soil**						**Elasto-Plastic Soil**					
E = 10 MPa	Permanent	-	-	-	9.9	14.0	18.3	-	-	-	9.9	14.0	18.3
	Sinusoidal	-	-	-	-	-	2.0	-	-	-	-	-	2.0
	Impulsive 1	-	-	-	-	-	0.2	-	-	-	-	-	0.2
	Impulsive 2	-	-	-	-	-	0.1	-	-	-	-	-	0.1
E = 50 MPa	Permanent	-	-	-	10.4	14.5	18.7	-	-	-	10.4	14.5	18.7
	Sinusoidal	-	-	-	1.1	2.2	3.3	-	-	-	1.1	2.2	3.3
	Impulsive 1	-	-	-	-	-	2.2	-	-	-	-	-	2.2
	Impulsive 2	-	-	-	-	-	0.2	-	-	-	-	-	0.2
E = 100 MPa	Permanent	-	-	-	10.4	14.6	18.7	-	-	-	10.4	14.5	18.7
	Sinusoidal	-	-	-	1.3	2.3	3.4	-	-	-	1.3	2.3	3.4
	Impulsive 1	-	-	-	-	1.4	2.5	-	-	-	-	1.4	2.5
	Impulsive 2	-	-	-	-	0.2	0.4	-	-	-	-	0.2	0.4
E = 200 MPa	Permanent	-	-	-	10.4	14.6	18.7	-	-	-	10.4	14.6	18.7
	Sinusoidal	-	-	-	1.4	2.4	3.5	-	-	-	1.4	2.4	3.5
	Impulsive 1	-	-	-	-	1.5	2.6	-	-	-	-	1.5	2.6
	Impulsive 2	-	-	-	-	0.3	0.5	-	-	-	-	0.3	0.5
E = 400 MPa	Permanent	-	-	-	10.4	14.6	18.7	-	-	-	10.4	14.6	18.7
	Sinusoidal	-	-	-	1.5	2.5	3.5	-	-	-	1.5	2.5	3.5
	Impulsive 1	-	-	-	0.6	1.6	2.6	-	-	-	0.5	1.5	2.6
	Impulsive 2	-	-	-	0.1	0.3	0.5	-	-	-	0.1	0.3	0.5
E = 27,000 MPa	Permanent	-	-	-	10.4	14.6	18.7						
	Sinusoidal	-	-	-	1.5	2.5	3.6						
	Impulsive 1	-	-	-	0.5	1.5	2.7						
	Impulsive 2	-	-	-	0.2	0.3	0.5						

However, the collapse of the soil did not occur in any of the simulations. Table 7 shows the results for the vertical deformation. It can be seen that the limit of deformation is not exceeded in any of the cases (See Table A2 in Appendix B).

Table 7. Vertical deformation in simulations under load state B.

		Sliding Failure (% of Vertical Deformation)											
Sliding Safety Coefficient		1.35	1.16	1.01	0.90	0.81	0.74	1.35	1.16	1.01	0.90	0.81	0.74
Young's Modulus	Signal	Elastic Soil						Elasto-Plastic Soil					
E = 10 MPa	Permanent	4.1%	5.9%	8.9%	8.4%	7.8%	7.8%	6.5%	9.7%	15.6%	14.5%	13.4%	13.3%
	Sinusoidal	3.5%	4.5%	5.8%	6.4%	6.6%	5.4%	5.4%	7.3%	9.6%	10.9%	11.1%	8.9%
	Impulsive 1	2.6%	3.4%	4.3%	5.3%	6.4%	7.2%	3.7%	5.0%	6.6%	8.5%	10.6%	12.1%
	Impulsive 2	1.5%	1.8%	2.1%	2.5%	3.0%	3.5%	2.6%	3.1%	3.7%	4.5%	5.4%	6.4%
E = 50 MPa	Permanent	0.7%	1.1%	1.5%	1.4%	1.3%	1.1%	1.3%	1.9%	2.8%	2.6%	2.2%	1.9%
	Sinusoidal	0.8%	1.1%	1.6%	1.6%	1.5%	1.4%	1.4%	2.0%	2.9%	2.9%	2.7%	2.6%
	Impulsive 1	0.7%	0.9%	1.2%	1.4%	1.6%	1.4%	1.0%	1.4%	1.9%	2.2%	2.6%	2.4%
	Impulsive 2	0.5%	0.7%	0.9%	1.1%	1.2%	1.1%	1.1%	1.4%	1.8%	2.2%	2.3%	2.1%
E = 100 MPa	Permanent	0.3%	0.5%	0.7%	0.7%	0.6%	0.5%	0.6%	0.9%	1.3%	1.3%	1.0%	0.9%
	Sinusoidal	0.3%	0.5%	0.7%	0.7%	0.7%	0.7%	0.7%	1.0%	1.5%	1.5%	1.4%	1.4%
	Impulsive 1	0.4%	0.5%	0.7%	0.8%	0.7%	0.7%	0.6%	0.8%	1.1%	1.4%	1.1%	1.1%
	Impulsive 2	0.3%	0.4%	0.5%	0.6%	0.6%	0.6%	0.7%	0.9%	1.2%	1.4%	1.2%	1.2%
E = 200 MPa	Permanent	0.1%	0.2%	0.3%	0.3%	0.3%	0.2%	0.3%	0.4%	0.7%	0.6%	0.5%	0.4%
	Sinusoidal	0.1%	0.2%	0.3%	0.3%	0.3%	0.3%	0.3%	0.5%	0.7%	0.7%	0.7%	0.7%
	Impulsive 1	0.2%	0.3%	0.4%	0.5%	0.4%	0.4%	0.3%	0.4%	0.6%	0.8%	0.6%	0.6%
	Impulsive 2	0.1%	0.2%	0.3%	0.3%	0.3%	0.3%	0.4%	0.5%	0.7%	0.8%	0.7%	0.7%
E = 400 MPa	Permanent	0.1%	0.1%	0.1%	0.1%	0.1%	0.1%	0.2%	0.2%	0.3%	0.3%	0.3%	0.2%
	Sinusoidal	0.1%	0.1%	0.1%	0.1%	0.1%	0.2%	0.2%	0.2%	0.4%	0.4%	0.4%	0.4%
	Impulsive 1	0.1%	0.2%	0.2%	0.2%	0.2%	0.1%	0.2%	0.2%	0.3%	0.3%	0.3%	0.3%
	Impulsive 2	0.1%	0.1%	0.1%	0.1%	0.1%	0.1%	0.2%	0.3%	0.4%	0.4%	0.4%	0.4%
E = 27,000 MPa	Permanent	0.0%	0.0%	0.0%	0.0%	0.0%	0.0%						
	Sinusoidal	0.0%	0.0%	0.0%	0.0%	0.0%	0.0%						
	Impulsive 1	0.0%	0.0%	0.0%	0.0%	0.0%	0.0%						
	Impulsive 2	0.0%	0.0%	0.0%	0.0%	0.0%	0.0%						

Table 8 shows the modes of failure produced in the simulations carried out with the load state B and the corresponding sliding and bearing capacity safety coefficients calculated at the time of failure. It can be seen that in the case of soils with a certain flexibility and different signal shape than permanent loads, sliding safety coefficients could be less than 1.0 without resulting in sliding. However, for rigid soils, failure always occurs with a sliding safety coefficient equal or less than unity (SSC $\leq$ 1.0).

4.3. Load State C

The influence of soil mechanics and the shape of the loading state in overturning were studied in the analysis of load state C.

A total of 360 simulations were carried out, varying the forces acting against the crown wall (9), the shape of the loading state signal (4 shapes), the elasticity of the soil (5 elasticity modules), and the model of the soil (2 models).

Figure 21 shows, for the load state C with each signal shape and type of soil, the turning of the crown wall in relation to the overturning safety coefficient (OSC) of the structure. It can be seen that, in the case of permanent loads and similar to static calculation, if the foundation has a very low Young's modulus, turning occurs in initial stages of the loading, reaching instability with lower loads. However, for very rigid soils, the instability occurs with OSC values close to 1.0 (OSC $\approx$ 1.0). It is also highlighted that for other signal shapes, the crown wall may have OSC values less than 1; however, the turning of the structure could be acceptable. For example, in the case of Impulsive 2, even for materials with E (MPa) = 50, the rotation produced with a OSC of 0.8 is less than 2°.

Table 8. Modes of failure in the simulations carried out under load state B and safety coefficients obtained at failure.

| Young's Modulus | Signal | Elastic Soil | | | Elasto-Plastic Soil | | |
| | | Safety Coefficient at the Time of Failure | | Mode of Failure | Safety Coefficient at the Time of Failure | | Mode of Failure |
		SSC	BCSC		SSC	BCSC	
E = 10 MPa	Permanent	1.01	1.14	SLIDING	1.01	1.14	SLIDING
	Sinusoidal	0.81	0.40	SLIDING	0.81	0.40	SLIDING
	Impulsive 1	0.81	0.40	SLIDING	0.81	0.40	SLIDING
	Impulsive 2	0.81	0.40	SLIDING	0.81	0.40	SLIDING
E = 50 MPa	Permanent	1.01	1.14	SLIDING	1.01	1.14	SLIDING
	Sinusoidal	1.01	1.14	SLIDING	1.01	1.14	SLIDING
	Impulsive 1	0.81	0.40	SLIDING	0.81	0.40	SLIDING
	Impulsive 2	0.81	0.40	SLIDING	0.81	0.40	SLIDING
E = 100 MPa	Permanent	1.01	1.14	SLIDING	1.01	1.14	SLIDING
	Sinusoidal	1.01	1.14	SLIDING	1.01	1.14	SLIDING
	Impulsive 1	0.90	0.69	SLIDING	0.90	0.69	SLIDING
	Impulsive 2	0.90	0.69	SLIDING	0.90	0.69	SLIDING
E = 200 MPa	Permanent	1.01	1.14	SLIDING	1.01	1.14	SLIDING
	Sinusoidal	1.01	1.14	SLIDING	1.01	1.14	SLIDING
	Impulsive 1	0.90	0.69	SLIDING	0.90	0.69	SLIDING
	Impulsive 2	0.90	0.69	SLIDING	0.90	0.69	SLIDING
E = 400 MPa	Permanent	1.01	1.14	SLIDING	1.01	1.14	SLIDING
	Sinusoidal	1.01	1.14	SLIDING	1.01	1.14	SLIDING
	Impulsive 1	1.01	1.14	SLIDING	1.01	1.14	SLIDING
	Impulsive 2	1.01	1.14	SLIDING	1.01	1.14	SLIDING
E = 27,000 MPa	Permanent	1.01		SLIDING			
	Sinusoidal	1.01		SLIDING			
	Impulsive 1	1.01		SLIDING			
	Impulsive 2	1.01		SLIDING			

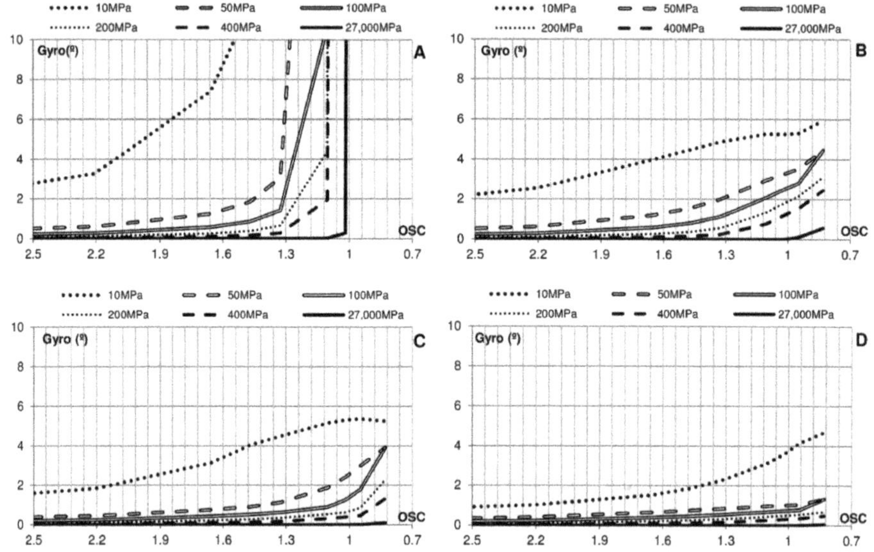

Figure 21. Wall rotation as a function of Overturning Safety coefficient (OSC) and Young's modulus, assuming elastic soil. (**A**) Permanent; (**B**) Sinusoidal; (**C**) Impulsive 1; (**D**) Impulsive 2.

The influence of the loading signal shape on the crown wall rotation has been analyzed (Figure 22). It can be verified that in the case of very rigid soil (E (MPa) = 27,000), until the OSC is less than 1.0 (OSC < 1.0), the crown wall does not begin to turn. At that time, in the case of permanent loads, the structure failure occurs immediately. However, in the other load types, the crown wall does not fail because the signal duration is not enough to cause the total instability of the structure. With deformable soils (e.g., E (MPa) = 10 and E (MPa) = 100), in the case of permanent loads, the failure of the structure occurs for values of OSC larger than 1.0. Also, if the soil has a certain flexibility in the case of impulsive loads, overturning may not occur even with overturning safety coefficients smaller than 1.0.

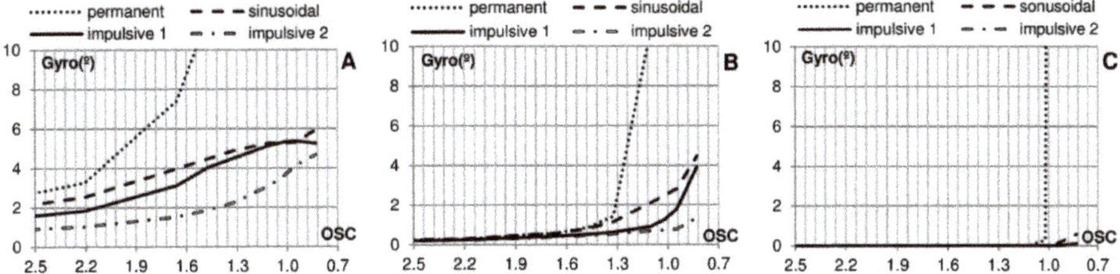

Figure 22. Crown wall rotation as a function of OSC and the shape of the signal, assuming elastic soil (**A**) E (MPa) = 10; (**B**) E (MPa) = 100; (**C**); E (MPa) = 27,000.

In Figure 23, the deformation of the foundation related to the bearing capacity safety coefficient (BCSC) is presented. For the four signal shapes and the two types of soils (elastic and elasto-plastic) studied, it can be seen that in the case of permanent load (static calculation), the maximum deformation of the foundation exceeds the maximum acceptable defined value for BCSC; i.e., greater than 1.0. Only in the case of E (MPa) = 27,000 (concrete), the maximum allowed deformation is not exceeded, even with safety coefficients close to 0.0. In this case, the section fails due to rigid overturning, not plastic. In impulsive loads with a bearing capacity safety coefficient lower than 1.0, the maximum acceptable deformation is not exceeded. On the right side of Figure 23, the deformations for elasto-plastic soils are presented, showing larger deformations than in the case of elastic soil.

Table 9 shows the turning of the crown wall in degrees for both elastic and elasto-plastic soils, indicating the overturning and bearing capacity safety coefficients. As expected, before the failure due to rigid overturning occurs, the structure fails due to bearing capacity of the foundation (except in the case of E (MPa) = 27,000, for which this has not been considered). In the case of E (MPa) = 27,000 and permanent loads (static calculation), the failure due to rigid overturning occurs when the overturning safety factor is <1.0. Also, it can be observed that the shape of the signal has a decisive influence on crown wall failure. In the case of impulsive pressures and elastic soil, the failure does not occur. Likewise, in the case, both of permanent and sinusoidal loads with elastic soil, the bearing capacity safety condition can be overcome (BCSC < 1.0) and the collapse of the foundation does not occur. Moreover, for elasto-plastic soil and impulsive signals, the deformations of the foundation may become unacceptable (See Table A2 in Appendix B) when the rigidity of the soil increases.

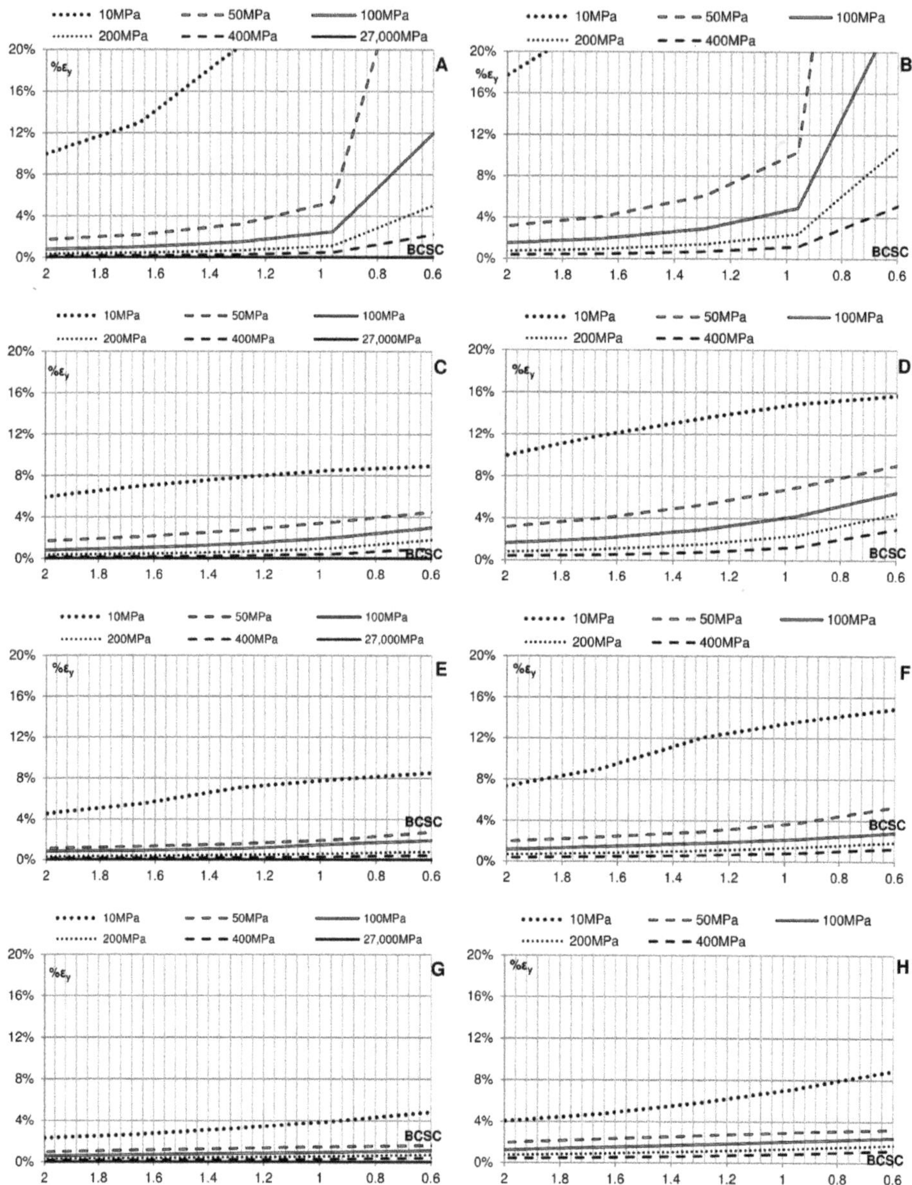

Figure 23. Soil deformation as a function of Bearing Capacity Safety Coefficient (BCSC) and Young's modulus. Permanent: (**A**) elastic soil; (**B**) elasto-plastic soil. Sinusoidal: (**C**) elastic soil; (**D**) elasto-plastic soil. Impulsive 1: (**E**) elastic soil; (**F**) elasto-plastic soil. Impulsive 2: (**G**) elastic soil; (**H**) elasto-plastic soil.

Table 9. Crown wall turning in simulations under load state C. [The simulations with failure of the structure are highlighted in dark gray (failure due to rigid overturning) and in light gray (failure due to bearing capacity of the foundation)].

		Bearing Capacity and Overturning Failure (Rotation in Degrees)												
Bearing Capacity Safety Coefficient		**2.49**	**1.66**	**1.29**	**0.96**	**0.35**	**0.07**	**0.01**	**2.49**	**1.66**	**1.29**	**0.96**	**0.35**	**0.07**	**0.00**
Overturning Safety Coefficient		**2.21**	**1.66**	**1.47**	**1.33**	**1.10**	**1.02**	**0.95**	**2.21**	**1.66**	**1.47**	**1.33**	**1.10**	**1.02**	**0.95**
Young's Modulus	**Signal**	**Elastic Soil**							**Elasto-Plastic Soil**						
10 MPa	Permanent	3.3	7.4	11.7	23.8	90.0	90.0	90.0	5.4	13.4	21.9	46.1	90.0	90.0	90.0
	Sinusoidal	2.5	4.0	4.5	4.9	5.3	5.3	5.3	4.1	6.8	7.8	8.5	9.3	9.3	9.4
	Impulsive 1	1.8	3.1	4.0	4.5	5.2	5.3	5.4	2.8	5.2	6.9	7.8	9.0	9.3	9.4
	Impulsive 2	1.0	1.5	1.9	2.2	3.1	3.6	4.1	1.8	2.7	3.3	4.1	5.8	6.7	7.8
50 MPa	Permanent	0.6	1.2	1.8	3.1	35.9	90.0	90.0	1.1	2.3	3.5	5.9	73.5	90.0	90.0
	Sinusoidal	0.6	1.2	1.6	2.0	3.0	3.3	3.5	1.1	2.3	3.0	4.0	6.0	6.7	7.2
	Impulsive 1	0.4	0.8	0.9	1.1	1.9	2.4	3.0	0.8	1.4	1.7	2.1	3.7	4.8	6.0
	Impulsive 2	0.4	0.6	0.7	0.8	1.0	1.0	1.0	0.8	1.3	1.5	1.7	2.0	2.1	2.1
100 MPa	Permanent	0.3	0.6	0.9	1.4	10.6	90.0	90.0	0.5	1.1	1.7	2.8	21.9	90.0	90.0
	Sinusoidal	0.3	0.6	0.8	1.1	2.1	2.5	2.8	0.6	1.2	1.7	2.4	4.6	5.5	6.3
	Impulsive 1	0.2	0.4	0.5	0.6	0.9	1.2	1.8	0.5	0.8	1.0	1.3	1.9	2.7	3.9
	Impulsive 2	0.2	0.4	0.5	0.5	0.7	0.7	0.8	0.5	0.9	1.0	1.2	1.5	1.6	1.7
200 MPa	Permanent	0.1	0.3	0.4	0.7	4.4	90.0	90.0	0.2	0.5	0.8	1.4	9.4	90.0	90.0
	Sinusoidal	0.1	0.3	0.4	0.6	1.4	1.8	2.2	0.3	0.6	0.9	1.4	3.3	4.5	5.4
	Impulsive 1	0.1	0.2	0.3	0.3	0.5	0.6	0.9	0.2	0.5	0.6	0.8	1.3	1.5	2.1
	Impulsive 2	0.1	0.2	0.2	0.3	0.4	0.5	0.5	0.3	0.5	0.6	0.8	1.1	1.2	1.4
400 MPa	Permanent	0.1	0.1	0.2	0.3	1.9	90.0	90.0	0.1	0.3	0.4	0.7	4.5	90.0	90.0
	Sinusoidal	0.0	0.1	0.2	0.2	0.8	1.2	1.5	0.1	0.3	0.4	0.7	2.4	3.7	4.8
	Impulsive 1	0.0	0.1	0.1	0.2	0.3	0.4	0.5	0.1	0.3	0.4	0.5	0.9	1.1	1.4
	Impulsive 2	0.0	0.1	0.1	0.2	0.3	0.3	0.4	0.2	0.3	0.4	0.5	0.8	0.9	1.1
27,000 MPa	Permanent	0.0	0.0	0.0	0.0	0.0	0.3	90.0							
	Sinusoidal	0.0	0.0	0.0	0.0	0.0	0.0	0.1							
	Impulsive 1	0.0	0.0	0.0	0.0	0.0	0.0	0.0							
	Impulsive 2	0.0	0.0	0.0	0.0	0.0	0.0	0.0							

Table 10 shows the modes of failure produced in the simulations carried out for load state C and the corresponding bearing capacity and overturning safety coefficients at the time of failure. It can be observed that the failure of the structure always occurs due to the foundation bearing capacity, except in the case of E (MPa) = 27,000 (concrete), in which the failure occurs due to overturning. In certain cases of impulsive signals, the failure occurs with bearing capacity safety coefficients below unity. In some cases, the crown wall did not fail. For example, in the case of E (MPa) = 100, with permanent loads, the maximum acceptable deformation is reached for OSC = 1.32 and BCSC = 0.95 (for elastic soil), and OSC = 1.50 and BCSC = 1.36 (for elasto-plastic soil). In the case of sinusoidal load, maximum acceptable deformation is reached for OSC = 1.22 and BCSC = 0.66 (for elastic soil), and OSC = 1.51 and BCSC = 1.36 (for elasto-plastic soil); and in the case of Impulsive 1 load, for OSC = 0.97 and BCSC = 0.01 (for elastic soil), and OSC = 1.20 and BCSC = 0.61 (for elasto-plastic soil). It is remarkable that the maximum acceptable deformation for Impulsive 2 load is not reached.

Table 10. Mode of failure in simulations under load state C and safety coefficients obtained at the time of failure of the crown wall.

			Failure Produced in Simulations under Load State C					
			Elastic Soil			Elasto-Plastic Soil		
Young's Modulus	Signal	Maximum Acceptable Deformation	Safety Coefficient at the Time of Failure		Mode of Failure	Safety Coefficient at the Time of Failure		Mode of Failure
			BCSC	OSC		BCSC	OSC	
E = 10 MPa	Permanent	19.8%	1.32	1.49	bearing capacity	1.88	1.80	bearing capacity
	Sinusoidal		no	no	no	no	no	no
	Impulsive 1		no	no	no	no	no	no
	Impulsive 2		no	no	no	no	no	no
E = 50 MPa	Permanent	4.5%	1.09	1.38	bearing capacity	1.58	1.62	bearing capacity
	Sinusoidal		0.59	1.19	bearing capacity	1.51	1.58	bearing capacity
	Impulsive 1		0.05	1.00	bearing capacity	0.79	1.26	bearing capacity
	Impulsive 2		no	no	no	0.00	0.83	bearing capacity
E = 100 MPa	Permanent	2.8%	0.95	1.32	bearing capacity	1.36	1.50	bearing capacity
	Sinusoidal		0.66	1.22	bearing capacity	1.36	1.51	bearing capacity
	Impulsive 1		0.03	0.97	bearing capacity	0.61	1.20	bearing capacity
	Impulsive 2		no	no	no	0.09	1.03	bearing capacity
E = 200 MPa	Permanent	1.7%	0.91	1.31	bearing capacity	1.20	1.43	bearing capacity
	Sinusoidal		0.64	1.21	bearing capacity	1.22	1.44	bearing capacity
	Impulsive 1		0.01	0.95	bearing capacity	0.73	1.24	bearing capacity
	Impulsive 2		no	no	no	0.58	1.17	bearing capacity
E = 400 MPa	Permanent	1.0%	0.85	1.29	bearing capacity	1.09	1.38	bearing capacity
	Sinusoidal		0.59	1.19	bearing capacity	1.14	1.40	bearing capacity
	Impulsive 1		0.01	0.84	bearing capacity	0.84	1.25	bearing capacity
	Impulsive 2		no	no	no	0.79	1.24	bearing capacity
E = 27,000 MPa	Permanent			1.00	overturning			
	Sinusoidal			no	no			
	Impulsive 1			no	no			
	Impulsive 2			no	no			

5. Conclusions

This study presents a simplified model for the dynamic calculation of breakwater crown walls, providing results of the relative movements expected based on hydrodynamic forces and soil characteristics (elastic, elasto-plastic). The proposed simplified model gives a first estimation of crown wall stability, comparing the displacement and rotation for certain designs, regardless of their structural stability, and whether they are compatible with the defined operating conditions.

The model has been validated in a real case, the main breakwater of the Outer Port of Punta Langosteira (A Coruña, Spain), by means of its comparison with the numerical model FLAC 2D 7.0. The good performance of the model proposed was verified with static calculation (permanent loads) and elastic soil in a range of Young's modulus values (E (MPa) = 10–27,000), producing sliding and rigid overturning whenever the safety coefficients were less than 1.0. The usual modes of failure of the crown walls were found to be the sliding and the bearing capacity of the foundation. However, rigid overturning only takes place in case of absolutely rigid foundations (E (MPa) = 27,000).

A fundamental aspect of the study was the decisive importance of the shape of the loading state signal. Four different shape of loads were analyzed. The results show that the impulsive forces of short duration may not transmit enough energy, regardless of its intensity, to cause a failure of the structure, even with safety coefficients less than unity obtained by static calculation.

The importance of the Young's modulus of the soil was also investigated. The analysis shows that with sufficiently flexible foundations, the soil becomes an absorber of the movement, allowing the structure to turn (and its subsequent recovery). However, if the crown wall is based on a very rigid soil or foundation (e.g., concrete), the impulsive forces can be decisive and causes the failure of the structure regardless of their limited duration. This is likely to be the cause of numerous recent crown wall breakages recently found in the upper parts which are not full protected by the main armor of the breakwater slope, in the north coast of Spain (Bermeo, A Garda, Cariño, Malpica). Therefore, it is recommended to carry out a soil-type sensitivity analysis.

The plasticity of the soil was found to be very important. The residual deformations reached in certain loading states can lead to the maximum deformation states in subsequent load stages being exceeded. It seems advisable to always carry out a double analysis: on one hand, assuming elastic soil, which provides information on the rhythmic movement of the structure and its possible resonance; and on the other hand, assuming elasto-plastic soil to determine the residual deformations of the soil and to incorporate the influence of the hysteresis of the foundation materials.

Another important aspect for the design of crown walls found in this study is that the truncation of the pressure signal registered in laboratory tests, [to eliminate a percentage of waves (e.g., 1‰) or a specific number of data registered (e.g., 0.1‰)], must be carried out with caution. This is relevant to avoid the omission of impulsive combinations potentially harmful to the structure. The proposed simplified model allows the use of the complete series of data and check the expected movements of the crown wall and its compatibility with the established design criteria.

Author Contributions: Conceptualization, E.M.; methodology, E.M.; formal analysis, E.M., E.P., J.S. and A.F.; writing—original draft preparation, E.M., E.P., J.S. and A.F.; writing—review and editing, E.M., E.P., J.S. and A.F.

Funding: This research received no external funding.

Acknowledgments: The authors give special thanks to the Port Authority of A. Coruña, Spain, for allowing access to data of the project of the New Breakwater at Punta Langosteira.

Conflicts of Interest: The authors declare no conflict of interest.

Appendix A. Simplified Problem Approach

The movement of the crown wall can be estimated by solving the general equation of the dynamics of the solid rigid (Equation (A1)), with six degrees of freedom (three movements and three turns):

$$M\ddot{x} + D\dot{x} + Kx = F(t) \tag{A1}$$

where M represents the matrix of the masses, D that of damping, and K that of stiffness. The resolution of this system requires, in addition to the knowledge of instantaneous acting forces, the data of the constituent materials of the breakwater, mainly its core (elasticity, stiffness, damping, permeability, heterogeneity, etc.). In an analytical form, the resolution becomes very complex and requires a simplification of the system in order to address it. On the other hand, it can be solved by means of the use of codes to calculate point-to-point iterations between the structure, the soil, and the acting mass of water in the time domain. It also must consider the flow inside the breakwater. However, the dynamic problems posed by interacting with a plastic soil are still not well solved, which is why in practice, there are proposed simplifications for solving them, both in materials and in forces.

This system of complex second-order differential equations can be simplified for a simple analytical resolution with the use of common software codes that allow iterative calculations.

First, a simplification of the movement system is assuming that there are no movements in the vertical and transverse directions and that the rotations are limited to those produced with respect to a transverse axis (Figure A1). The simplified system is limited to the problem of the motion of translation and that of the oscillation, becoming a system of two second-order differential equations with two unknowns. The terms involving "d" represent the damping of the soil, and the terms involving "k" the soil stiffness [2] (Equation (A2)):

$$\begin{bmatrix} m & 0 \\ 0 & I_{CG} \end{bmatrix}\begin{bmatrix} \ddot{x}_{CG} \\ \ddot{\theta} \end{bmatrix} + \begin{bmatrix} d_x & -d_x r_{CG} \\ -d_x r_{CG} & d_\theta + d_x r_{CG}^2 \end{bmatrix}\begin{bmatrix} \dot{x}_{CG} \\ \dot{\theta} \end{bmatrix} + \begin{bmatrix} k_x & -k_x r_{CG} \\ -k_x r_{CG} & k_\theta + k_x r_{CG}^2 \end{bmatrix}\begin{bmatrix} x_{CG} \\ \theta \end{bmatrix} = \begin{bmatrix} F_x \\ M_{CG} \end{bmatrix} \tag{A2}$$

where m—crown wall mass, x_{CG}—movement of the centre of gravity, M_{CG}—turning moment related the centre of gravity, I_{CG}—inertia moment related to the centre of gravity, r_{CG}—distance from the centre of gravity to the turning point, d_θ and d_x—damping of the foundation, and k_θ and k_x—stiffness of the foundation.

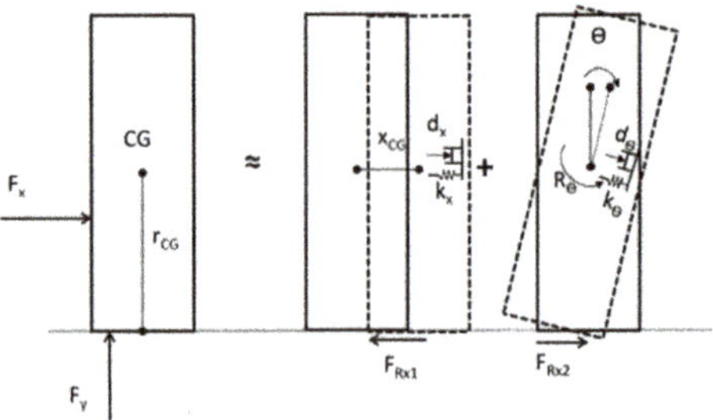

Figure A1. Crown wall movements. Adapted from the system proposed by Pedersen [2].

In a second approach, considering that the crown wall rotates with respect to an "O" point of its contact with the soil (Figures A2 and A3), the system can be analysed in a more simplified form (Equation (A3)). Thus, the equations can be expressed as:

$$\begin{bmatrix} m & 0 \\ 0 & I_O \end{bmatrix}\begin{bmatrix} \ddot{x} \\ \ddot{\theta} \end{bmatrix} + \begin{bmatrix} d_x & 0 \\ 0 & d_\theta \end{bmatrix}\begin{bmatrix} \dot{x} \\ \dot{\theta} \end{bmatrix} + \begin{bmatrix} k_x & 0 \\ 0 & k_\theta \end{bmatrix}\begin{bmatrix} x \\ \theta \end{bmatrix} = \begin{bmatrix} F_x \\ M_O \end{bmatrix} \tag{A3}$$

where M_O—overturning moment related the point "O" of turn, I_O—inertia moment related to the point "O" of turn, x—movement of the crown wall, and θ—rotation of the crown wall.

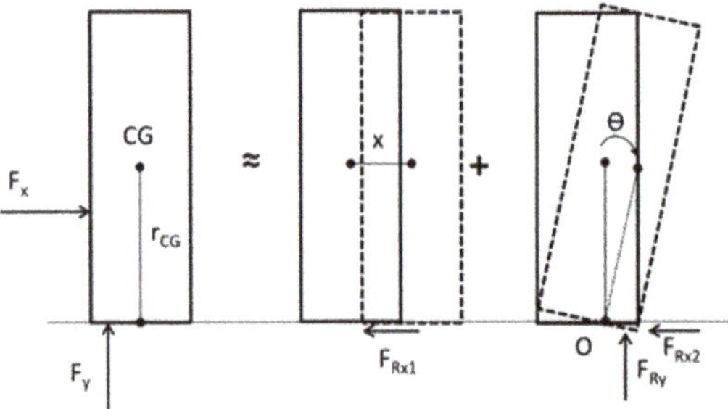

Figure A2. Scheme proposed of the dynamic system. (CG—centre of gravity; r_{CG}— radio of rotation of the center of gravity, θ—gyro; and F_R—resistance of the soil to the movement).

Appendix A.1. Soil Stiffness

It is proposed to estimate the soil stiffness as a function of the deformation through the application of the loads.

Appendix A.1.1. Stiffness against Turning

In the previous simplification, it was considered that the crown wall turned around an "O" point located at the contact with the soil. As an approximation, it is suggested that the distance from said point "O" to the inner corner of the crown wall, "C", which is variable over time, is equal to the active area of contact between the crown wall and foundation (B*) and double that of the distance between the point of application of the resultant forces in that contact plane and the inner corner "C" (called the equivalent foundation breath [5,18]; see Figure A3).

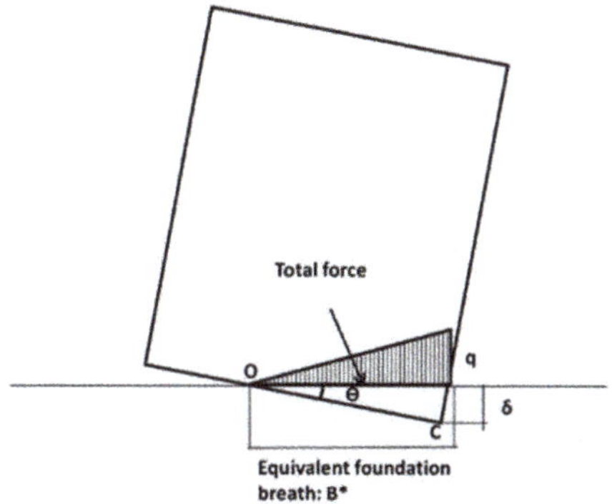

Figure A3. Definition of equivalent foundation breath.

On the other hand, considering the soil as an elastic medium and that the distribution of the loads transmitted to the soil (which depends on its stress-deformational state) is known and has a triangular shape, a vertical movement δC is generated at the point "C". This parameter is a function of the Young's modulus of elasticity (E) and the Poisson's coefficient of the soil (v), which with a permanent load, can be considered as follows [17] (Equation (A4)):

$$\delta_C = \left(q\left(1 - \upsilon^2\right) B^* \right)/\pi E = B^* \theta \; \rightarrow \; qB^* = \pi E \theta B^* / \left(1 - v^2\right) \tag{A4}$$

Therefore, given that $M_{Def_v} = k_\theta \theta = \left(qB^{*2}\right)/3 = \left[\left(\pi E B^{*2}\right)/\left(3\left(1 - v^2\right)\right)\right]\theta$, the soil stiffness for a permanent load, depending on the angle of rotation of the structure, could be assimilated as such:

$$k_\theta = \left(\pi E B^{*2}\right)/\left(3\left(1 - v^2\right)\right) \tag{A5}$$

Gazetas [19] proposed a lower stiffness in the case of strip foundations with a layer of incompressible soil at a depth "D" under cyclic loads (Equation (A6)):

$$k_\theta = \left[\left(\pi E B^{*2}\right)/\left(4\left(1 - v^2\right)\right)\right]\left(1 + 0.1 B^*/D\right) \tag{A6}$$

Appendix A.1.2. Stiffness against Movement

The same assumptions made in the previous case can be applied here, considering the soil as an elastic medium and that the distribution of the loads transmitted to the soil is known. So, the movement of the soil, δ_x, as a consequence of the application of a permanent horizontal load in the previous point "C" can be considered [17] (Equation (A7)):

$$\delta_x = [(Bq(1+v))/\pi G](\ln B/2 - \ln B) \tag{A7}$$

where B—width of the crown wall, G—shear modulus of the soil

Therefore, given that $q = k_x\delta_x = (\pi G)/[B(1+v)(\ln B/2 - \ln B)]\delta_x$, the soil stiffness for a permanent horizontal load, as a function of horizontal movement, can be expressed as:

$$k_x = \pi G/[B(1+v)(\ln B/2 - \ln B)] \tag{A8}$$

where the shear modulus of the soil "G", related to the elasticity Young's modulus and Poisson coefficient, is considered with the following expression (Equation (A9)):

$$G = E/(2(1+v)) \tag{A9}$$

Gazetas [19] proposed a larger stiffness in the case of strip foundations with a layer of incompressible soil at a depth "D" under cyclic loads (Equation (A10)):

$$k_x = [(2.1G)/(2-v)](1+B/D) \tag{A10}$$

Appendix A.2. Consideration of the Damping of the Movement

The damping of the movement due to the stiffness and viscous characteristics of the foundation can be approached as a response to the instantaneous deformation of the foundation (as a consequence of the punctual stresses in the crown wall at the instant immediately before).

Appendix A.2.1. Damping Corresponding to the Turn

The damping is introduced as an opposite turn to that produced by the turning moment caused by the soil's reaction to its deformation up until the previous moment (Figure A4).

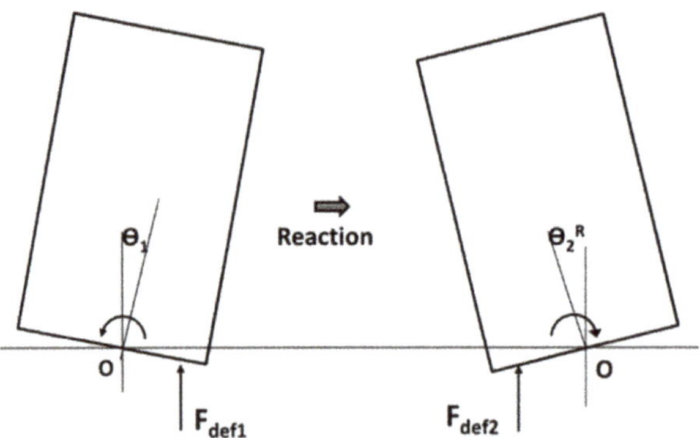

Figure A4. Damping of the turn due to soil reaction. (F_{def}—resistance of the soil to deformation produced by the rotation).

Since $M_{Def_{V1}} = I_O\ddot{\theta}_2^R + M_{Def_{V2}}$, then, as an a approximation, $\theta_2^R \sim (k_{\theta_1}\theta_1 t^2/2)/(I_O + k_{\theta_2}t^2/12)$. So, the net rotation at instant "i" is the turn of the loading state in the instant "i" minus the rotation produced by the soil reaction in the instant "i − 1":

$$\theta_{i_{net}} = \theta_i - \theta_{i-1}^R \tag{A11}$$

Appendix A.2.2. Inertial Damping

In addition, another damping of the turning movement is produced as a consequence of the stabilizing moment. This is generated as a consequence of the location of the centre of gravity of the crown wall with respect to the point of turning. If the projection is located towards the seaward side of point "O" (Figure A5), it is necessary to introduce it in the iterative process.

$$M_{stab} = W \times \overline{x_{CGi}O_i} \quad \text{where,} \quad \overline{x_{CGi}O_i} = \overline{x_{CG0}O_i} - y_{CG}tg(\theta_i) \tag{A12}$$

The consequent damping in the turn is equal to a stabilizing turn:

$$\theta_i^{estab} = \iint M_{stab_i}dtdt/I_O \tag{A13}$$

Therefore, adding this stabilizing turn as well as the net rotation obtained in Equation (A11) at the instant "i" of load application, the following total net rotation occurs (Equation (A14)):

$$\theta_{i\,net}^{total} = \theta_{inet} - \theta_{i-1}^{stab} \tag{A14}$$

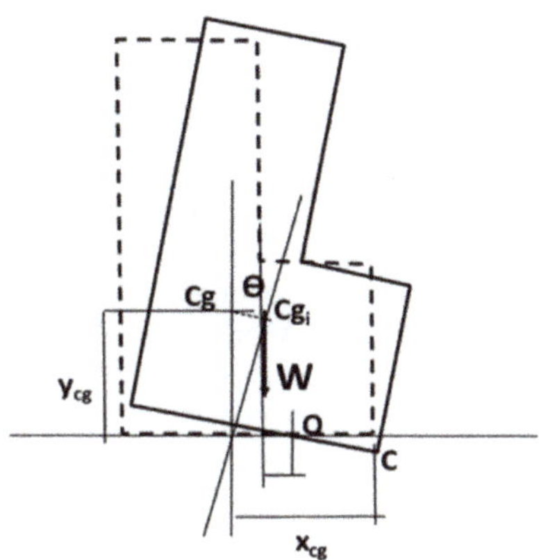

Figure A5. Damping of the rotation as a result of the stabilizing moment.

Appendix A.2.3. Damping Corresponding to Movement

The same simplification done for the turn can be done for displacements, and the movement of the crown wall along the *x*-axis and the soil reaction due to its stiffness before its breakage could be estimated. However, in the case of movements, it is considered, according to Coulomb's hypothesis [20], that the resistance of the soil to the movement is proportional to the normal force applied against the soil and independent of the velocity.

Appendix A.3. Simplified Model

Therefore, the system of differential equations and its resolution in the time domain are considerably simplified. The system of equations can be written in a simpler matrix form (Equation (A15)):

$$\begin{bmatrix} m & mr_{CG} \\ mr_{CG} & I_O \end{bmatrix}\begin{bmatrix} \ddot{x} \\ \ddot{\theta} \end{bmatrix} + \begin{bmatrix} 0 & 0 \\ 0 & k_\theta \end{bmatrix}\begin{bmatrix} x \\ \theta \end{bmatrix} = \begin{bmatrix} G_{Displ} \\ M_O \end{bmatrix} \tag{A15}$$

where x—crown wall movement, θ—crown wall turn, m—crown wall mass, r_{CG}—distance of the centre of gravity of the crown wall to the turning point "O", M_O—turning moment related to the point "O", and I_O—inertial moment with respect to point "O".

Furthermore, $k_\theta = \left[\left(\pi E B^{*2}\right)/\left(4\left(1-v^2\right)\right)\right](1+0.1B^*/D)$, where B*—equivalent foundation breath, D—depth of non-deformable soil related to the crown wall foundation, E—Young's modulus, and v—Poisson coefficient. Finally, $G_{Displ} = (1-p)\left[F_x - \mu\left(W - F_y\right)\right]$, where p—percentage of reduction of F_x as a function of the water mass that returns in the opposite direction of the movement and the reduced relative velocity of the wave and the moveable section compared to a fixed structure [20] (it is considered tp = 0, μ—friction coefficient, F_x—horizontal forces, F_y—vertical forces, W—m·g, and g—gravity.

According to this simplification, the transversal deformation of the soil is now introduced in the calculation, taking into account that until the slip condition is exceeded, there is no movement. So, this horizontal movement of the soil can be limited better by means of the following expression (Equation (A16)):

$$\varepsilon_{x\%max} = \max\left(q_h/G_{tang}\right) \tag{A16}$$

where $q_h = F_x/B$ is the horizontal shear pressure and G_{tang} is the tangent shear modulus.

The movement estimation presented with this simplification (Equation (A15)) is considered to be more precise than that presented by Burcharth et al. [21] (Equation (A17)) for the analysis of the movement of a caisson subjected to a horizontal force when introducing the effect of the turning movement of the crown wall:

$$F(t) = F_x(t) - \left(W - F_y(t)\right)\mu_t = (m_{caisson} + m_{added})\ddot{x} \tag{A17}$$

where W—weight of the caisson, $m_{caisson}$—mass of the caisson, and m_{added}—water mass that overtops the caisson and contributes to its stability.

The formula proposed by Burcharth et al. [21] (Equation (A17)) was checked against a numerical model of finite elements, in the case of a vertical caisson [22], underestimating the movements as a function of the permeability characteristics of the foundation, and in the case of crown walls [20], being on the safe side. However, it can be valid as an approximation if the movement of the core of the breakwater is not taken into account and we consider the relative movements to be absolute with respect to a specific point. This approach is useful for many of the usual problems in port engineering.

Appendix B. Elasto-Plastic Model of the Soil Proposed

The study was developed considering the hyperbolic elasto-plastic model of soil response proposed by Schanz et al. [10] and Kondner [11], considering the hysteresis of the materials, introducing a variable stress–strain relation according to the following expression (Equation (A18)):

$$E_{tang} = dQ_H/d\varepsilon_y = (1/E_0)/\left[\left(1/E_0 + \varepsilon_a/q_a\right)^2\right] \tag{A18}$$

where $Q_H = \varepsilon_a/\left(1/E_0 + \varepsilon_a/q_a\right)$, E_0—initial Young's modulus of elasticity, E_{tang}—tangent Young's modulus for a specific deformation $\%\varepsilon_a$, q_a—deviator in the asymptote of the hyperbola (difference between total vertical load and horizontal strain of the soil) and ε_a—vertical deformation of the soil (%).

The following adjustment parameters are proposed (Table A1):

Table A1. Adjustment parameters of the soil considered (initial Young module and Deviator).

Type of Soil	E_0 (MPa)	q_a (KN)
1	10	350
2	50	400
3	100	500
4	200	600
5	400	700

The analysis is presented in the following curves, with stress–strain relation and variation of the elasticity modulus as a function of deformation (Figure A6):

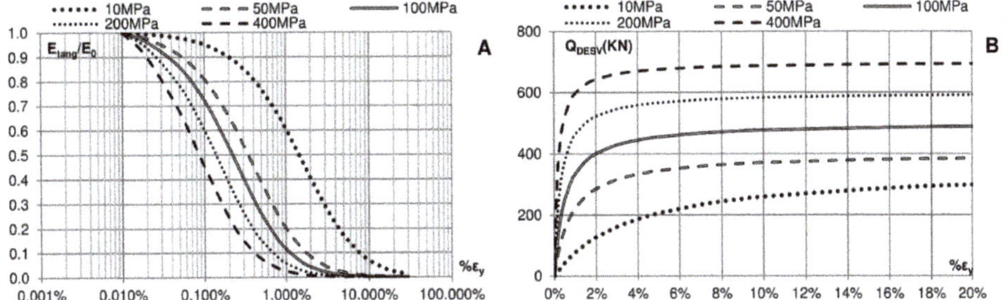

Figure A6. (**A**) Stress–strain relation and (**B**) Young's modulus variation with deformation of the soil.

To analyze the failure of the system, it is proposed to consider as the admissible deformation of each material the value that corresponds to the application of 85% of the deviator in the asymptote (Table A2); e.g., approximately 5% of E_0.

Table A2. Admissible deformation definition.

E (MPa)	Deviator in the Asymtote (KN)	Admissible Deviator (KN)	$\%\varepsilon_y$ Admissible
10	350	280	19.8%
50	400	320	4.5%
100	500	400	2.8%
200	600	480	1.7%
400	700	560	1.0%

On the other hand, to take into account that the stiffness of the soil increases in the load reduction process, it is proposed to consider that during the recovery process, Young's modulus increases up to $E_{rec} = E_0 = 2E_{50}$. Therefore, the stress–strain diagram considered in the loading–unloading process has the following form (Figure A7).

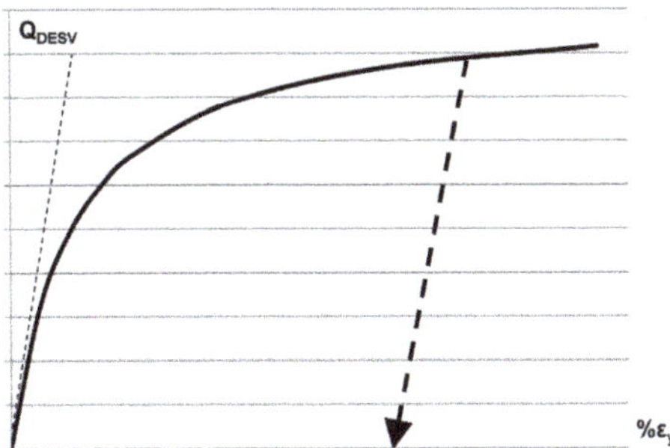

Figure A7. Stress–strain diagram of the loading–unloading process.

To consider the hysteresis of the materials, the model proposed by Atkinson [16] is applied. When the stress exceeds the yield point, residual deformations accumulate to the elastic deformations. Once the load has been reduced, according the stress–strain diagram of Figure A7, in a new cycle of reloading, the yield point moves upward in the stress–strain curve. It follows a process of hardening by deformation (Figure A8), up to the deformation that is considered as breakage, as suggested by Atkinson [16].

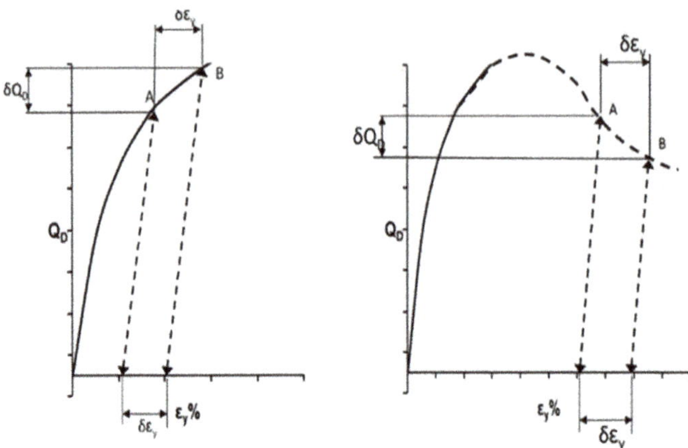

Figure A8. Process of hardening and softening produced by deformation, adapted from Atkinson [14].

References

1. US Army Corps of Engineers. *Coastal Engineering Manual*; US Army Corps of Engineers: Washington, DC, USA, 2002; Part 6, Chapter VI-5—Fundamentals of Design.
2. Pedersen, J. *Wave Forces and Overtopping on Crown Walls of Rubble Mound Breakwaters: An Experimental Study*; Aalborg University: Aalborg, Denmark, 1996; ISSN 0909-4296.
3. Kortenhaus, A.; Oumeraci, H. Clasification of wave loading on monolithic coastal structures. In Proceedings of the 26th Conference on Coastal Engineering, Copenhagen, Denmark, 22–26 June 1998; pp. 1–14. [CrossRef]
4. Cuomo, G.; Allsop, W.; Bruce, T.; Pearson, J. Breaking wave loads at vertical walls and breakwaters. *Coast. Eng.* **2010**, *57*, 424–439. [CrossRef]
5. Gomez-Barquín, G. *Análisis Dinámico de un Dique de Abrigo Vertical*, 1st ed.; Asociación Técnica de Puertos y Costas (Spanish Section of PIANC): Madrid, Spain, 1998.
6. Lamberti, A.; Martinelli, L.; de Groot, M.B. Dynamics. In *Mast III/Proverbs, Volume IIb (Geotechnical Aspects)*; de Groot, M.B., Ed.; Technical University of Braunschweig: Braunschweig, Germany, 1999.
7. Plaxis. *Plaxis 2D Scientific Manual*; Delft University of Technology and Plaxis: Delft, The Netherlands, 2017.
8. Itasca. *An Introduction to Flac 8 and a Guide to Its Practical Application in Geotechnical Engineering*; Itasca: Minneapolis, MN, USA, 2015.
9. Kvalstad, T. Degradation and residual pore pressures. In *Mast III/Proverbs, Volume IIb (Geotechnical Aspects)*; de Groot, M.B., Ed.; Technical University of Braunschweig: Braunschweig, Germany, 1999.
10. Schanz, T.; Vermeer, A.; Bonnier, P. The Hardening Soil Model: Formulation and Verification. In Proceedings of the Beyond 2000 Computational Geotechnics 10 Years PLAXIS, Amsterdam, The Netherlands, March 1999.
11. Kondner, R.L. Hyperbolic stress-strain response: Cohesive soils. *J. Soil Mech. Found. Div.* **1963**, *89*, 115–144. [CrossRef]
12. CEDEX. *Ensayo en Modelo Físico 2D a Gran esCala Sobre la Sección Tipo del Dique del Nuevo Puerto Exterior de La Coruña en Punta Langosteira*; CEDEX: Madrid, Spain, 2005.
13. Soriano, A. *Movimientos del Espaldón Durante un Temporal*; Dique de Langosteira, Autoridad Portuaria de A Coruña: A Coruña, Spain, 2016. (In Spanish)
14. CEDEX. *Informe de Laboratorio para UTE Galería Langosteira, Proyecto Constructivo de galería para protección de tuberías en el dique de las nuevas instalaciones portuarias en Punta Langosteir*; CEDEX: Madrid, Spain, 2016.
15. Hansen, J.B. *A Revised and Extended Formula for Bearing Capacity*; Bulletin No. 28; Danish Geotechnical Institute: Copenhagen, Denmark, 1970; pp. 5–11.
16. Atkinson, J. *The Mechanics of Soils and Foundations*, 2nd ed.; Taylor & Francis: Abingdon, UK, 2007; 475p, ISBN 9780415362566.
17. Jimenez Salas, J.A.; de Justo, J.L.; Serrano, A.A. *Geotecnia y Cimientos II. Mecánica del suelo y de las Rocas*, 2nd ed.; Rueda: Madrid, Spain, 1981; p. 1188, ISBN 84-7207-021-2.

18. *Recomendaciones Geotécnicas para Obras Marítimas y Portuarias*, 1st ed.; ROM 0.5.05; Puertos del Estado: Madrid, Spain, 2005; 546p, ISBN 84-88975-52-X.
19. Gazetas, G. Analysis of machine foundation vibrations: State of the art. *Int. J. Soil Dyn. Earthq. Eng.* **1983**, *2*, 2–42. [CrossRef]
20. Nørgaard, J.Q.H.; Andersen, L.V.; Andersen, T.L.; Burcharth, H.F. Displacement of monolithic rubble-mound breakwater crown-walls. In Proceedings of the 33rd International Conference on Coastal Engineering, Santander, Spain, 29–30 June 2012; Volume 2, pp. 966–981.
21. Burcharth, H.F.; Andersen, L.; Andersen, T.L. Analyses of stability of caisson breakwaters on rubble foundation exposed to impulsive loads. In *Coastal Engineering 2008*; World Scientific Publishing Company: Singapore, 2009; pp. 3606–3618. [CrossRef]
22. Andersen, L.; Burcharth, H.F.; Lykke Andersen, T. Validity of Simplified Analysis of Stability of Caisson Breakwaters on Rubble Foundation Exposed to Impulsive Loads. In Proceedings of the 32nd International Conference on Coastal Engineering, Shanghai, China, 30 June–5 July 2010; Volume 46, pp. 1–14. [CrossRef]

Article

Stability of Rubble Mound Breakwaters—A Study of the Notional Permeability Factor, Based on Physical Model Tests

Mads Røge Eldrup *, Thomas Lykke Andersen and Hans Falk Burcharth

Department of Civil Engineering, Aalborg University, 9220 Aalborg, Denmark; tla@civil.aau.dk (T.L.A.); hansburcharth@gmail.com (H.F.B.)
* Correspondence: mre@civil.aau.dk

Received: 3 April 2019; Accepted: 30 April 2019; Published: 2 May 2019

Abstract: The Van der Meer formulae for quarry rock armor stability are commonly used in breakwater design. The formulae describe the stability as a function of the wave characteristics, number of waves, front slope angle and rock material properties. The latter includes a so-called notional permeability factor characterizing the permeability of the structure. Based on armor stability model tests with three armor layer compositions, Van der Meer determined three values of the notional permeability. Based on numerical model results he added for a typical layer composition one more value. Based on physical model tests, the present paper provides notional permeability factors for seven layer compositions of which two correspond to the compositions tested by Van der Meer. The results of these two layer compositions are within the scatter of the results by Van der Meer. To help determination of the notional permeability for non-tested layer compositions, a simple empirical formula is presented.

Keywords: rock armor stability; breakwater; damage; notional permeability factor

---

1. Introduction

The rock armor stability of rubble mound breakwaters has been estimated with the formulae by Van der Meer [1] in the last decades. The formulae are still used worldwide even though the study was performed approximately 30 years ago. An alternative to the stability formulae by Van der Meer [1] could be a numerical model. However, computational fluid models like the volume of fluid (VOF) and smoothed-particle hydrodynamics (SPH) are still computationally demanding and need to be coupled to a solid state model like a discrete element method (DEM). Furthermore, the numerical models rely on parameters found in physical model tests, as for example, the porosity parameters used to describe the water flow inside the rubble mound breakwater. Thus numerical models cannot be used as a standalone but need input parameters based on physical model tests. Sarfaraz and Pak [2] used a coupled SPH-DEM model to test the stability of cube armored rubble mound breakwaters. They compared the numerical results to empirical formulae and the numerical results were not far from the empirical estimations. Numerical models can be a supplement to empirical formulae used to solve complex problems but, in most situations, empirical formulae are still highly relevant.

Van der Meer [1] performed a large number of model tests with rubble mound breakwaters exposed to irregular mainly non-breaking Rayleigh distributed waves corresponding to $H_{1/3}/h \leq 0.2$ in which $H_{1/3}$ is the significant incident wave height and h the water depth. The tests included cross-sections with five different front slopes in the range of $\cot(\alpha) = 1.5$–6, and three different layer compositions. The three compositions were: An armor layer on a thin filter layer on an impermeable core, an armor layer on a coarse permeable core and a homogeneous structure, see layer compositions A, H and M in Figure 1. Van der Meer used the work by Thompson and Shuttler [3] as a starting point, finding three notational permeability values (P = 0.1, 0.5, 0.6) for the tested layer compositions. The tests included the impermeable core composition tested by Thompson and Shuttler.

The notional permeability parameter has no physical meaning but was introduced to ensure that the effect of permeability was taken into account. For the very typical layer composition consisting of a permeable core, underlayer and armor layer, Van der Meer [1] estimated the value P = 0.4 on the basis of above given P-values and the numerical HADEER model by Hölscher and Barends [4], which models the wave introduced flow in the porous structure of rubble mounds.

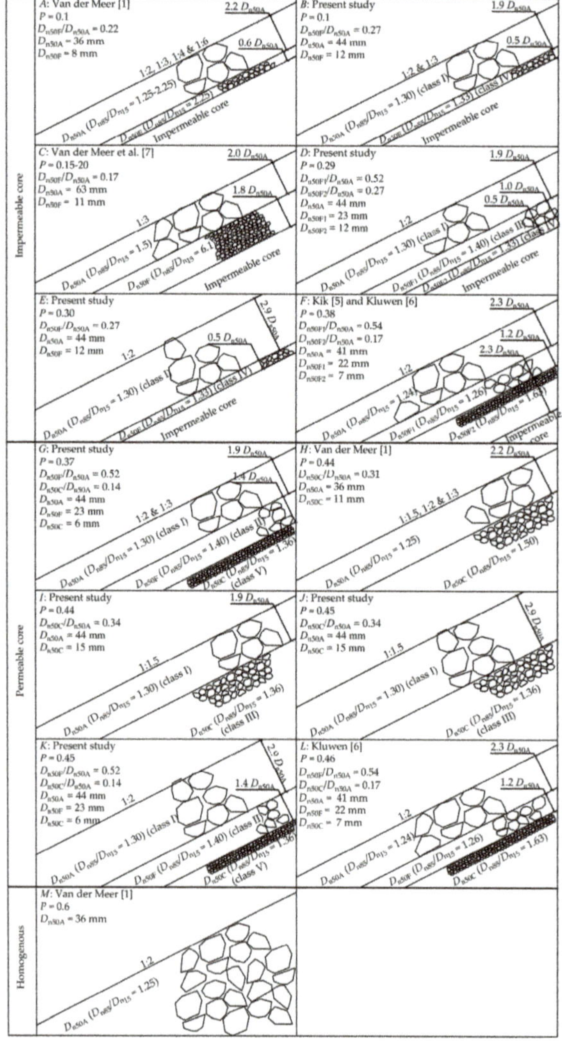

Figure 1. Fitted notional permeability factor of different layer compositions, for which *A*, *H* and *M* are given by Van der Meer [1], *F* is given by Kik [5] and Kluwen [6], *L* is given by Kluwen [6], *C* is given by Van der Meer et al. [7] and *B*, *D*, *E*, *G*, *I*, *J* and *K* are tested in the present study. D_{n50A} is the nominal stone size of the armor, D_{n50F} is the nominal stone size of the filter/underlayer material, and D_{n50C} is the nominal stone size of the core material.

Kik [5] tested a layer composition with an impermeable geo-textile placed underneath a relatively thick second underlayer, cf. composition F in Figure 1, and found $P = 0.37$, but recommend a design value of $P = 0.35$ due to limited tests. Kluwen [6] tested the same structure under similar wave conditions but extended the number of data. Based on all the tests, Kluwen fitted a notional permeability of $P = 0.38$. Kluwen [6] also tested layer composition L in Figure 1 and determined $P = 0.46$. This layer composition is similar to the layer composition for which Van der Meer [1] estimated $P = 0.4$, but the composition by Kluwen [6] had a thicker armor layer, a thinner underlayer and coarser material in both underlayer and core.

Recently, Van der Meer et al. [7] studied the influence of grading and thickness of the underlayer/filter layer for a structure with an impermeable core. They observed that an underlayer with a thickness of $0.5D_{n50A}$ of the armor stone size gave complete failure, while a thickness of $1.75D_{n50A}$ reduced the armor layer damage by 50%. For the layer composition with a thick underlayer, they estimated $P = 0.15$–0.2, see layer composition C in Figure 1. Furthermore, they observed that a very wide-graded underlayer material (including fine material) gave as expected more damage than a narrow graded underlayer material with the same D_{n50}. Since only two wave steepnesses were tested for each composition, no final recommendations on P were given.

In addition to the above given existing notional permeability factors, Figure 1 also presents the obtained results from the present study on seven layer compositions (B, D, E, G, I, J and K). A more detailed description of these layer compositions and the analysis of the obtained notional permeability factors are given later in the present paper.

The influence of the notional permeability value on armor stability is demonstrated in Table 1. Based on the notional permeability factors given in Figure 1, the related required rock armor masses are given as calculated from the Van der Meer [1] formulae for some typical conditions including three different deep water wave steepnesses, significant wave height $H_{1/3} = 4$ m, front slope $\cot(\alpha) = 2$, damage $S_d = 2$, rock mass density of 2650 kg/m^3, water mass density of 1025 kg/m^3 and number of waves $N = 1000$ waves. Table 1 shows that changing the notional permeability from $P = 0.46$ to $P = 0.38$ demands an increase in armor unit mass of approximately 10–35%, depending on the breaker parameter $\xi_{0m} = \tan\alpha/s_{0m}^{0.5}$ in which $s_{0m} = 2\pi H_{1/3}/(gT_m^2)$ and T_m is the mean wave period. Changing the notional permeability from $P = 0.38$ to $P = 0.17$ demands an increase in armor unit mass of approximately 50–100%. This large sensitivity of the armor mass to the notional permeability motivates the determination of more notional permeability values.

Table 1. Estimated rock armor weight in tonnes with the use of the Van der Meer [1] formulae.

P	s_{0m} (-) 0.05	T_m (s) 7.2	ξ_{0m} (-) 2.2	s_{0m} (-) 0.02	T_m (s) 11.3	ξ_{0m} (-) 3.5	s_{0m} (-) 0.01	T_m (s) 16.0	ξ_{0m} (-) 5.0
0.10		10.9			21.6			19.8	
0.17		8.1			16.2			17.4	
0.38		5.3			10.5			8.6	
0.46		4.8			9.5			6.3	
0.50		4.6			9.0			5.4	

A method to estimate the notional permeability was proposed by Jumelet [8]. He developed a numerical volume exchange model, which couples the external processes with the internal processes. The external process is described by the wave run-up, and the internal process by the Forchheimer equation for flow through porous media. The model was calibrated with the tests by Van der Meer [1]. The model determines the notional permeability factor based on the breaker parameter ξ, the ratio between the armor and core material size, and the relation between the wave run-up for a rubble mound with an impermeable core and a rubble mound with a permeable core. The wave run-up at the armor surface is for a permeable core dependent on the water infiltration into the core. The model assumes that the surface roughness reduces the wave run-up on the armor layer with a roughness factor of $\gamma_f = 0.75$ compared to a smooth slope while the run-up at the core was considered to be $\gamma_{Ru} = 0.5$ of the run-up at the surface.

Van Broekhoven [9] conducted a range of experimental model test data to further investigate these assumptions by Jumelet. He tested layer compositions with permeable and impermeable cores and placed the armor material directly on the core material surface. He found that the wave run-up at the armor surface was not influenced by the permeability of the core, but a clear influence from the permeability was observed for the wave run-up at the core surface. Van Broekhoven [9] concluded that the wave run-up below the armor layer is better correlated to the notional permeability factor than the wave run-up at the armor surface. Based on that observation he determined the notional permeability factor from the breaker parameter and the relation between the wave run-up under the armor layer for a permeable core and an impermeable core. Van Broekhoven [9] did not test layer compositions with filter layers.

Van der Neut [10] used the volume of fluid method by IH Cantabria (IH2VOF) to estimate the notional permeability on the layer compositions tested by Van der Meer [1]. He calibrated the numerical model against small-scale stability tests and found relations between the notional permeability and four different dimensionless parameters determined from the numerical model. Thus the model is not a simulation of the stability, but is a coupling between rock armor stability tests and some dimensionless parameters describing the notional permeability factor.

The aim of the present paper is to get estimates of P for a wider range of layer compositions. For this purpose, new rock armor stability model tests were carried out with seven different layer compositions having different permeabilities.

Following a short presentation in Section 2 of the stability formulae by Van der Meer [1] the model setup and the model materials are presented in Section 3. Wave generation and wave analysis are explained in Section 4. Following a description in Section 5 of the applied damage measuring technique, the test program and the test procedure are given in Section 6. A comparison of the model test results with the results of Van der Meer is presented in Section 7 followed by a presentation and a discussion in Section 8 of the notional permeability factors determined for the new layer compositions. Finally, in Section 9, a discussion of possible methods to estimate the notional permeability is given and a simple empirical method for the estimation of the notional permeability is presented.

2. Stability Formulae by Van der Meer

The Van der Meer [1] formulae for the stability of rock armored non-overtopped breakwaters is as follows, Equation (1):

Plunging waves $(\xi_{0m} < \xi_{0m,cr}$ or $\cot(\alpha) \geq 4)$:

$$\frac{H_{1/3}}{\Delta D_{n50A}} = 6.2 P^{0.18} \left(\frac{S_d}{\sqrt{N}}\right)^{0.2} \xi_{0m}^{-0.5}$$

Surging waves $(\xi_{0m} \geq \xi_{0m,cr}$ and $\cot(\alpha) < 4)$:

$$\frac{H_{1/3}}{\Delta D_{n50A}} = P^{-0.13} \left(\frac{S_d}{\sqrt{N}}\right)^{0.2} \sqrt{\cot(\alpha)}\ \xi_{0m}^{P} \tag{1}$$

Transition between plunging and surging formula :

$$\xi_{0m,cr} = \left(6.2\ P^{0.31}\ \sqrt{\tan(\alpha)}\right)^{\frac{1}{P+0.5}}$$

Here $\Delta = \rho_{armor}/\rho_{water} - 1$ is the reduced relative density of the armor stones. $D_{n50A} = \sqrt[3]{W_{50A}/\rho_{armor}}$ is the nominal size of the armor stones based on the median armor stone mass W_{50A} as described in the Rock Manual [11]. P is the notional permeability factor. α is the angle of the seaward slope of the structure. $\xi_{0m} = \tan(\alpha)/s_{0m}^{0.5}$ is the surf similarity parameter where the wave steepness ($s_{0m} = H_{1/3}/L_{0m}$) is calculated based on the significant wave height ($H_{1/3}$) and the mean wave period (T_m) at the toe, using deep water wavelength formulae ($L_{0m} = T_m^2 g/2\pi$). The tested range of ξ_{0m} was 0.7–7. N is the number of waves (no more than 8500 waves should be used). The waves in the present tests deviated to some extend from Rayleigh distributed waves in that $H_{2\%}/H_{1/3} = 1.19–1.47$. For such cases, Van der Meer recommends $H_{1/3}$ in Equation (1) replaced by $H_{2\%}/1.4$.

3. Model Test Setup and Model Materials

The new tests were carried out in a wave flume at Aalborg University with dimensions of $25.0 \times 1.5 \times 1.0$ m (l × w × h). For the present tests a 1:100 concrete foreshore was used in order to generate depth-limited waves without wave breaking at the wavemaker. Figure 2 illustrates the wave flume.

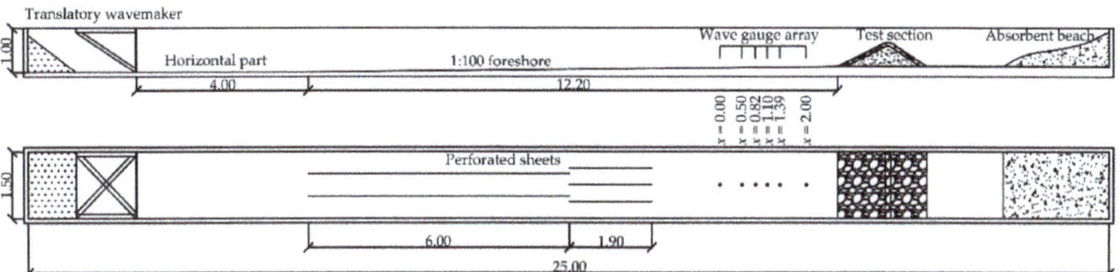

Figure 2. Experimental setup of the flume. Measurements are in meters.

The water depth at the toe of the model breakwater was 0.5 m in all tests. Five different rock materials were used for the tested layer compositions. Table 2 lists the properties of the materials. Figure 3 shows typical shapes of the tested armor rocks. The shapes are of importance for the armor stability. Figure 4 shows the grading curves of the materials listed in Table 2.

Table 2. Test materials used for all layer compositions.

Rock Class	Median Weight W_{50} (g)	Mass Density ρ (kg/m^3)	Nominal Diameter D_{n50} (m)	Grading Ratio $f_g = D_{n85}/D_{n15}$
I	221.0	2620	0.044	1.30
II	32.2	2618	0.023	1.40
III	9.0	2768	0.015	1.36
IV	4.0	2485	0.012	1.33
V	0.7	2936	0.006	1.36

Figure 3. Class I rocks used in the armor layer for the present tests.

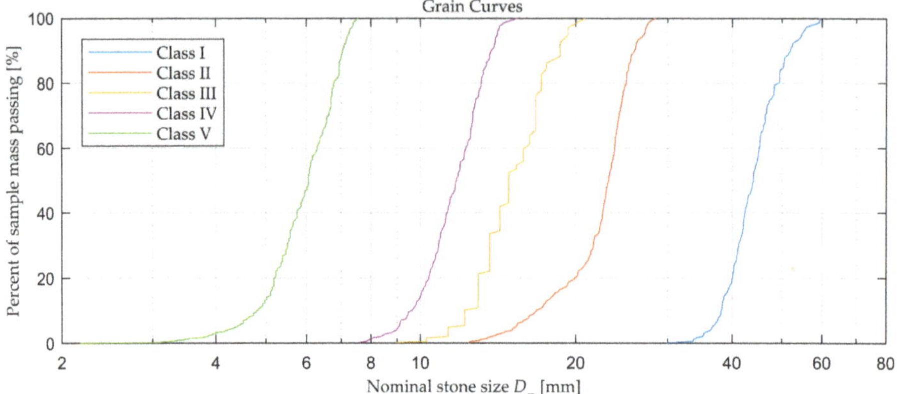

Figure 4. Grain curves of the rock material used in the present tests.

The layer composition with an impermeable core proved to be difficult to be modeled. In the first tests, the impermeable membrane was made of a plywood plate on which the underlayer was directly placed. Unfortunately, sliding of the underlayer was observed when exposed to low steepness waves. An attempt to increase the roughness between the underlayer and the impermeable membrane was made by replacing the plywood plate by concrete slabs with an impermeable membrane below. However, sliding still occurred. Finally, a solution with Class V rocks glued to the plywood plate was found acceptable, see Figure 5. This shows that the interface between the rock material and the core is important. If not modeled correctly this could lead to incorrect stability results. To ensure that the plate was stable and no displacement of the plate could occur, the plywood plate was placed on top of Class V rocks, see Figure 6.

Figure 5. Plywood plate with glued Class V rocks. Used to increase the roughness of the impermeable interface.

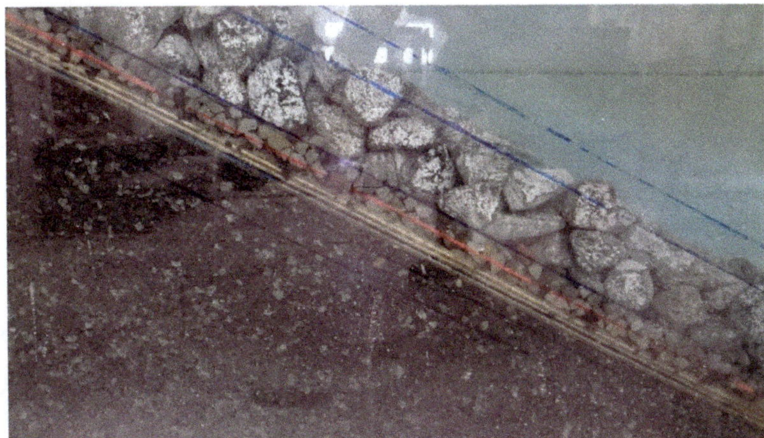

Figure 6. One of the layer compositions with the impermeable core. The figure shows the plywood plate placed on the Class V rocks.

Seven layer compositions as given in Figure 1 were tested. Two of them are similar (compositions *B* and *I*) to what Van der Meer [1] tested (compositions *A* and *H*). The first composition (*B*) consists of an armor layer, underlayer and an impermeable core. The composition has an armor thickness of $1.9D_{n50A}$ and an underlayer thickness of $0.5D_{n50A}$ with a rock size of $0.27D_{n50A}$. The second composition (*I*) consists of an armor layer and a permeable core. The composition has an armor thickness of $1.9D_{n50A}$ with a core rock size of $0.34D_{n50A}$. The third layer composition, *D*, has an armor layer with two underlayers and an impermeable core. The armor layer has a thickness of $1.9D_{n50A}$ and the first underlayer a thickness of $1D_{n50A}$ and a rock size of $0.52D_{n50A}$. The second underlayer has a thickness of $0.5D_{n50A}$ and has a rock size of $0.27D_{n50A}$. This composition will show how sensitive the notional permeability is to the underlayer thickness when an impermeable core is present and thus provide additional insight to the study by Van der Meer et al. [7], Kik [5] and Kluwen [6]. The fourth composition, *G*, consists of an armor layer, an underlayer and a permeable core. The armor layer has a thickness of $1.9D_{n50A}$, and the underlayer has a thickness of $1.4D_{n50A}$ with a rock size of $0.52D_{n50A}$. The core has a rock size of $0.14D_{n50A}$. This layer composition is similar to the non-tested ($P = 0.4$) layer composition by Van der Meer [1]. Furthermore, it is also similar to what Kluwen [6] tested, but she had thicker armor layer and underlayer with a slightly coarser material in the core. Finally, three additional compositions were tested, see layer compositions *E*, *J* and *K*. These layer compositions have an armor layer thickness of $2.9D_{n50A}$ compared to $1.9D_{n50A}$, which is used for layer compositions *B*, *I* and *G*. This will give additional information of the notional permeability and the influence of the layer thickness for the armor material.

4. Wave Generation and Wave Analysis

The waves were generated by the software AwaSys 7 by Aalborg University [12], which includes the used wave generation theories by Eldrup and Lykke Andersen [13] and Zhang et al. [14]. The second-order wave generation by Eldrup and Lykke Andersen [13] was used when free unwanted waves were of acceptable small amplitude. When not acceptable (shallow water cases), the wave generation method by Zhang et al. [14] was used. The method by Zhang et al. [14] uses a depth-averaged velocity as input, which for the present study was generated by MIKE 21 BW by propagating waves from deep to shallow water by a 1:100 foreshore. During all tests, active absorption of reflected waves was used based on wave gauges at the paddle face using the Lykke Andersen et al. [15] method, which has been proven effective also for nonlinear irregular waves, cf. Lykke Andersen et al. [16]. JONSWAP spectra with peak enhancement factor $\gamma = 3.3$ were used in all tests.

To measure and separate incident and reflected waves, six resistant type wave gauges placed in front of the structure with distances between gauges of 0.50, 0.82, 1.10, 1.39 and 2.00 m as shown in Figure 2 were used. The distance from the breakwater to the nearest wave gauge was approximately $0.4L_P$ (peak wavelength) based on the recommendation given by Klopman and Van der Meer [17]. The water depth in the middle of the array was approximately 1.7 cm higher than at the toe, and due to that, depth-limited waves would be slightly smaller at the toe than at the wave gauge array. The difference in $H_{2\%}$ in the middle of the array compared to the toe is estimated by linear shoaling and Battjes and Groendjik [18] to be maximum 1%, which is judged acceptable compared to the scatter in the stability results. In case a steeper foreshore was used the difference would have been significantly larger. In such a case it would be recommended to also measure the waves at the toe without the structure in place. The nonlinear method by Eldrup and Lykke Andersen [19] was used to separate the incident and reflected waves. As opposed to the methods of Goda [20] and Mansard and Funke [21] this separation method includes both bound and free components and amplitude dispersion, which is essential for accurate determination of low exceedance wave parameters, for example $H_{2\%}$, in nonlinear sea states. The method is included in the software package WaveLab 3 by Aalborg University [22].

5. Damage Measurement

After each test, the reshaped profile was measured by a computer controlled non-contact laser profiler run by the software EPro by Aalborg University [23], cf. Figure 7. The measurement grid had a spacing of 10 mm in length and 5 mm in width. The eroded area A_e and the damage $S_d = A_e/D_{n50A}^2$ given in the present paper were based on average values (averaged over the measurement grid) where 20 cm on each side of the flume was disregarded to minimize effects from the walls. Furthermore, only the part of the eroded area where clear erosion was observed was evaluated, which means that small settlements on the upper part of the slope were not included in the eroded area. This is in agreement with the procedure used by Van der Meer [1].

Figure 7. Profiler used to measure the eroded area.

To get more exact measurements, the flume was emptied before laser profiling. Figure 8 shows an example of the averaged measured profile after two consecutive tests.

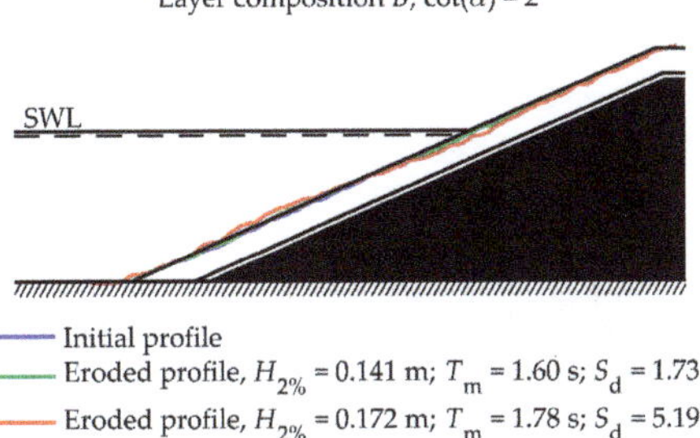

Figure 8. Example of averaged measured profiles showing the damage development in two consecutive tests.

6. Test Program and Test Procedure

In total 149 model tests were performed to non-breaking and slightly breaking wave attack on the seven different permeabilities. Table 3 shows the parameter ranges covered by the tests. To ensure that viscous scale effects are negligible, the Reynolds number, given in Equation (2), should be larger than a critical value typically taken as $Re_{\mathrm{crit}} = 3 \times 10^4$ (Dai and Kamel [24]).

$$Re = \frac{\sqrt{g\,H_{1/3}}\,D_{\mathrm{n50A}}}{\nu} > Re_{\mathrm{crit}} \tag{2}$$

where ν is the kinematic viscosity, D_{n50A} the nominal armor stone size and $\sqrt{g\,H_{1/3}}$ is the characteristic velocity. This is fulfilled for all layer compositions, cf. Table 3.

Table 3. Main characteristics of tests.

Layer Compositions (Figure 7)	B	E	I	J	G	K	D
Number of tests	28	14	20	17	34	23	13
Breaker parameter, ζ_{0m}	1.57–6.86	2.29–6.90	2.82–9.34	3.50–6.91	1.56–7.04	2.26–5.27	2.29–6.91
Wave steepness, s_{0m}	0.005–0.048	0.005–0.048	0.005–0.056	0.009–0.036	0.005–0.049	0.009–0.049	0.005–0.048
Relative wave height, $H_{1/3}/h$	0.20–0.34	0.21–0.33	0.23–0.40	0.23–0.40	0.24–0.51	0.23–0.41	0.22–0.34
Relative wave breaking, $H_{2\%}/H_{1/3}$	1.30–1.46	1.32–1.44	1.29–1.44	1.29–1.43	1.19–1.44	1.29–1.47	1.30–1.41
Relative wave length, L_{0m}/h	6.00–59.26	6.11–59.15	5.05–59.47	7.35–38.03	6.46–56.38	6.42–37.20	6.40–59.50
Relative freeboard, $A_c/H_{1/3}$	1.57–2.66	1.65–2.59	1.37–2.33	1.57–2.66	1.07–2.30	1.33–2.36	1.60–2.49
Stability number, $H_0 = H_{1/3}/\Delta D_{n50}$	1.43–2.42	1.47–2.30	1.63–2.78	1.65–2.80	1.65–3.56	1.61–2.85	1.52–2.38
Reynolds number for armor layer stones, $Re \times 10^{-4}$	3.4–4.4	3.4–4.3	3.6–4.7	3.6–4.7	3.6–5.3	3.6–4.7	3.5–4.3

The 149 tests consisted of different tests series where the wave height was increased in steps, while the wave steepness remained constant. Accumulated damage was measured after each test in the series. Test series were terminated when the underlayer was visible and as such exposed and, after that, the structure was rebuilt for a new test series. In each test, 1000 waves were used. Van der Meer [1] assumed all structures to be non-overtopped when the dimensionless freeboard ($A_c/H_{1/3}$) > 1–2, which is valid for all the present tests. A_c is the freeboard, $H_{1/3}$ is the average of the highest 1/3 of the waves.

The present test procedure was not identical to that of Van der Meer [1]. He did not measure accumulated damage, but instead the damage after 1000 waves and 3000 waves was measured. Afterwards, the breakwater was rebuilt, and a new sea state was tested. His wave series had a constant wave period with different wave heights. Based on these wave series, he fitted a damage curve for a constant wave period from which he extracted the relation between the wave heights and the damage values in the interval of $S_d = 2$–17. Thus his stability formulae were established on fitting to the damage curves. The present tests were made with 1–5 wave heights for each wave steepness. Damage curves for each of the present wave series were fitted in the present work. However, because most of the data corresponded to accumulated damage, a conversion to non-accumulated damage with the use of Equation (3) was made in order to comply with the basis of the Van der Meer formulae. The conversion is based on the relations found by Van der Meer [1] between damage, wave height and number of waves. The remaining parameters in the Van der Meer formulae are kept constant in the test series and are thus included in A. A is the slope of the continuous line seen in Figure 9, which describes the relation between the damage, the number of waves and the wave height. For the accumulated test series ($i > 1$) an extra number of waves $N_{extra,i}$ were added to the number of waves N_i used in the test. Since the extra number of waves is a function of A, an iterative procedure was applied to Equation (3) until convergence of A was found.

$$\frac{H_{2\%,i}}{\Delta D_{n50A}} = A\left(\frac{S_{d,i}}{\sqrt{N_i}}\right)^{0.2}$$

$$N_{extra,i} = \frac{A^{10}S_{d,i-1}^2}{\left(\dfrac{H_{2\%,i}}{\Delta D_{n50A}}\right)^{10}}$$

$$N_{total,i} = N_i + N_{extra,i}$$

(3)

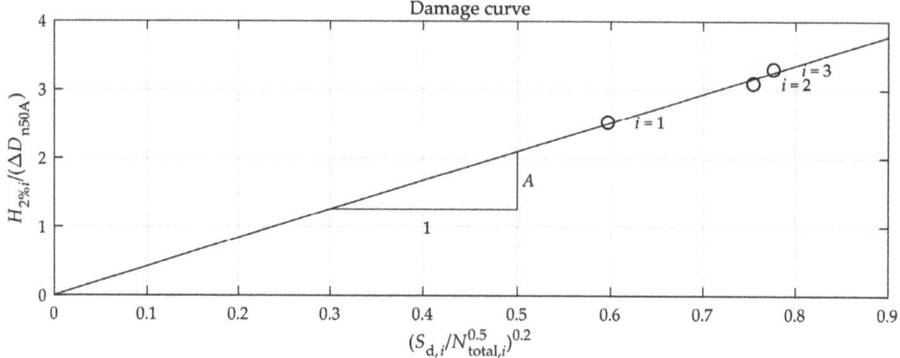

Figure 9. Damage curve for accumulated tests. The tests are shown with markers and the test number in the wave series is given by i. A and $N_{total,i}$ are found by iterating Equation (3) until convergence of A is obtained.

In the tests series of Van der Meer [1] and Thompson and Shuttler [3] the breaker parameter was not kept constant. Consequently, their tests series do not correspond to a constant value of A. However, as their tests already represent non-accumulated damage the raw data were instead plotted in Figure 11.

7. Comparison with Physical Tests of Van der Meer

All results of the previously tested layer compositions and the results of Van der Meer [1] and Thompson and Shuttler [3] are shown in Figure 10. The estimation of the notional permeability of the present layer compositions is based on minimization of the root mean square error (RMSE) on $H_{2\%}/(\Delta D_{n50A}(S_d/N^{0.5})^{0.2})$ in the context of the Van der Meer formulae. The waves in the tests by Van der Meer [1] and Thompson and Shuttler [3] are all in deeper water for which wave heights can be assumed Rayleigh distributed, i.e., $H_{2\%} = 1.4H_{1/3}$. The results are shown for measured damage levels in the range $2 \leq S_d \leq 8$ for $\cot(\alpha) = 1.5$ and 2, and damage levels in the range $2 \leq S_d \leq 12$ for $\cot(\alpha) = 3$.

Figure 10a,b show the results of the composition with an impermeable core for slopes $\cot(\alpha) = 2$ and $\cot(\alpha) = 3$ for compositions A and B. While the results for $\cot(\alpha) = 3$ seemed in fair agreement with a fitted $P = 0.10$, the results for the steeper $\cot(\alpha) = 2$ disagreed in the surging wave domain. Quite higher stability in the surging domain and $P = 0.21$ were found for $\cot(\alpha) = 2$.

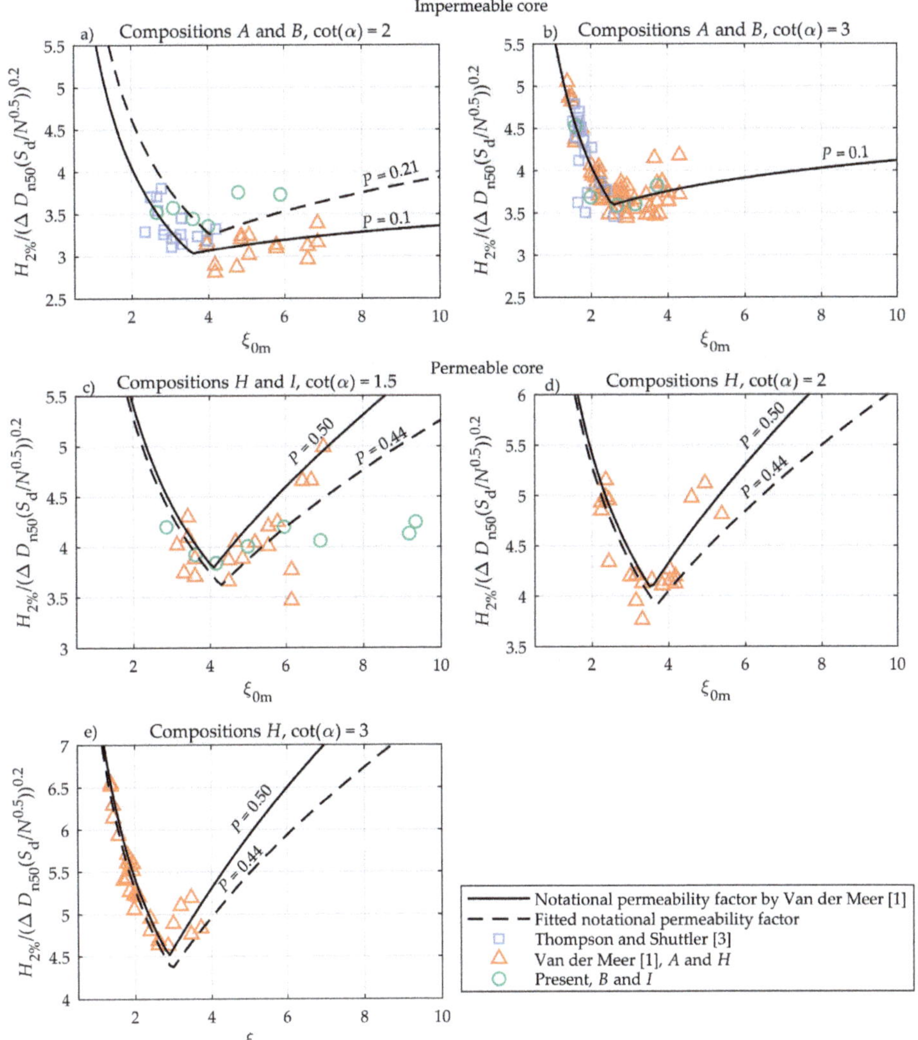

Figure 10. Tests with layer compositions A, B, H and I and front slopes $\cot(\alpha) = 1.5$, 2 and 3. Continuous lines indicate the given permeability (P) by Van der Meer [1]. Dashed lines indicate fitted values of P derived from the Van der Meer formulae. Markers indicate the results.

In the present layer composition B, the grading ratio of the filter layer is $D_{n85}/D_{n15} = 1.33$ compared to $D_{n85}/D_{n15} = 2.25$ used in the tests by Van der Meer [1]. This difference makes the present filter layer more permeable due to a larger porosity, which increases the stability. Furthermore, Van der

Meer [1] tested two grain size distributions of the armor layer for the impermeable layer composition. $D_{n85}/D_{n15} = 2.25$ was tested with $\cot(\alpha) = 2, 3, 4$ and 6 while $D_{n85}/D_{n15} = 1.25$ was tested with $\cot(\alpha) = 3$ and 4. Thus he did not test the impermeable layer composition with $D_{n85}/D_{n15} = 1.25$ for $\xi_{0m} >> 4$, see Figure 10b. Comparing the armor layer for compositions A and B for $\cot(\alpha) = 2$ shows that the present compositions has a more narrow gradation and thus a more permeable armor layer. The combination of a more permeable armor layer and filter layer might be the reason for the increase in armor stability of the present tests for $\xi_{0m} > 4$ for $\cot(\alpha) = 2$. The results were too few to prove a significant change to the notional permeability factor and it was kept to $P = 0.1$ to be on the safe side.

Figure 10c shows the results of layer compositions H and I. The present data were for most of the tests within the scatter of the data by Van der Meer [1]. Two of the tests ($\xi_{0m} \approx 9$) were significantly more unstable than predicted by the formula with $P = 0.5$. These two tests with very low steepness waves were outside the applicability range ($0.7 < \xi_{0m} < 7$) of the Van der Meer [1] formulae and therefore not used in the fitting of P. However, the reason for the two tests deviating might be related to the lower water depth used in the present tests as the wavelength of the large wave period was significantly more affected by the water depth compared to smaller wave periods. It should also be noted that Van der Meer had two significant outliers at $\xi_{0m} \approx 6$. The present tests had for composition H with slopes $\cot(\alpha) = 1.5, 2.0$ and 3.0 a fitted $P = 0.44$, which was lower than the value $P = 0.5$ given by Van der Meer [1], see Figure 10c–e. However, it should be noted that the Van der Meer formulae with $P = 0.5$ overpredicted the stability for all of his own data for $\cot(\alpha) = 1.5$.

Considering the differences in layer compositions, there seems to be a fair agreement between the present results and the results of Van der Meer. Therefore, new layer compositions can be tested and a notional permeability factor can be fitted with use of the formulae by Van der Meer [1].

8. Notional Permeability for New Layer Compositions

Figures 11 and 12 show the fitted notional permeability factors for the new layer compositions. The results of the layer compositions with an armor layer thickness of two rocks are shown in Figure 11 and results of the three layered armor thickness are shown in Figure 12. Since compositions with an armor layer thickness of three rocks could suffer more damage before failure than an armor layer thickness of two rocks, a wide damage level range of $2 \leq S_d \leq 12$ was included in the analysis.

Figure 11a,b shows the results of the layer compositions G with a permeable core, one underlayer and an armor layer. By fitting the present results of layer composition G to the formulae by Van der Meer, a notional permeability of $P = 0.37$ was found for the lowest RMSE for both front slopes. This notional permeability was significantly lower than $P = 0.46$ found by Kluwen [6] (composition L) for an almost identical composition. The reason for the differences in the P value is not clear.

Figure 11c shows the results of layer composition D with an impermeable core, two underlayers and an armor layer. The lowest RMSE was found for a notional permeability of 0.29. This notional permeability was significantly larger than $P = 0.1$ given for layer composition A and B, which also has an impermeable core, cf. Figure 10a. Even though the fitted notional permeability factor for layer composition B was $P = 0.21$ the increase of the notional permeability factor for layer composition D was significant. The notional permeability factor was significantly influenced by the layer thickness of the permeable layers for compositions with an impermeable core. The results with the fitted P were in good agreement with the Van der Meer formulae, but a slight underprediction was observed for the data in the plunging regime and overprediction in the surging regime.

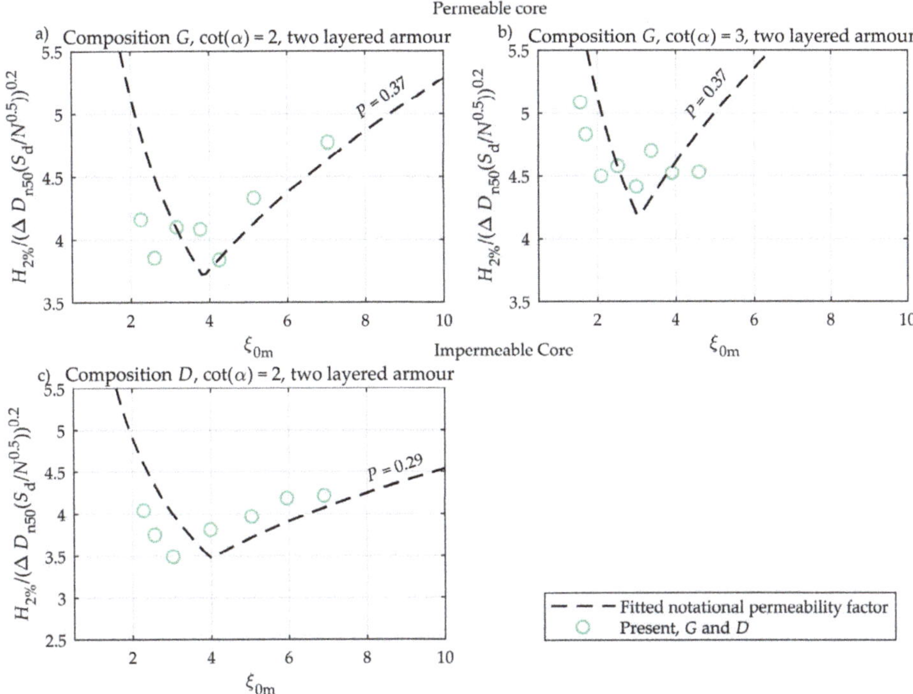

Figure 11. Tests with layer compositions G and D and front slope of $\cot(\alpha) = 2$ and 3. Dashed lines indicate fitted values of P derived from the Van der Meer [1] formulae. Markers indicate the results.

Figure 12a shows the results of layer compositions J with a permeable core and an armor layer. The best fitted notional permeability factor was found for $P = 0.45$. The results were a close match to the formulae by Van der Meer [1]. Comparing the results to layer composition I in Figure 10c showed that an increase in the armor layer thickness had no significant change to the notional permeability factor.

Figure 12b shows the results of the layer composition K with a permeable core, one underlayer and an armor layer. The composition had a fitted notional permeability factor of 0.45. Comparing the results to layer composition G in Figure 11a shows that the increase in the armor layer thickness has a small influence to the notional permeability factor. Thus for the finer core material, the influence of the armor layer thickness was higher compared to a coarse permeable core. The scatter of the stability results was significant but it was clear that the stability was increased when comparing Figures 11a and 12b. Therefore, an increase of the notional permeability factor was also expected when using the formulae by Van der Meer.

Figure 12c show the results of layer composition E with an impermeable core, thin filter layer and an armor layer. Layer composition E had a best fitted $P = 0.30$ and the results were only having small deviations with the Van der Meer formulae at the transition between the plunging and surging formulae. Comparing the results to layer composition B in Figure 10a shows that an increase in the armor layer thickness increased the notional permeability factor significantly. Thus the notional permeability for structures with an impermeable core seemed very sensitive to the layer thickness and material size of the permeable layers. This was also observed by Van der Meer et al. [7]. Moreover, the results show that the effect of armor layer thickness for a coarse permeable core was far from being as significant as found for an impermeable core.

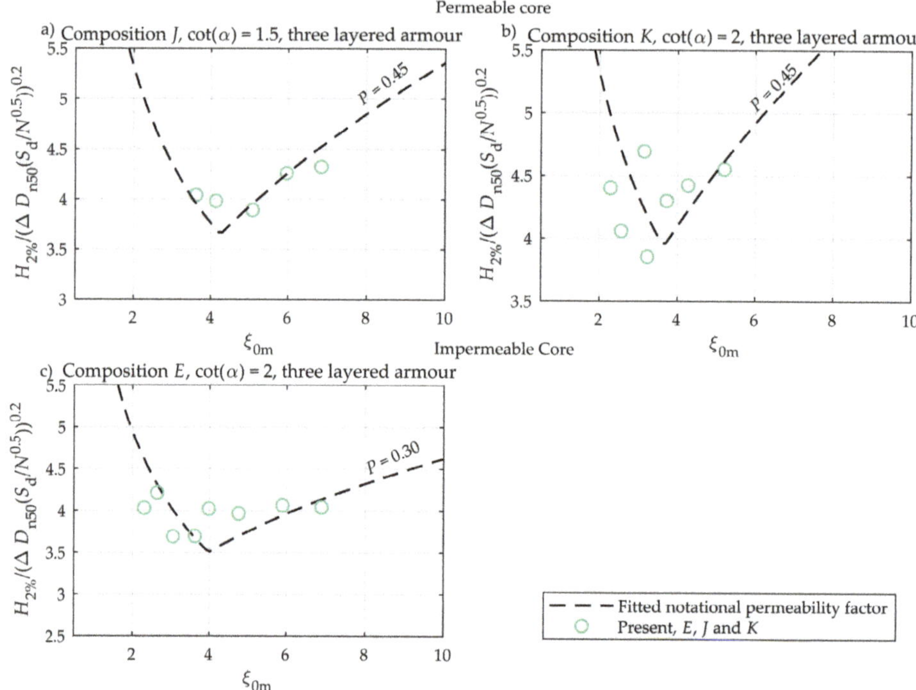

Figure 12. Tests with layer compositions *J*, *K* and *E* and front slope of cot(α) = 1.5 and 2. Dashed lines indicate fitted values of *P* derived from the Van der Meer [1] formulae. Markers indicate the results.

The experimental results allowed the effect on the notional permeability of upgrading two layer rock armor to three layer rock armor to be extracted. Table 4 shows that the influence of the thickness of the armor layer on the notional permeability factor was largest for impermeable layer compositions like *B* and *E*, and smallest for permeable layer compositions like *G* and *K*.

Table 4. Influence of armor layer thickness.

Rock Class	Layer Composition	Notional Permeability Factor Two Layers	Three Layers	Influence
Impermeable core	*B* and *E* for cot(α) = 2	0.10	0.30	Significant
Permeable core no filter	*I* and *J* for cot(α) = 1.5	0.44	0.45	Insignificant
Conventional layer composition	*G* and *K* for cot(α) = 2	0.37	0.45	Moderate

9. New Method for the Estimation of the Notional Permeability Factor

Application of the Van der Meer rock armor stability formulae for desk study design of new rock armor layer compositions demands knowledge of the notional permeability factor *P*. In the introduction we explained that application of numerical models might help estimating the notional permeability. However, such approach has some difficulties. First of all the notional permeability has no physical meaning as also explained by Van der Meer [1]. It is a parameter fitted to a complex formula fitted to physical model tests of armor stability, and could be indirectly related to phenomena as run-up and porous flow resistance/dissipation. Secondly, because numerical models are partly based on parameters fitted to results of physical model tests, direct determination of *P* from basic physical principles is not possible. Moreover, while the value of the notional permeability is fixed for a specific layer composition, the other phenomena vary with the wave conditions.

A pragmatic approach to obtain a tool for the prediction of the notional permeability would be to fit a formula to all the parameter values obtained in model tests for all tested layer configurations. Such a formula is presented in the following.

The new formula will include the known physical processes in an empirical way. It is well known that a homogenous structure has the largest P factor, and when introducing a core with smaller size material the wave run-up and the loads on the armor units will increase corresponding to a reduction in the P factor. This is partly due to higher porous flow resistance and a decrease in the buffer capacity of the permeable layers. This effect on the P factor is clearly seen when comparing composition M with I and G in Figure 1. Furthermore, the structure stability is influenced by the thickness of the layers. Comparing the tests with an armor layer thickness of two rocks with the compositions with a thickness of three rocks, it can be seen that the increase in P is largest for compositions where the impermeable layer is closest to the armor layer. This shows that the effect on the P factor from the material size is decreasing with increasing distance into the breakwater. To describe the relative distance from each layer to the surface of the armor layer, a relative distance $z^* = z/D_{n50A}$ is used. The distance z is perpendicular to the front slope, see Figure 13.

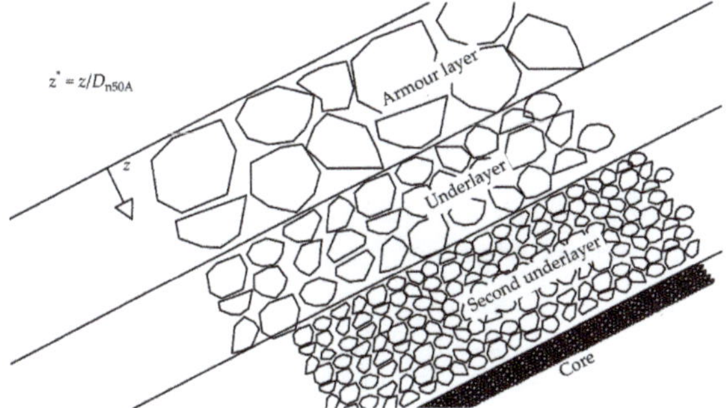

Figure 13. Definition of the relative depth z^* as a function of z and the nominal size of the armor stones D_{n50A}.

The grading of the materials plays a role because a very wide grading has small porosity. However, this effect cannot be studied based on the present tests as all gradings were narrow with $D_{n85}/D_{n15} < 2.25$. Therefore, the developed empirical formula is limited to narrow graded materials with grain size distributions within the ranges of the materials tested in the laboratory. Based on the above considerations, the empirical formula for the P factor can be expressed as a function of rock size and the relative depth z^*. The functions f and g defined in Equation (4) and plotted in Figure 14 are empirically fitted to model the influence of the rock size and the relative depths, respectively.

$$f = 0.79\left(1 - \exp\left(-4.1\frac{D_{n50,z^*}}{D_{n50,A}}\right)\right) \text{ for } \frac{D_{n85}}{D_{n15}} < 2.5 \quad g = \exp(-0.62\, z^*) \quad (4)$$

here D_{n50A} is the nominal size of the armor units, and D_{n50,z^*} is the nominal size of the units in the given layer at relative depth z^*.

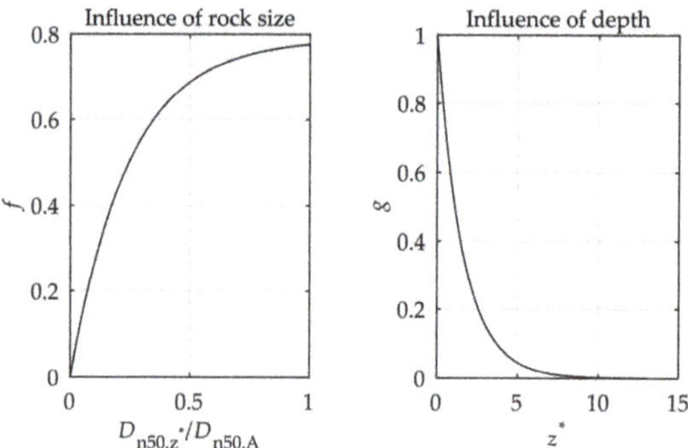

Figure 14. Influence of the relative rock size on (f) and influence of relative layer depth on (g). The subplots are plotted with the use of Equation (4).

The g function implies that the layers with the largest influence on the permeability are those close to the slope surface, whereas the f function implies that large material is more permeable than the fine material. Analysis has shown that an integration function, k, giving the influence from the relative rock size and the relative depth, can be used to estimate the notional permeability factor, $P(k)$. The integration function k is given by Equation (5).

$$k = \int_0^{z^*_{max}} f(z^*)g(z^*)dz^* \tag{5}$$

z^*_{max} is the value of z^* for the impermeable layer, but has a maximum value of 13. For a layer composition consisting of N permeable layers the integration function Equation (5) can be rewritten into a closed form as:

$$k = \sum_{i=1}^{N}\left(0.79 - 0.79\exp\left(-4.1\frac{D_{n50,i}}{D_{n50,A}}\right)\right)\left(\frac{\exp(-0.62z^*_1) - \exp(-0.62z^*_2)}{0.62}\right) \tag{6}$$

where $D_{n50,i}$ is the nominal size of the units in the given layer. Figure 15 shows the definition of z^*_1 and z^*_2 for $i = 2$ in Equation (6). z^*_2 should stop at the impermeable layer or at a maximum value of 13.

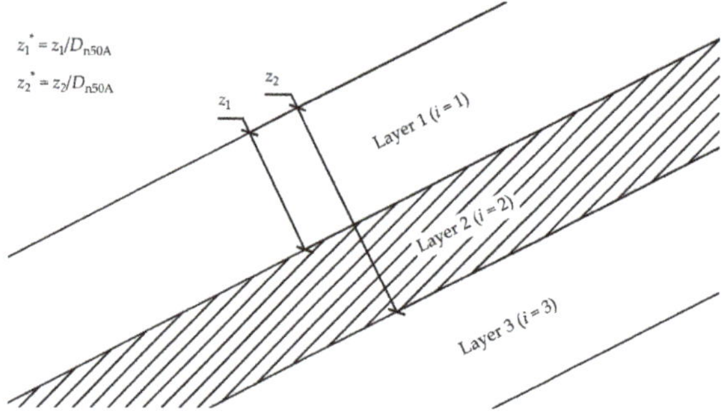

Figure 15. Example of the definition of the relative depth z^*_1 and z^*_2 for $i = 2$ in Equation (6) with D_{n50A} as the nominal size of the armor stones.

Based on the integration function k the new empirical formula for estimating P can be given as

$$P = \max\begin{cases} 0.1 \\ 1.72k - 1.58 \end{cases} \tag{7}$$

Equation (7) is limited to compositions in which the material size decreases from the armor layer to the core. For example, if a layer composition with an identical core and armor layer is used, but a thin and almost impermeable layer is separating these layers, the integration function k should stop at the impermeable layer, and the material size should never increase with z.

Figure 16 shows the estimated P factors from Equation (7) compared with the fitted P factors given in Figure 1. Good correlation was found for all layer compositions having a typical deviation of ± 0.03 between the estimated and the fitted notional permeability factor.

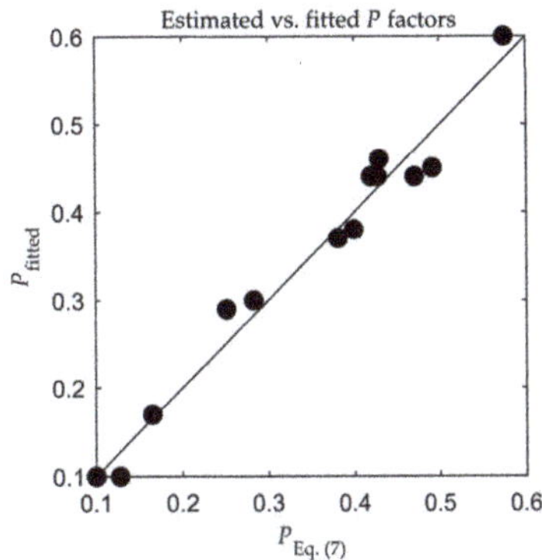

Figure 16. Comparison between the fitted P factors shown in Figure 1 and the estimated P factors calculated by Equation (7).

10. Conclusions

The present paper presents notional permeabilities for various rock armored layer compositions based on hydraulic model tests. The model test program included two layer compositions similar to those previously tested by Van der Meer [1].

Notional permeability factors were determined for two new layer compositions, one with an impermeable core and one with a permeable core. Furthermore, three additional compositions similar to previously tested compositions but having a 50% thicker armor layer were tested. In total seven layer compositions were added to the database with known notional permeability factors.

Based on notional permeability factors for all 13 layer compositions, an empirical formula for the estimation of the notional permeability was established. Given typical deviations of 0.03, the formula shows good agreement with the known P factors determined from model tests.

Increasing the armor layer thickness from two to three layers of rocks, the notional permeability factor and the related armor stability were increased. For compositions with impermeable core, the increase was significant. For conventional layer compositions with filter layer(s) and quarry rock core, the increase was moderate. For compositions with armor layer placed directly on a very permeable core, the increase was insignificant.

Author Contributions: M.R.E. and T.L.A. planned the test campaign for the paper in cooperation. The experimental tests and the analysis of the results were performed by the M.R.E. A common discussion on the outline of the analysis was done in a shared effort between the authors. The outline of the paper was written by M.R.E. T.L.A. and H.F.B. performed a detailed review on the draft paper and contributed with valuable suggestions.

Funding: This research received no external funding.

References

1.   Van der Meer, J.W. Rock Slopes and Gravel Beaches under Wave Attack. Ph.D. Thesis, Delft University of Technology, Delft, The Netherlands, 1988.

2. Sarfaraz, M.; Pak, A. An integrated SPH-polyhedral DEM algorithm to investigate hydraulic stability of rock and concrete blocks: Application to cubic armours in breakwaters. *Eng. Anal. Bound. Elem.* **2017**, *84*, 1–18. [CrossRef]

3. Thompson, D.M.; Shuttler, R.M. *Riprap Design for Wind-Wave Attack, a Laboratory Study in Random Waves*; HR Wallingford: Oxfordshire, UK, 1975.

4. Hölscher, P.; Barends, F.B.J. *Transport in Porous Media*; Delft Geotechnics: Delft, The Netherlands, 1986.

5. Kik, R. The Notional Permeability of Breakwaters: Experimental Research on the Permeability Factor P. Mater's Thesis, Delft University of Technology, Delft, The Netherlands, 2011.

6. Kluwen, J.G.M. Physical Model Tests of the Notional Permeability on Breakwaters. Mater's Thesis, Delft University of Technology, Delft, The Netherlands, 2012.

7. Van der Meer, J.W.; Van Gent, M.R.A.; Wolters, G.; Heineke, D. *New Design Guidance for Underlayers and Filter Layers for Rock Armour under Wave Attack*; ICE Publishing: London, UK, 2018.

8. Jumelet, H.D. The Influence of Core Permeability on Armour Layer Stability. Mater's Thesis, Delft University of Technology, Delft, The Netherlands, 2010.

9. Van Broekhoven, P.J.M. The Influence of Armour Layer and Core Permeability on the Wave Run-Up. Mater's Thesis, Delft University of Technology, Delft, The Netherlands, 2010.

10. Van der Neut, E.M. Analysis of the Notional Permeability of Rubble Mound Breakwaters by Means of a VOF Model. Mater's Thesis, Delft University of Technology, Delft, The Netherlands, 2015.

11. CIRIA/CUR/CETMEF. The Rock Manual. The Use of Rock in Hydraulic Engineering. 2007. Available online: www.kennisbank-waterbouw.nl/DesignCodes/rockmanual/introduction.pdf (accessed on 2 May 2019).

12. Aalborg University AwaSys—Software for Wave Laboratories. Available online: www.hydrosoft.civil.aau.dk/awasys (accessed on 2 May 2019).

13. Eldrup, M.R.; Andersen, T.L. Applicability of Nonlinear Wavemaker Theory. *J. Mar. Sci. Eng.* **2019**, *7*, 14. [CrossRef]

14. Zhang, H.; Schäffer, H.A.; Jakobsen, K.P. Deterministic combination of numerical and physical coastal wave models. *Coast. Eng.* **2007**, *54*, 171–186. [CrossRef]

15. Lykke Andersen, T.; Clavero, M.; Frigaard, P.; Losada, M.; Puyol, J.I. A new active absorption system and its performance to linear and non-linear waves. *Coast. Eng.* **2016**, *114*, 47–60. [CrossRef]

16. Lykke Andersen, T.; Clavero, M.; Eldrup, M.R.; Frigaard, P.; Losada, M. Active Absorption of Nonlinear Irregular Waves. In Proceedings of the Coastal Engineering Conference, Baltimore, MD, USA, 30 July–3 August 2018.

17. Klopman, G.; van der Meer, J.W. Random Wave Measurements in Front of Reflective Structures. *J. Waterw. Port Coast. Ocean Eng.* **1999**, *125*, 39–45. [CrossRef]

18. Battjes, J.A.; Groenendijk, H.W. Wave height distributions on shallow foreshores. *Coast. Eng.* **2000**, *40*, 161–182. [CrossRef]

19. Eldrup, M.R.; Lykke Andersen, T. Estimation of Incident and Reflected Wave Trains in Highly Nonlinear Two-Dimensional Irregular Waves. *J. Waterw. Port Coast. Ocean Eng.* **2019**, *145*, 04018038. [CrossRef]

20. Goda, Y.; Suzuki, T. Estimation of Incident and Reflected Waves in Random Wave Experiment. In Proceedings of the 15th Coastal Engineering Conference, Honolulu, HI, USA, 1–17 July 1976; ASCE: Reston, VA, USA, 1976; pp. 828–845.

21. Mansard, E.P.D.; Funke, E.R. The Measurement of Incident and Reflected Spectra Using a Least Squares Method. In Proceedings of the 17th Conference on Coastal Engineering, Sydney, Australia, 23–28 March 1980; ASCE: Reston, VA, USA, 1980; pp. 154–172.

22. Aalborg University WaveLab—Software for Wave Laboratories. Available online: www.hydrosoft.civil.aau.dk/wavelab (accessed on 27 April 2019).

23. Aalborg University EPro—Software for Wave Laboratories. Available online: www.hydrosoft.civil.aau.dk/epro/ (accessed on 27 April 2019).

24. Dai, Y.B.; Kamel, A.M. *Scale Effect Tests for Rubble-Mound Breakwaters: Hydraulic Model Investigation*; U. S. Army Engineer Waterways Experiment Station: Vicksburg, MS, USA, 1969.

 *water*

Correction

Correction: Eldrup, M.R., et al. Stability of Rubble Mound Breakwaters—A Study of the Notional Permeability Factor, based on Physical Model Tests. *Water* 2019, *11*, 934

Mads Røge Eldrup *, Thomas Lykke Andersen and Hans Falk Burcharth

Department of Civil Engineering, Aalborg University, 9220 Aalborg, Denmark; tla@civil.aau.dk (T.L.A.); hansburcharth@gmail.com (H.F.B.)
* Correspondence: mre@civil.aau.dk

Received: 6 November 2019; Accepted: 7 November 2019; Published: 25 November 2019

The authors wish to make the following corrections to this paper [1]:
Replace Equation (6)

$$k = \sum_{i=1}^{N} 0.79 - 0.79 \exp\left(-4.1 \frac{D_{n50,i}}{D_{n50,\,A}}\right)\left(\frac{\exp\left(-0.62 z_1^*\right) - \exp\left(-0.62 z_2^*\right)}{0.62}\right)$$

with:

$$k = \sum_{i=1}^{N} \left(0.79 - 0.79 \exp\left(-4.1 \frac{D_{n50,i}}{D_{n50,\,A}}\right)\right)\left(\frac{\exp\left(-0.62 z_1^*\right) - \exp\left(-0.62 z_2^*\right)}{0.62}\right)$$

The authors would like to apologize for any inconvenience caused to the readers by these changes. The manuscript will be updated, and the original will remain online on the article webpage, with a reference to this Correction.

Reference

1.   Eldrup, M.R.; Lykke Andersen, T.; Burcharth, H.F. Stability of Rubble Mound Breakwaters—A Study of the Notional Permeability Factor, based on Physical Model Tests. *Water* **2019**, *11*, 934. [CrossRef]

MDPI

St. Alban-Anlage 66

4052 Basel

Switzerland

Tel. +41 61 683 77 34

Fax +41 61 302 89 18

www.mdpi.com

Water Editorial Office

E-mail: water@mdpi.com

www.mdpi.com/journal/water

www.ingramcontent.com/pod-product-compliance
Lightning Source LLC
LaVergne TN
LVHW070157170726
843515LV00007B/1940